Networks Illustrated

8 Principles Without Calculus

CHRISTOPHER BRINTON AND MUNG CHIANG
PRINCETON UNIVERSITY

EdWiser Scholastic Press

EdWiser Scholastic Press
422 Executive Drive, Princeton, NJ 08540
USA

Published in the United States of America by EdWiser Scholastic Press

First published 2013

Printed in the United States
ISBN 978-0-9895430-1-9 Paperback

Networks Illustrated

Are you curious about the social and technological networks that shape our lives, like Amazon, the Internet, and WiFi? But are you worried about a Calculus or Linear Algebra prerequisite that typically comes along with such courses? If so, then this book is for you. You will learn the key principles behind the networks that surround us every day, such as "sharing is hard" and "crowds are wise", and you will see how these simple principles are reflected in how these networks operate. Plus, all the mathematics you need is adding and multiplying numbers, as this book relies on pictures, anecdotes, and analogies as instructional tools.

This book can also be used for an introductory course in college or high school, offered to those interested in networking from any majors and disciplines. Instructional resources, including lecture videos and slides, as well as comprehensive questions for homework assignments, can found at the continuously updated course website `www.networksillustrated.org`, or by emailing the authors at `networksillustrated@gmail.com`.

Christopher Brinton is a Ph.D. Candidate in the Department of Electrical Engineering at Princeton University. He received his Masters Degree in EE from Princeton in May 2013, and his BSEE from The College of New Jersey (summa cum laude) in May 2011. He is a co-instructor for the MOOC based on this book.

Mung Chiang is a Professor of Electrical Engineering at Princeton University and the Director of Princeton EDGE Lab that has commercialized several technologies. His research on networking receives the Waterman Award and the IEEE Tomiyasu Award, his education activities received the ASEE Terman Award, and his inventions received a TR35 Award.

"Its broad coverage of network technology and algorithms, with lucid examples and diagrams, make this text the quintessential network primer."

–Keith Cambron, author of *Global Networks* and former President of AT&T Labs.

"*Networks Illustrated* is the definitive book about networks. It is an intelligent, comprehensive overview of the networks that shape our lives: social networks, crowd-based networks, auction-based networks, cellular networks, and the Internet, the network of networks. This is a must read for anyone who wants to understand how these networks operate and attempt to serve the greater good."

–Lynda Clarizio, former President of Advertising.com

To Mom, Dad, Eric, and Kate
–Christopher Brinton

To Novia and Augustan
–Mung Chiang

Contents

Preface

The motivation to write this book has an interesting story, bringing us back to September of 2012. At that time, co-author Mung had just published his first textbook, *Networked Life: 20 Questions and Answers*, using the just-in-time, case-study-based pedagogical approach, and written in a language geared towards junior or senior level undergraduates in science and engineering.

Subsequently, we began to offer the corresponding course, **Networks: Friends, Money, and Bytes**, during the Fall 2012 semester, as a Massive Open Online Course (MOOC). The Princeton students had the opportunity to experience a flipped classroom – where the homework becomes watching the lectures and the class time is used for discussion – while online students were able to get course content for free online.

Driven by intellectual curiosity rather than a certificate or practical skills for a new job, over 65,000 people enrolled. But as the course progressed, we were getting emails, posts on the discussion forums, and Facebook messages from people saying they didn't have the mathematical prerequisite of multivariable calculus and linear algebra. They had either never before seen vectors, matrices, derivatives and so on, or didn't care to see it.

So we began thinking, can we explain this material *without* using all the equations in the "Long Answer" section of the Networked Life textbook? How much do you really have to know about eigenvalues and eigenvectors to get the gist of PageRank? Do we really have to take derivatives to explain the notions of supply and demand that underly data pricing?

As we thought more, we realized that we could teach the fundamentals of networking using math that is not much more complicated than $2 + 3 = 5$, as long as there are sufficient pictures, analogies, and anecdotes to compliment the text. And that lead us to write this book, which explains the social and technological networks that surround us each day by focusing on *8 principles*. We still wanted this book to be more detailed than a popular science text, while keeping it at a level such that anyone with middle school mathematics can access the material.

Now, what are those principles? They are simple phrases that summarize a wide range of ideas behind networks:

1. *Sharing is hard.*
   Whether you're making a phone call or connecting to the Internet on WiFi, you have to share the network media (e.g., the air) that you're using with many other people. We will look at the techniques that have been designed to allocate the network's resources fairly.
2. *Consensus is hard.*
   Many websites today have to resolve large banks of raw data to find a consensus. This could take the form of allocating ad spaces from bids, ranking web pages from relationships between pages, or resolving disputes when conflicts of opinion arise. Such tasks become increasingly convoluted as the size of the input increases.
3. *Crowds are wise.*
   Today, online retail and rental businesses commonly allow customers to rate the items they purchase. We can leverage the opinions of large "crowds" of ratings to make product ranking and recommendation more accurate and useful. But we have to make certain assumptions about the way the ratings were made in the first place.
4. *Crowds are not so wise.*
   Unfortunately, the assumptions we need to make about "crowds" in order for their opinions to be useful are not always valid. For instance, on social networks like Facebook or YouTube, many people can readily be influenced to watch videos or purchase products, since they will naturally follow either the "crowd" or the influential members of the community.
5. *Network is expensive.*
   Our behavior in networks is often driven by prices, shaped by the balance between demand and supply. We will look at reasons why usage-based pricing sends more useful pricing signals to customers.
6. *Network of networks.*
   The Internet itself is a huge, ubiquitous network that is comprised of many inter-connected networks. We will first look at the principles of packet switching, and then how the Internet routes packets from one location to another.
7. *Layers on layers.*
   One of the fundamental principles behind the Internet's architecture is the notion of layering. These layers range from the physical media over which our data flows, to the applications that we run on our smart phones. We will see how load is controlled and multicast is achieved in some of these layers.
8. *Bigger and bigger.*
   The size of the Internet is hard to comprehend. We will explore why

there can still be small worlds, and how the "cloud" can keep getting bigger and bigger.

There are other resources available for this book when used as a textbook for colleges (or even high school) students *outside* of science and engineering.

- *Students* can have access to homework problems by going to `www.networksillustrated.org` and registering. This will not include the solutions.
- *Instructors* can obtain any of the following:
  - Homework problems and solutions
  - Power point presentations
  - Student project ideas

  by emailing us at `networksillustrated@gmail.com`. Tell us what school you're from and we'll be happy to provide you with this.

By the way, there's also a free online course based on this book, which will likely be offered every other semester. Tune into the course website above for more details. On that note, we should also mention that this book is part of a much larger effort on our part to help redraw the boundaries of education: both online with MOOCs, and in person with flipped classroom. We have many thoughts on this matter, and have compiled a list of ten questions (with no definitive answers): `http://scenic.princeton.edu/files/MOOC_FLIP_thoughts.pdf`.

There are a few ongoing research projects we've undertaken to explore if teaching can be made both massively scalable and effective. A personalized learning platform is one of them. Why should everyone read the same book? How about creating "parallel universes" of different paths of learning? Rather than cutting our feet to fit the same shoes, the shoes should feel the feet and adapt itself. Together with a team of engineers, we have been developing such a technology, which we call the Individualized and Integrated BookApp (IIB). We unveiled the first version of IIB back on Valentine's Day 2013, and the 100 students who used the IIB liked this new mode of learning. More IIB updates are on the way.

Back to this non-personalized version of the book, in hardcopy or in kindle. No calculus, just 8 principles illustrating the networks we live in. We had so much fun writing it. Hope you'll enjoy reading it!

–Chris Brinton and Mung Chiang, June 2013

Acknowledgements

Co-authoring this book was one of the most enriching, and simply put, *fun* experiences that I've ever had. I am thankful to so many people who have played a large part in defining who I have become as a person today. I would like to acknowledge all of my friends, family, former professors and school teachers, as well as colleagues at all the institutions and internships I've been part of. I must point out a few of these individuals specifically.

- My undergraduate advisor, Dr. Allen Katz, who provided me with invaluable guidance through my studies, and helped me decide to pursue graduate studies in the first place.
- Five undergraduate students at Princeton – Pranav Gokhale, Ethan Berl, Harvest Zhang, Mo Luo, and Rohan Sharma – who volunteered to proofread and/or help make examples for this book.
- A few of my closest friends – Rob, Rachael, Dave, and Matt – who have always stood by my side, even when I dropped off the face of the earth for months at a time to write this book.
- My family – Mom, Dad, and Eric – who have provided me with invaluable advice and moral support throughout my life. I could not have asked for a more caring and compassionate family. Dad spent about 30 – 40 hours of his time helping to proofread the technicalities of this material.
- My girlfriend, Kate, whom I have known closely for just under two years but has already become my best friend. She kept me company countless nights when I was writing this book, while preparing her lesson plans to educate the younger generation.
- My PhD advisor and co-author of this book, Dr. Mung Chiang, who has provided me with unbounded opportunities to excel, and without whom I would have never written this book in the first place.

A big thank you to each of these outstanding individuals.

–Chris Brinton, June 2013

In 2011–12, I wrote my first textbook *Networked Life: 20 Questions and Answers*. In the Acknowledgement section there, I thanked many who made that book possible.

In addition to all of them, in preparation for this second textbook, I am grateful to those who helped with our Massive Open Online Course (MOOC) experiment since 2012, including many colleagues at Princeton University, supportive alumni such as Dr. Bill Ju '78, outstanding TAs: Mike Wang, Arpan Ghosh, and my coauthor Chris Brinton who kept pushing the envelope of education innovation tirelessly, as well as many online students who participated at our Virtual Office Hours, discussion forums, and course wiki.

The biggest and most joyful difference the past 9 months have made was the arrival of Augustan Chiang. There is no better inspiration than the little August and his big sister Novia.

–Mung Chiang, June 2013

1 Sharing Is Hard

For a few seconds, sit back and imagine how your life would be without a cell phone or a computer. This may be difficult, since it would require you to revert to landlines every time you have to contact your boss, family, or have any conversation with anyone remotely. But let's go back even further, to even before you could have a telephone in your home. At this time, information could only travel as fast as you or a mailman could walk from street to street, as a train could travel from city to city, or as a freighter could cross the sea from continent to continent. Though hard to imagine, this is how people lived for thousands of years before revolutionaries laid the foundation of wireline and wireless communications.

The main focus of this chapter will be two types of wireless technologies: cellular and WiFi networks. Brief histories of both will be given and the major engineering marvels explained. In doing so, we will see two ways in which we "share" the resources (in this case the air) by which we communicate. Managing interference is really the key: it requires methods to figure out when you communicate and how loudly you communicate, completely behind the scenes.

1.1 Power control in cellular networks

Cell phones are now part of our daily lives. Take Figure 1.1, which shows the **mobile penetration** for selected countries in late 2012. This is the average number of cellular subscriptions per person in the country. Notice that the leftmost six are over 100%, meaning there are *more cell phone subscriptions than people* in these countries.

The ten countries in Figure 1.1 were the ones in late 2012 with over 100 million subscriptions. And if we took the worldwide total number of purchased cell phone subscriptions at this time, it was over 6,000,000,000.

With these enormous numbers, you might ask, how is it possible that we can all communicate effectively with one another without disrupting each other's calls, messages, or Internet usage?

The modern mobile **cellular** system is the end result of decades of

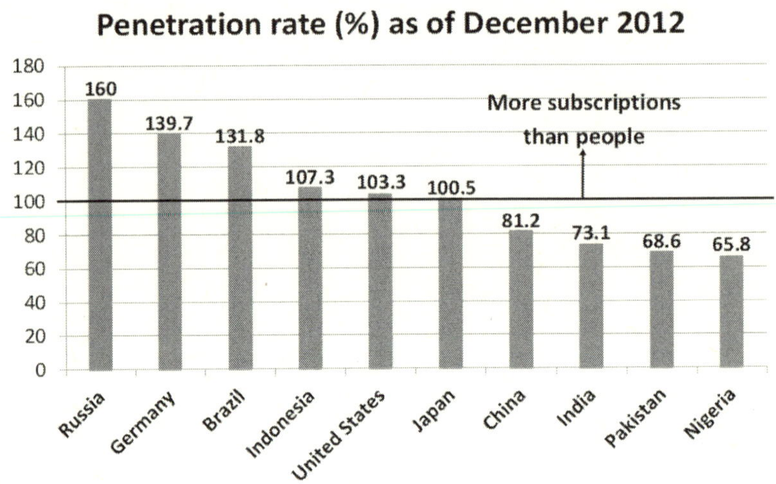

Figure 1.1 Mobile penetration, or percentage of cellular subscriptions per person, in selected countries as of December 2012. These are the ten countries that had over 100 million total subscriptions at that time. Six countries have a penetration of over 100%, meaning there are more subscriptions than people. In particular, Russia has an average of 1.6 phones per person.

technological innovation. And as mobile devices moved from being a luxury item in the 1940s–80s to an utter necessity by the 21st century, engineers had to come up with different ways of letting people share the air.

1.1.1 From telephone to cell phone

Before the advent of the Internet and the cell phone, communication networking was really all about **wireline** phone calls. Wireline is simply communication using wires, as opposed to the more commonly used term **wireless**. Though what we conceive of as a "cell phone" was barely available before the 1990s, the first phone call was made all the way back on October 9, 1876, from the famous Alexander Graham Bell to his associate Thomas A. Watson, over a two-mile wired stretch from Boston to Cambridge. Bell is widely credited with the invention of the telephone. People of Bell's time understood that this was a significant achievement in world history, but nobody could have imagined its impact; the starting point of telecommunications innovation stretching to the present day.

In 1877, Bell founded the Bell Telephone Company, the first company

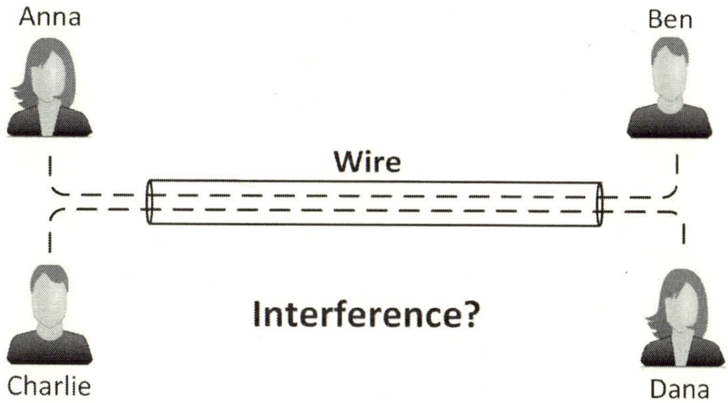

Figure 1.2 Suppose Anna and Ben are having a phone conversation, as are Charlie and Dana. How is it possible for both pairs to use the same wire without interfering with one another? We will look at a few ways this can be accomplished in this chapter.

that provided a **public switched telephone network** (PSTN) service (we commonly refer to PSTN as simply "landline"). In the same year, it was providing service to just about one thousand homes. Running the network initially required a lot of manual work, as operators acted as human "switches", completing the physical connections between each end of each call.

Even before Bell designed the telephone, he was experimenting with the telegraph, an earlier invention. He designed the "multiple telegraph," where multiple **transmitters** (senders of messages) and **receivers** (receivers of messages) could operate over a single wire.

Stop there for a second: how is it possible for us to have many different people sharing the *same* wire, as shown in Figure 1.2? If Anna and Ben are trying to talk to each other, as are Charlie and Dana, wouldn't this cause them to disrupt each other's conversations?

Not necessarily. Even though they are sharing the same *space* (the wire), we can separate them along some other dimension. The most intuitive one is perhaps *time*: let Anna and Ben use the wire for a bit, then Charlie and Dana, then back to Anna and Ben, and so on. But that would be really inefficient, especially as we kept adding more people to the same wire. We'd need some way to allow people to talk at the same time. But is that even possible?

Well, what about separating by *pitch* instead? Have Anna and Ben talk in a really high tone of voice, and Charlie and Dana talk in a really

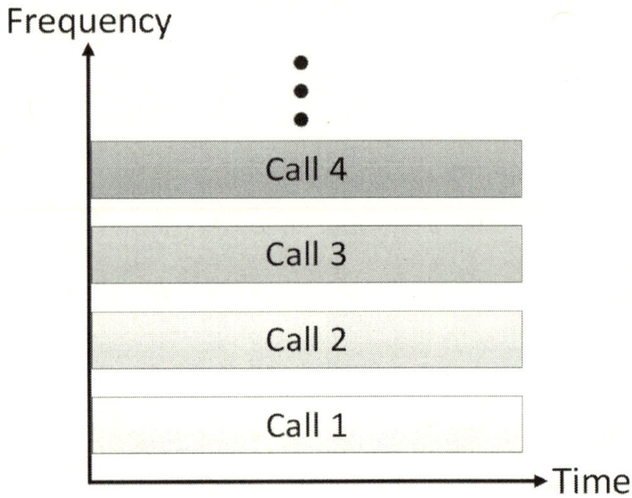

Figure 1.3 The basic concept underlying FDMA. Calls are disginguished by their designated frequency channels: the users on Call 1 get one channel, those on Call 2 a different one, and so on. FDMA was the only multiple access technology available for mobile networks until the 1980s.

low one. If we could somehow separate these pitches far enough, to the point where each pair couldn't hear the other, we'd be in good shape. Both parties could talk at the same time, without a problem.

But human vocal chords (transmitters) and ears (receivers) cannot actually operate in separate-enough pitch ranges that would enable sharing the air. So what about separating by *language*? Let one pair talk in English, and the other in Spanish. Then, they can just listen for their own language while still talking at the same time.

Using these three dimensions – time, pitch, and language – as such, we allow multiple users to share the same transmission medium (e.g., wire or air). In fact, each of these cases is a simplified example of a different **multiple access** technology, which we will investigate in more detail throughout this chapter.

FDMA

Now, back to the multiple telegraph. With this device conversations were separated along different *frequencies*, in what is known as **frequency division multiple access** (FDMA). FDMA allocates each transmitter-receiver pair, also known as a **link**, a separate frequency by which they can communicate. We illustrate this in Figure 1.3.

FDMA is how AM and FM radio operate. You can picture it if you

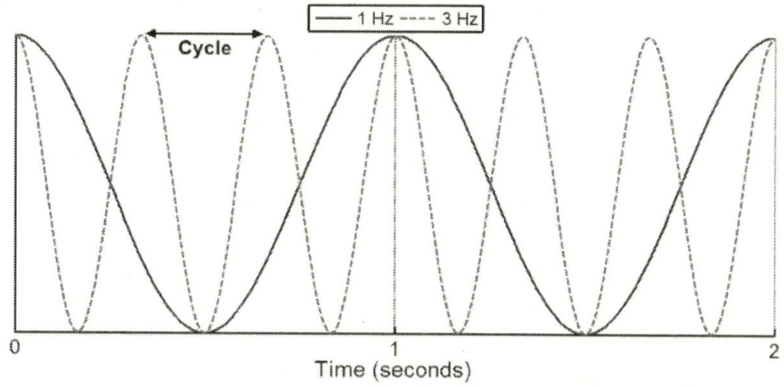

Figure 1.4 Two waves of different frequencies. The frequency is the number of cycles the wave completes in one second. One "cycle" implies returning back to the same point, for instance, from high to high, as we show here. The solid wave completes one cycle in one second, and hence is 1 Hz, while the dashed completes three, and is 3 Hz.

imagine yourself tuning your car radio (the receiver) to different frequencies in order to switch radio stations. Each radio station is a different wireless transmitter, all of which may be broadcasting simultaneously. The way they are distinguished is through their designated frequency.

So what is a "frequency?" For the ones we can hear, we distinguish them as different pitches of sound. The standard unit of measurement for frequency is the **Hertz** (Hz), which indicates the number of cycles per second in a wave. So 10 Hz means that a wave completes 10 cycles per second, 20 Hz implies 20 cycles per second, and so on. We illustrate this in Figure 1.4.

The unit of frequency will be mentioned many times in this book, but the actual ranges we will deal with are many orders of magnitude higher. Wireless frequency bands are typically in millions of Hertz, known as Megahertz (MHz), or billions of Hertz, known as Gigahertz (GHz). To give you an idea of just how large this is, the highest frequency a human can hear is about 20 Kilohertz (kHz), or 20 thousand Hertz.

As we said, with FDMA, frequency is the resource split among different users. Each connection uses a different **frequency channel**, which encapsulates a range of frequencies, defining the **channel width**. For instance, if the channel for our call was between 40 MHz and 60 MHz, its width would be $60 - 40 = 20$ MHz. But we only have a certain

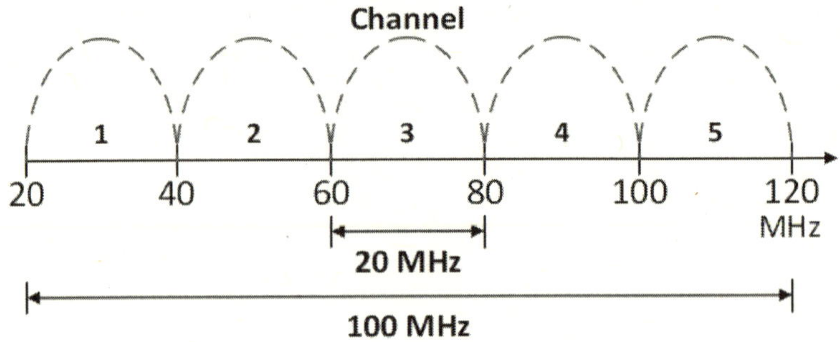

Figure 1.5 Each frequency channel has a certain width that it needs to operate. Here, we show the spectrum between 20 and 120 MHz, or 100 MHz total. If each channel requires a width of 20 MHz, this means we can have at most $100/20 = 5$ channels, as shown.

amount of frequency that we can divide into channels to start with. As such, there is a limit to the number of links the network can support at one time. We illustrate this idea in Figure 1.5.

The first mobile phones, dating all the way back to the 1920-30s, used FDMA. They were completely **analog** in nature (as was the original telephone network, and as is radio), meaning that their signals traversed the air in their exact electrical forms. But while they were "mobile," they were nowhere near "handheld:" Weighing several pounds, they could only be installed in automobiles. Additionally, they cost a fortune and weren't affordable by the average person.

Phones are great, but pretty useless unless we have a *network* to connect them. In 1946, the first "mobile-phone" network, called the Mobile Telephone Service (MTS), was introduced by Bell Telephone. This was an FDMA system, which linked to the wireline telephone network. Then, in 1964, it was replaced by the Improved MTS (IMTS). This was still using FDMA, but added a new feature: direct dialing.

Both MTS and its successor used 32 channels, each in the frequency range of 40 to 460 MHz. And both are considered zero generation, or **0G**, technologies, as opposed to the **3G** and **4G** technologies we use today.

The first handheld

Back in the 1970s, the head of Motorola's car phone division, Martin Cooper, was convinced that handhelds were the wave of the future.

To this end, in 1973, he and his team spent a mere 90 days making the first-ever mobile handheld phone: the DynaTAC. And on April 3

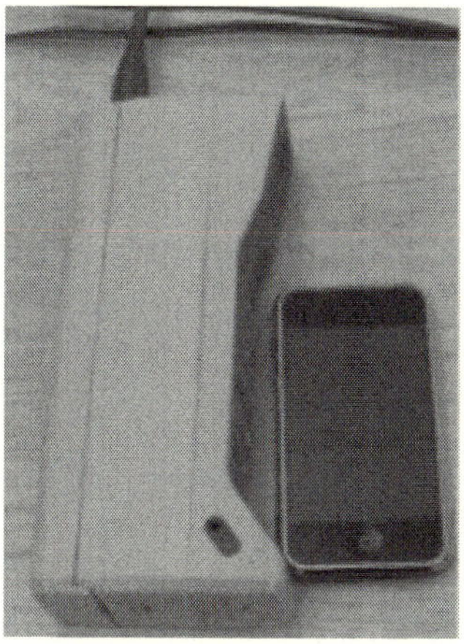

Figure 1.6 Visual comparison between Motorola's DynaTAC 8000X, made first in 1973, and Apple's iPhone, made first in 2007. Weighing close to 2 lbs, the DynaTAC cost almost $3000 (in 1973 dollars) and offered only 30 minutes of call time before charging was necessary. By comparison, an iPhone in 2012 weighed less than one-third of a pound, cost as low as $150 (depending on the model and the wireless contract), and offered many hours of voice and data applications after each recharge.

of that year, he made the first call from his phone to his counterpart Joel Engel at Bell Labs.

But the DynaTAC was not what we view today as a "handheld." It was quite hefty, and weighed almost 2 lbs. We show a comparison between the DynaTAC and the first iPhone in Figure 1.6.

In fact, it wouldn't be until the mid-1990s that the industry would really start switching from car phones. This was due to the weight, short battery life, and sheer cost of handhelds: even in the 1980s, a car phone still cost over $3,000. Similar to digital networks, palm-sized phones only became practical once the size and cost of electronic components began to reduce drastically, which is in turn partially driven by the volume of demand, which is in turn partially driven by the applications enabled by such technology.

The (non-biological) cell

By the 1970s, mobile networks were bursting at the seams. In 1976, New York City alone had about 500 mobile subscribers, and more than six times this number on the waiting list. Obviously, there was a dire need to increase network **capacity**, or the maximum number of users that can be subscribed to a network. So what could network opera-

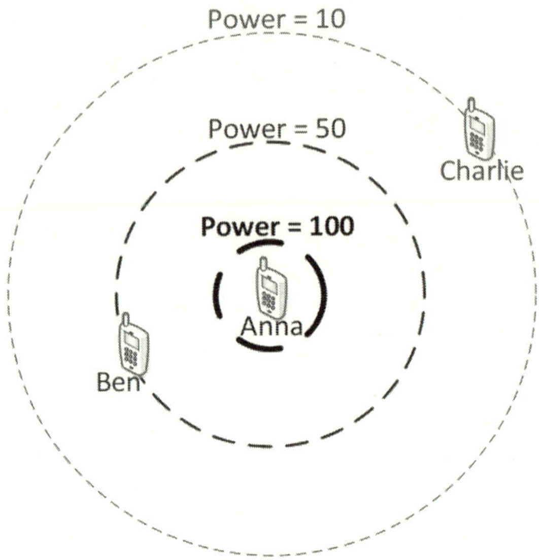

Figure 1.7 As signals propagate through space, their power levels attenuate. The farther away two antennae are from one another, the more dramatic the attenuation will be. This diagram illustrates a transmitter, Anna, sending her signal, and two receivers, Ben and Charlie. Around Anna's phone, the power level is 100. By the time the signal gets to Ben, it's 50, and once at Charlie, it's 10. You could think of attenuation as happening in this radial fashion: on any given circle, the power levels will be the same.

tors do? Well, there were really only two options: petition the Federal Communications Commission (FCC) for more spectrum, or figure out a way to fit more users into the same spectrum.

The FCC was established in 1934 to oversee the development of communications technologies in the US. As such, it has been in charge of *licensing* spectra to service providers. Some portions of the spectrum, like the part WiFi uses, is unlicensed and can be used by anyone. But most portions, including that for cellular, require a license to use. Operators need to go through a long, grueling process with the FCC to get these licenses. Depending on the number of petitioning operators and the availability of additional spectra, this process can take years and be quite expensive.

So, maybe the operators petition the FCC at this point, and wait a few years. But in the mean time, it's necessary to explore the second option.

How can we put more users in the same spectrum? Is it possible to *reuse* some of the channels? That seems kind of farfetched: if we have two links right next to each other, and they are using the same channel, certainly they would interfere.

But what if they *aren't* right next to each other? If they are far enough away, can we reuse the same channel?

The answer is yes. When signals propagate through the air (as well as through wires), their power levels **attenuate**. This means that they diminish with increasing distance, as shown in Figure 1.7. The precise *amount* of decrease is dependent upon the environment: outdoors, it's much less than indoors, since there's much less "stuff" to bump into.

Typically, attenuation is looked upon as a negative quality. It causes a signal to weaken, making it harder to transmit over long distances. But it's exactly what we need here: if you and I are far enough away, we can each make a call without interfering with one another. In fact, without attenuation, you would be able to hear other people's conversations miles away, which is a clearly undesirable scenario.

The concept of attenuation led engineers to begin dividing mobile regions geographically into **cells**, often represented as "hexagons" in diagrams (and for good reason, beyond our scope here). The idea is that any given cell can be assigned a set of frequencies that are *not* being used by a cell adjacent to it. In this way, cells that are using the same channel will be far enough away from each other that it won't matter.

We show an example in Figure 1.8. Any of the hexagons with the same color will be assigned the same frequencies, since they are not adjacent. So maybe green/darkest gets channels 1 - 4, blue/second darkest gets 5 - 8, red/second lightest gets 9 - 12, and white/lightest gets 13 - 16 (color vs. shade depends on whether you're reading the color version or not). So, suppose Robert is in a red (second lightest) cell and is on channel 2. If Rachael is in a different red (second lightest) cell, she can also be assigned channel 2, since she is far enough away. Of course, someone like Ben in a blue cell cannot get channel 2, because they are too close to the red cells. But still, this technique saves a lot of space, allowing us to reuse channels.

This "coloring" problem will always have a combination that uses the least amount of colors (frequency bands), which is naturally the best scenario. But finding such a combination can be quite difficult, especially as the number of cells gets very large. The minimum number of frequency bands that is possible is known as the **frequency reuse factor**. In Figure 1.8, the frequency reuse factor is 4.

So, what's in a cell? Each one has six **base stations** (BSs), each of

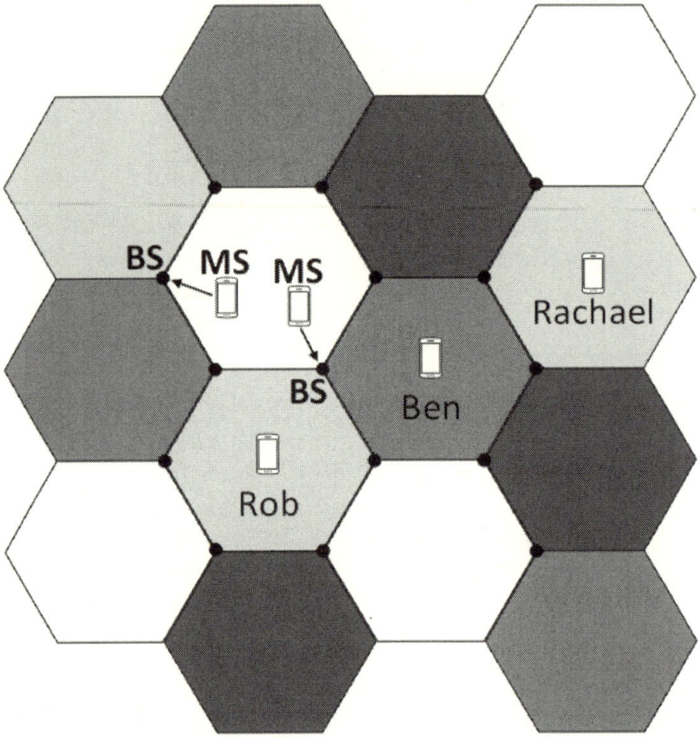

Figure 1.8 Cellular network with a frequency reuse factor of 4 (color-coded/shaded accordingly). Each cell is a hexagon, with multiple mobile stations, and six base stations that each sit at the intersection of and serve three cells. Some MSs are close in proximity to the BS they are connected to, while others are further away.

which sits at the intersection of three cells. Each BS is connected on one side to the wired core network and the Internet, and on the other to each **mobile station** (MS) assigned to it. An MS could be a cell phone, tablet, or any device that can transmit and receive according to a cellular standard. There are a few other names for them; for example, in some industry standards an MS is also called a User Equipment (UE) and a BS called a Node B (NB). Our focus in this chapter will be on the air-interface between the BSs and MSs.

Cells were first used in the Advanced Mobile Phone System (AMPS), which marked the beginning of **1G** technology in the US. It was an analog network that operated in the 800 MHz frequency range, with over 400 channels. Bell System's first field trial of AMPS in 1978 consisted of ten cells in Chicago, which covered 21,000 square miles and had 90 participating employees. This trial demonstrated the feasibility of the cellular concept.

Under AMPS, the number of mobile subscribers skyrocketed. By the 1990s, there were 25 million cellular subscribers in the US alone. Unfortunately, this was a double-edged sword: due to its low capacity and high usage rates, analog just could not cut it anymore.

1.1.2 Digital cellular

With analog networks once again bursting at the seams, the US and other countries began experimenting with an alternative: **digital** systems. An analog signal is "digitized" by converting it into a sequence of **bits**, or 1s and 0s, before transmission (see Figure 1.9).

Digital systems can offer enormous advantages in terms of capacity, as they enable use of two other multiple access technologies that we will discuss shortly. Prior to the late 1980s, the small scale electronics necessary to realize these networks were just not yet available.

TDMA

The transfer from analog to digital cellular marked the migration from 1G to **2G**. The first 2G standard was the Global System for Mobile Communications (GSM), which began in 1982, and by 1987, it was confirmed to have three times the previous analog capacity.

So, where did this three-fold increase in capacity come from? Well,

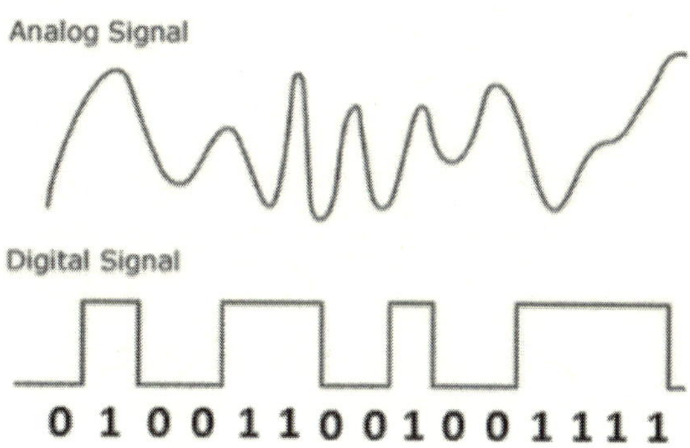

Figure 1.9 Illustration of the difference between analog (top) and digital (bottom) signals. An analog signal is one that varies continuously in time. A digital signal on the other hand is a series of bits, or 1s and 0s. Analog signals are converted to digital form by a process known as sampling.

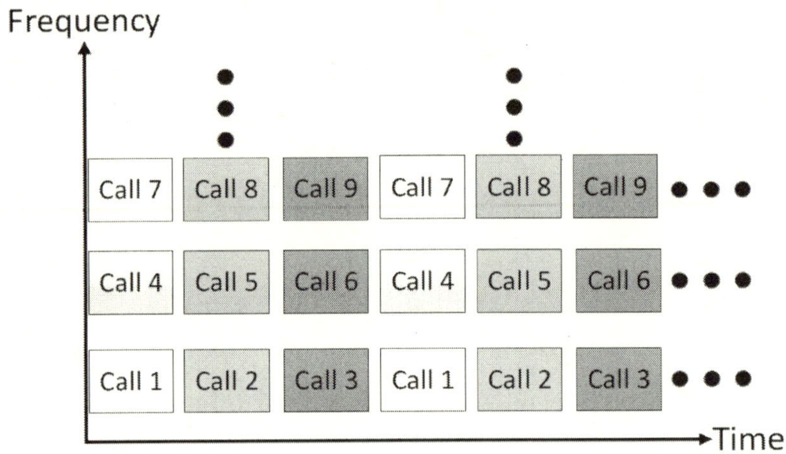

Figure 1.10 With TDMA, a certain number of calls (in this case three) can share the same frequency channel. For instance, calls 1, 2, and 3 are assigned the same channel, but are separated in time.

digital coding has another great advantage: it allows us to *compress* multiple conversations into one band. So, even within one cell, we could have a number of people sharing the *same frequency channel.* We just have to add another one of our dimensions into the mix, to make sure we distinguish between them. In this case, the dimension chosen was *time.*

The idea is as follows. We can pack multiple users onto the same frequency channel, but they all have to take turns: each is allocated a different *timeslot*. This scheme is known as **time division multiple access** (TDMA), in which time is resource that's shared, by dividing it into slots. TDMA was not possible in purely analog networks, because analog signals cannot be compressed onto the same channel directly. But with digital technology, GSM was able to accomplish this.

We illustrate TDMA in Figure 1.10. Clearly, TDMA provides capacity improvements over FDMA, since there can be multiple users on each frequency. GSM, in particular, allows up to eight users to share a single frequency channel.

GSM was adopted in much of Europe rather quickly, as the European Union favored the development of one common standard. A single standard guarantees that the networks will be *interoperable*. In fact, GSM is still used in parts of the world today, operating mainly in the 900 and 1800 MHz bands. It drove down the cost of phones and marked

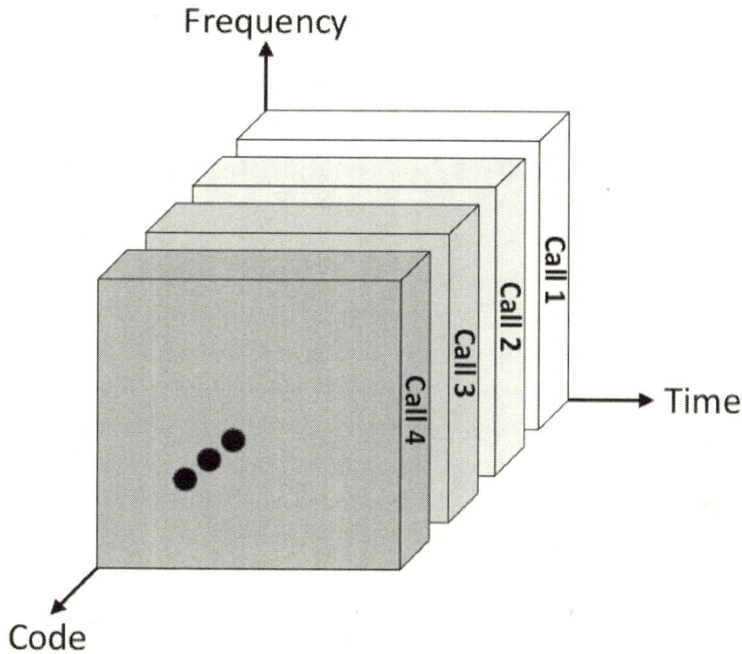

Figure 1.11 Illustration of how CDMA distinguishes between calls. While all calls may operate over the same frequencies and at the same times, they are distinguished along the "code" dimension. Each transmission in the network is assigned a unique code.

the transformation to handhelds offering texting, gaming, and other entertainment.

CDMA

The story of 2G standard adoption in the US is even more interesting. With the knowledge of increasing capacity demands, the US's Cellular Telecommunications Industry Association (CTIA) posted a set of performance requirements in 1988 that the industry should aim to meet with the first digital cellular standard. The main one was a ten-fold improvement in capacity over legacy analog networks.

By this time, virtually every network operator and device manufacturer in the US was at least familiar with TDMA. To them, it seemed like the best way to go.

But not to the company Qualcomm. They championed yet another technology, **code division multiple access**. Qualcomm had originally

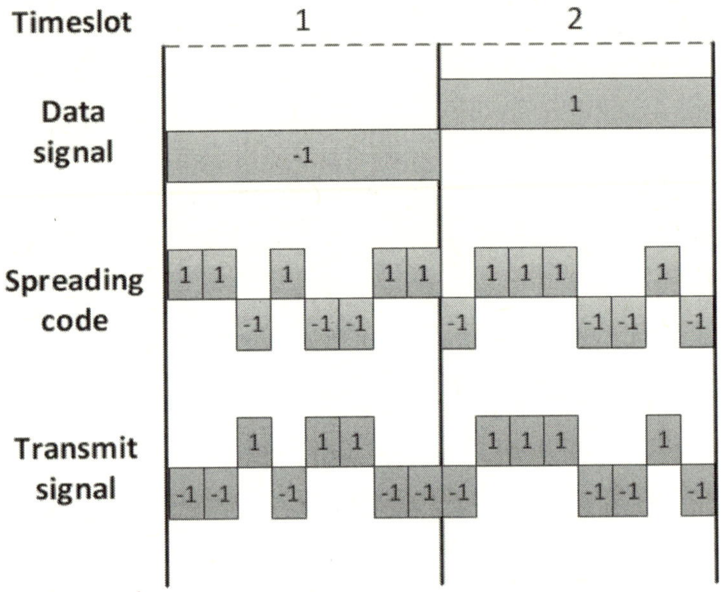

Figure 1.12 An example of how CDMA converts a data signal into the transmit signal using a spreading code. The 0 bit is represented as -1 for mathematical convenience (1 and -1 can also be thought of as "high" and "low"). In this example, the spreading code changes eight times more rapidly than the data signal. During the second timeslot, the data signal is 1: each value of the spreading code is multiplied by 1, which gives us the spreading code back. Conversely, in the first timeslot the data signal is -1: the spreading code is multiplied by -1, which makes the transmit signal the inversion of the spreading signal.

used CDMA for satellite communications, and was now pioneering its penetration into the cellular industry.

With CDMA, users are separated over the "code" dimension, and are undistinguished in both time and frequency. We show this in Figure 1.11. The best analogy to a "code" is perhaps language: it's as if you gave each link a different one to speak.

The idea behind CDMA is as follows. The transmitter multiplies the signals with a **spreading code**, which is simply a sequence of 1's and -1's. Each bit in the transmission is multiplied separately and sent as a much larger, composite stream. We illustrate this in Figure 1.12. Once the "coded" message arrives, the receiver uses the spreading code to recover the original message.

Each code is like a key. The sender locks the message, sends it out,

Figure 1.13 With CDMA, each transmitter-receiver pair uses a different code to communicate. So, frequencies can be re-used in each cell, making the frequency reuse factor 1.

and only gives the key to the receiver. The difficulty in designing such codes is that only one "key" should be able to "unlock" any given signal. If another receiver tries to descramble this message using its own, it should appear as noise. A collection of codes possessing this property, where each one "cancels" the other, is referred to as a family of **orthogonal codes**.

What did Qualcomm find so attractive about CDMA?

- Unlike TDMA, code-based wireless has a frequency reuse of *one*. Different cells no longer need to be spread along the frequency dimension. This is illustrated in Figure 1.13.
- Further, the company had an early prediction that CDMA could provide a capacity improvement of *forty times* over legacy analog networks. This far surpassed TDMA's potential, as well as the CTIA's expectations.

Despite Qualcomm's claims, most of the engineers, manufacturers, and operators at the time resisted CDMA. For one, code-based wireless was a radical shift from the more intuitive, time-based technology. Further, there had not yet been a demonstration of CDMA in even a prototype cellular network. So nobody really believed the forty-fold claim.

And in 1989, the CTIA voted and approved *TDMA* as the first 2G digital standard in the US. Most were relieved with this decision, as many had started to design and even manufacture time-based wireless phones by then. But Qualcomm was far from content. In the years that followed the ruling, they would work relentlessly to promote CDMA and demonstrate that it was a better alternative.

1.1.3 The cocktail party analogy

Having introduced three multiple access technologies up to now, let's consider a simple analogy to illustrate them more conceptually. Imagine a cocktail party taking place in a large mansion, in which there are many independent conversations occurring. Given that there are a lot of people at this party, if everyone was crammed into the same room and talking simultaneously, it would be difficult to hear what was going on in your own conversation. We leave it up to the host to determine the best way to manage this.

In this large mansion, there are many rooms. So the host decides that there can be one couple having a conversation in each room. Each couple has their room until their conversation is finished, and no other can enter until this couple has left. In this way, each person can speak as loud as he wants without overpowering another conversation, or being overpowered himself. This is an example of FDMA, where each room represents a separate frequency band. But given that there are probably many more guests than rooms, this would not be desirable to many of the couples who were not assigned a room (i.e., not assigned to a frequency channel).

To deal with this capacity issue, the host decides to allow many couples to share each room (i.e., many people per frequency band). In

Figure 1.14 The cocktail party analogy in the case of CDMA. Everyone at the party can share the same space and speak simultaneously since each group speaks a different language, hence using a different "code."

order to deal with inevitable conversation collision, has decided that each person will speak at separate times. So, in any given room, the first conversation has thirty seconds to talk, while everyone else remains silent, followed by the next group, and so on, and this repeats. Again, each person can speak as loud as he wants, because conversations will not overpower one another. This is an example of TDMA, where in each room every conversation is assigned a separate timeslot.

Rather than separating everyone into various rooms and assigning timeslots, suppose the host asks each pair to use a separate language to communicate. Then, everyone can speak simultaneously, since each pair is listening for one language in particular. This is an example of a CDMA system, where each language represents a different code (see Figure 1.14). But human languages were not designed as perfect codes. Further, volume control becomes an issue here, because everyone can hear all the other conversations. We would require some type of coordination where each person would adjust based upon their relative distances from one another.

1.1.4 Power control and handoffs

CDMA undoubtedly had its share of complications. We will now explore some of the major problems that Qualcomm had to overcome in the early 1990s.

Near-far problem

As we've discussed, whenever signals are traveling at the same time there is inevitable **interference**. It goes along with the saying that "one man's music is another man's noise". You only want to hear your conversation, and any others sharing the same space as you is undesirable, hindering your ability to decode the message meant for you.

And this problem becomes even more severe when you take distance from the base station into consideration. How can someone who may be many miles from the BS make a call without it being ruined by someone only a few meters from it? Not only does this person face more attenuation, but there are likely more objects (like trees) to obstruct its path. This leads to differing levels of **channel quality**, as illustrated in Figure 1.15.

This concept became known as the classic **near-far problem**. Our phones need some mechanism by which we can adjust our transmit powers to compensate for differences in channel quality.

Qualcomm's initially proposed solution to mitigate this was the **trans-**

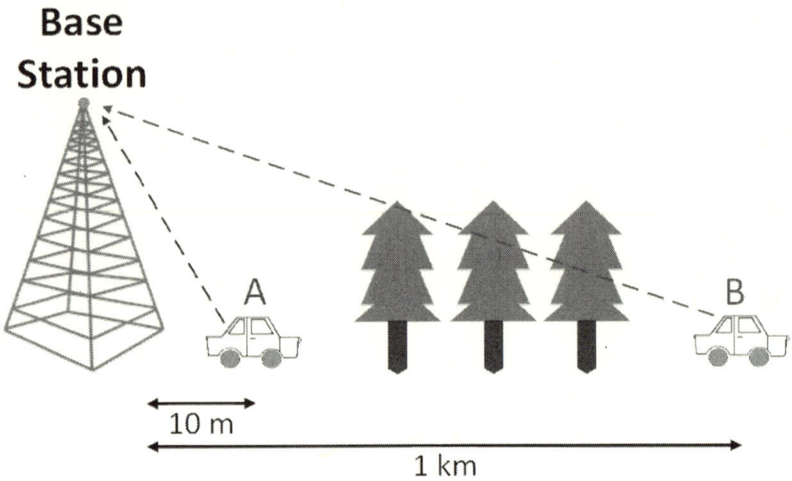

Figure 1.15 The farther a transmitter is from its receiver, the higher the attenuation, and the more objects there are to obstruct the path. As an example, we show two transmitters, A and B, who are 10 m and 1 km from the base station (the receiver), respectively. While A has a short, clear path to the tower, B has a long path which is obstructed by objects (*i.e.* trees). The initial transmit power control algorithm proposed by Qualcomm compensated for this by adjusting the transmit powers accordingly.

mit power control (TPC) algorithm. It was an attempt to equalize *received* signal powers. The base station would measure what it was receiving from each transmitter, compare this with what it wanted to receive, and send a feedback message to each device telling them to adjust accordingly.

How is power measured at the receiver then? We need a way to quantify it. Well, the standard unit of measurement for power is the **Watt** (W), which measures the amount of energy transmitted per second. So, 5 W means that five units of energy are transferred per second. In this book, we will typically be dealing with power levels that are many fractions of a Watt, either a Milliwatt (mW), which is a thousandth of a Watt, or a Microwatt (μW), which is a millionth of a Watt.

Back to the TPC algorithm. Suppose the desired power level at the tower is 10 mW. Two cell phones, A and B, start sending at this power, and the base station receives them at 5 mW and 1 mW, respectively. We see that channel degradation has led A's power to be halved, and B's power to be reduced by a factor of 10.

To "reverse" this, we want the transmitters to send at twice and ten times their current transmit powers, respectively. That means A should

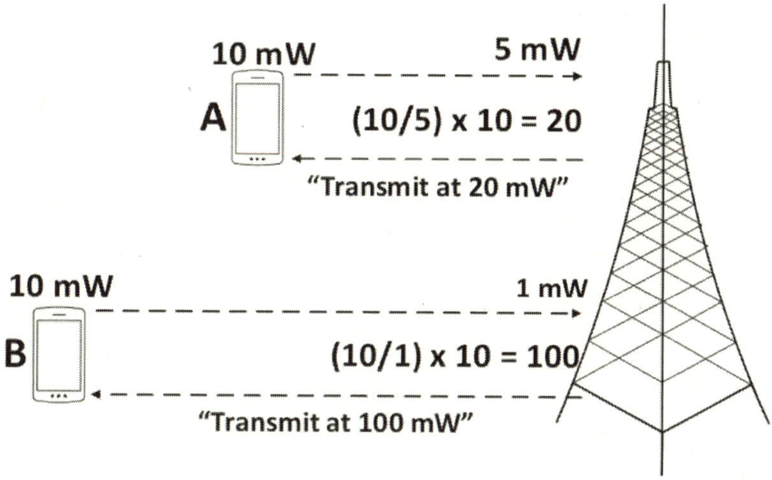

Figure 1.16 Cell phones A and B are each initially sending at 10 mW, the receiver's desired power. By the time the signals get there, they have diminished to 5 and 1 mW, respectively. Hence, the receiver tells them to instead send at double and ten times their transmit powers, or 20 mW and 100 mW.

send at 2×10 mW = 20 mW, and B should send at 10×10 mW = 100 mW. We illustrate this process in Figure 1.16.

More generally, the TPC algorithm uses the following equation when telling the transmitters what to send at:

$$\text{Next Power} = \text{“The Ratio”} \times \text{Current Power}, \tag{1.1}$$

where "The Ratio" is the desired power (10 mW) divided by the received power (5 and 1 mW).

Distributed power control

With TPC, our objective is to equalize the received signal *power*, by boosting the transmissions accordingly. Simple.

But voice calls typically need another parameter equalized: received signal *quality*. Quality is different than power because it also takes interference from other phones as well as receiver noise into consideration. This is important because while received power scales with transmit power, so does received *interference*, as shown in Figure 1.17.

How do we determine signal quality? Well, we can view it as a combination of three factors:

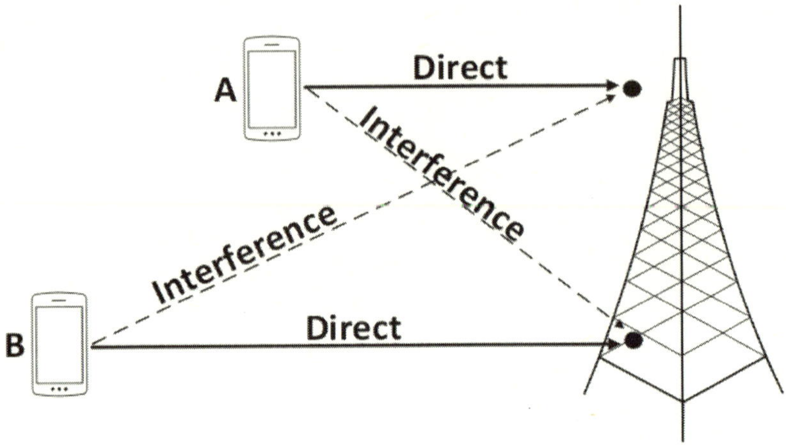

Figure 1.17 Here we depict the fact that there's always interference between channels, which affects the signal quality. We can think of the base station as having two receivers collocated. Ideally, we would like all the power from the transmitters to be "direct" to the right receiver, but this is not the case. Further, the level of interference scales with the transmit power: as A increases, this causes more interference for B, and vice versa. As a result, we need to look at equalizing signal quality rather than power.

- The received signal power from the *desired* transmitter. This is the person we are trying to listen to.
- The received signal power from the *undesired* (i.e., interfering) transmitters. These are the people we are trying to avoid.
- The receiver noise power. This is something inherent in every receiver.

The *quality* is the good (item 1) divided by the bad (items 2 and 3), which gives us the **signal to interference ratio** (SIR). We will look at how this computation is made soon.

Achieving target SIRs is more complicated than achieving target powers, because we cannot simply increase transmit powers to achieve all the desired SIRs simultaneously. After all, increasing Transmitter A's power will cause the SIRs of all other transmitters to decrease. We would then have to increase all the other transmitter powers to deal with this, which would affect everyone else, causing them to increase, and so on. The result would be an inevitable "arms race" ending in whoever could transmit the highest winning.

So, if each mobile device fixes a desired SIR, is it possible to find transmit powers for each device such that all of the SIRs are met simul-

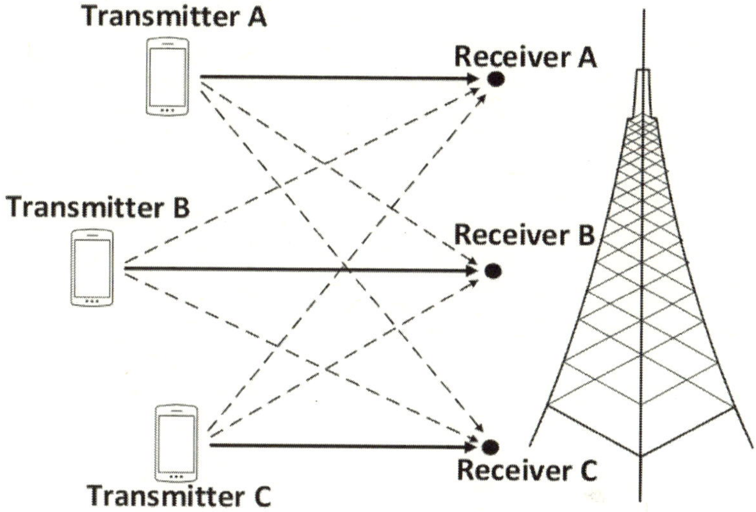

Figure 1.18 Channels between three mobile stations and the base station. The solid lines denote direct channel gains for transmitter-receiver pairs A, B, and C respectively. The higher these values, the better. On the other hand, the dashed lines represent the interference channel gains. These are generally much smaller, and the lower the better.

taneously? The answer is yes, given that the desired SIRs are **feasible.** What this means is that there is not a collective desire for "too much": the SIRs are *mutually compatible.* Users with transmission paths that are less susceptible to interference can have higher target SIRs, while users who have poorer channel quality probably need lower target values.

The solution to this SIR problem is known as **distributed power control** (DPC), and its operation is relatively simple.

1. The devices each start with some initial transmit power.
2. The receiver measures the SIR for each transmitter.
3. Based upon the ratio between the target and the measured SIRs, each transmitter adjusts its power level.
4. Repeat 2 - 3 as needed.

This process is an **iterative** algorithm that repeats over and over, unlike the single step, near-far TPC algorithm. But given that the target SIRs are feasible, the DPC algorithm has been proven to **converge**, meaning that the SIRs will be met by power levels that will stop updating. In fact, these convergent power levels will also be **optimal** ones, in the sense that they will satisfy the objective of having the lowest possible sum, thus using the least amount of total power.

Consider three MSs, A, B, and C in a single cell, as in Figure 1.18. Here, the solid lines represent the direct **channel gains** uplink for each of the transmitter-receiver pairs. Channel gain is a measure of how much the power is **amplified**, or for all practical purposes, multiplied, from source to destination (as these are fractional quantities, it's really attenuation). Direct channel gain should be as high as possible since it represents an intended conversation. On the other hand, the dashed lines represent the interference channel gains, or by what factor the unintended signals will couple to each receiver. We show the channel gains for this scenario in Table 1.1, and the target SIRs and receiver noise in Table 1.2. Also, suppose we set the initial transmit power, at time "zero", of each of the receivers to be 2 mW. Let's illustrate the DPC algorithm's computation for its first few iterations.

To compute the transmit powers for the next iteration, the DPC algorithm employs an intuitive equation similar to the near-far TPC algorithm, but involving SIRs rather than channel quality. For each transmitter, the update is given as

$$\text{Next Power} = \text{“The Ratio”} \times \text{Current Power}, \tag{1.2}$$

where "The Ratio" is the desired SIR divided by the measured SIR. This update is quite logical.

- If the measured SIR is lower than the desired SIR, then "The Ratio" will be larger than one. As a result, the transmit power will *increase*, in an attempt to equalize them.
- On the other hand, if the measured SIR is higher than the desired one, "The Ratio" will be lower than one. As a result, the transmit power will *decrease*, for two reasons. First, this transmitter can afford

Transmitter of Link	**Receiver of Link** A	B	C
A	0.9	0.1	0.2
B	0.1	0.8	0.2
C	0.2	0.1	0.9

Table 1.1 Channel gains in an example of DPC. The entries are for illustrating the algorithm. They do not represent actual numerical values typically observed in real cellular networks.

Link	Target SIR	Noise (mW)
A	1.8	0.1
B	2.0	0.2
C	2.2	0.3

Table 1.2 Desired SIR and noise parameters for each link in the DPC example. They do not represent actual numerical values typically observed in real cellular networks.

to use less power. Second, this will help improve the SIRs for the other transmitters.

- Finally, if the measured and desired SIRs are the same, "The Ratio" will be one. So, the transmit power will *stay the same*, since the target is already met.

Why is such an update necessary? Well, within a cell, each device imposes **negative externality** on all of the others, by causing them interference. In other words, while achieving its own benefit, a device does some "damage" to the rest of the network as well. The purpose of the update is to keep the devices in check: whenever the SIR is higher than it needs to be, the power level is dropped, and whenever it's too low, the power level is raised. The process of "message passing" between the base station and the devices to correct for such deviations is a form of **negative feedback**, where we try to correct for errors, as illustrated in Figure 1.19. This negative feedback forces the transmitters to *internalize* their negative externalities (e.g., pay for the "pain" they cause) by following the rules to make up for their added interference. The concepts of negative externality and negative feedback will be recurring throughout this book, in many contexts.

Now, back to our example. Out of all the quantities relevant to Equation (1.2), we know the desired SIR values from Table 1.2, and the current transmit powers. But we don't know the measured SIR, which we need to update the power. For each link, this is given by

$$\text{Measured SIR} = \frac{\text{Signal}}{\text{Interference} + \text{Noise}}. \tag{1.3}$$

How do we get "Signal," "Interference," and "Noise?" Let's start with Link A:

- *Signal*: This is just the *direct* gain from Transmitter A to Receiver A, multiplied by the transmit power. From Table 1.1, it is 0.9×2 mW $=$ 1.8 mW.

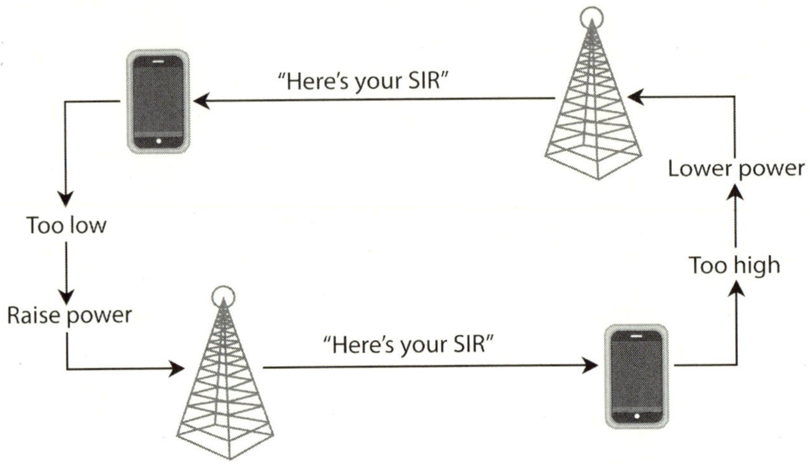

Figure 1.19 In a given cell, the tower tells each device its current, received SIR, which serves as a negative feedback signal. Using this, each device can update its transmit power independently. If it's lower than desired, the power will be raised, if it's higher, it will be lowered, and if it's the same, it won't change.

- *Interference*: This is the sum of the *indirect* gains from other transmitters to Receiver A, multiplied by their transmit powers. From Table 1.1, we look at the first column: from B to A, the gain is 0.1, and from C to A it is 0.2. The result is 0.1×2 mW $+ 0.2 \times 2$ mW $=$ 0.6 mW.
- *Noise*: This is just the receiver noise, which is given by Table 1.2 to be 0.1 mW for Link A.

So, the measured SIR for Link A is

$$\frac{0.9 \times 2}{0.1 \times 2 + 0.2 \times 2 \ + 0.1} = \frac{1.8}{0.7} = 2.57.$$

Now, let's look at Link B:

- *Signal*: From Transmitter B to Receiver B, the direct gain is 0.8. Hence, the signal power is 0.8×2 mW $=$ 1.6 mW.
- *Interference*: The indirect gains come from Transmitters A and C. From A to B, the gain is 0.1, and from C to B it is also 0.1. Hence, the interfering power is 0.1×2 mW $+ 0.1 \times 2$ mW $=$ 0.4 mW.
- *Noise*: The receiver noise for Link B is 0.2 mW.

It follows that the measured SIR for Link B is

$$\frac{0.8 \times 2}{0.1 \times 2 + 0.1 \times 2 \; + 0.2} = \frac{1.6}{0.6} = 2.67.$$

Through the exact same procedure, we can find the SIR for Link C to be

$$\frac{0.9 \times 2}{0.2 \times 2 + 0.2 \times 2 \; + 0.3} = \frac{1.8}{1.1} = 1.64.$$

Let's compare these values to the desired SIRs from Table 1.2. For Link A, the desired is 1.8, and hence the measured SIR is too *high* by $2.57 - 1.8 = 0.77$. Similarly, the measured SIR for Link B is too high by $2.67 - 2.0 = 0.67$, while that of Link C is too *low* by $2.2 - 1.64 = 0.56$.

We will now apply Equation (1.2) to calculate the power levels for the first iteration. What are the ratios? We divide the target by the measured SIRs for each link to get these: $1.8/2.57 = 0.700$, $2.0/2.67 = 0.75$, and $2.2/1.64 = 1.34$. As expected, the ratio is *less than one* for Links A and B (measured is too high), and *greater than one* for Link C (measured is too low). With these values, the next power levels will be

$$\begin{aligned} 0.70 \times 2 \text{ mW} &= 1.40 \text{ mW}, \\ 0.75 \times 2 \text{ mW} &= 1.50 \text{ mW}, \\ 1.34 \times 2 \text{ mW} &= 2.68 \text{ mW}, \end{aligned}$$

respectively. The power levels of the first two links have decreased, while that of the third link has increased, as we expected.

Now, what are the SIRs at these new power levels? Let's look at Link A:

- *Signal*: As before, the direct gain is 0.9. With the updated transmit power of 1.40 mW, the received signal power is 0.9×1.40 mW $=$ 1.26 mW.
- *Interference*: Again, the indirect gains from Transmitters B and C are 0.1 and 0.2, respectively. With the updated transmit powers of 1.50 mW and 2.68 mW, the interfering power becomes 0.1×1.50 mW $+ 0.2 \times 2.68$ mW $= 0.686$ mW.
- *Noise*: This is 0.1 mW, as before.

The updated measured SIR value for Link A is then

$$\frac{0.9 \times 1.40}{0.1 \times 1.50 + 0.2 \times 2.68 \; + 0.1} = \frac{1.26}{0.786} = 1.60.$$

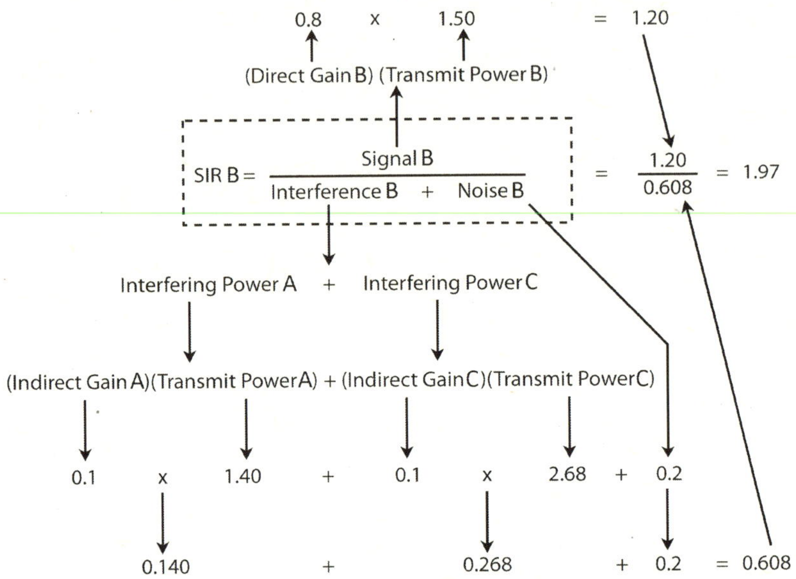

Figure 1.20 Depiction of the steps involved in computing SIR. Here, we use the values for Link B in the first iteration.

We can similarly find the new SIR values for Links B and C:

$$\frac{0.8 \times 1.50}{0.1 \times 1.40 + 0.1 \times 2.68 \ + 0.2} = \frac{1.20}{0.608} = 1.97,$$

$$\frac{0.9 \times 2.68}{0.2 \times 1.40 + 0.2 \times 1.50 \ + 0.3} = \frac{2.412}{0.880} = 2.75.$$

We illustrate the computations for Link B visually in Figure 1.20.

Each of the SIRs are closer to their target values than they were initially: Links A and B are now lower by $1.8 - 1.6 = 0.2$ and $2.0 - 1.97 = 0.03$ than the desired values, and Link C is higher by $2.75 - 2.2 = 0.55$. Initially, we saw that Link C was in need of higher SIR, while the other two links had an abundance, but now their statuses have switched. This characteristic of overshooting and undershooting the target values will likely occur in many of the iterations until convergence.

We continue now to the second iteration, this time using the transmit powers and measured SIRs from the first iteration in our calculations. The new transmit powers are

$$\frac{1.80}{1.60} \times 1.40 \text{ mW} = 1.58 \text{ mW},$$

$$\frac{2.0}{1.97} \times 1.50 \text{ mW} = 1.52 \text{ mW},$$

$$\frac{2.2}{2.75} \times 2.69 \text{ mW} = 2.15 \text{ mW},$$

resulting in measured SIRs of

$$\frac{0.9 \times 1.58}{0.1 \times 1.52 + 0.2 \times 2.15 \ + 0.1} = \frac{1.422}{0.682} = 2.08,$$

$$\frac{0.8 \times 1.52}{0.1 \times 1.58 + 0.1 \times 2.15 \ + 0.2} = \frac{1.216}{0.573} = 2.13,$$

$$\frac{0.9 \times 2.15}{0.2 \times 1.58 + 0.2 \times 1.52 \ + 0.3} = \frac{1.935}{0.920} = 2.10.$$

As before, the overshoot/undershoot pattern has reversed. Now Links A and B are too high by 0.28 and 0.13, and Link C is too low by 0.1.

The calculations for future iterations are all carried out in the same manner. Rather than continue with these computations, we graph the transmit powers and SIR levels for 30 iterations of the algorithm in Figure 1.21. After roughly 10 iterations, we can no longer see any noticeable changes in either quantity, indicating that DPC, roughly speaking, has converged to an **equilibrium** of this iterative algorithm. The measured SIRs have reached their target values of 1.8, 2.0, and 2.2, with power levels of 1.26, 1.31, and 1.99 mW.

Why does Link C have a much higher power level than the other two? This is for three reasons.

- First, it has the *highest noise component*: 0.3 mW.
- Second, it has the *highest interference gains* from other links: both 0.2.
- Third, it has the *highest target SIR value* of any link: 2.2.

Link C needs a higher transmit power to overcome these disadvantages.

We can verify that the power levels 1.26, 1.31, and 1.99 mW indeed produce the target SIRs, and that the DPC algorithm will no longer update. We apply Equation (1.3) to compute the SIRs:

$$\frac{0.9 \times 1.26}{0.1 \times 1.31 + 0.2 \times 1.99 \ + 0.1} = 1.8,$$

$$\frac{0.8 \times 1.31}{0.1 \times 1.26 + 0.1 \times 1.99 \ + 0.2} = 2.0,$$

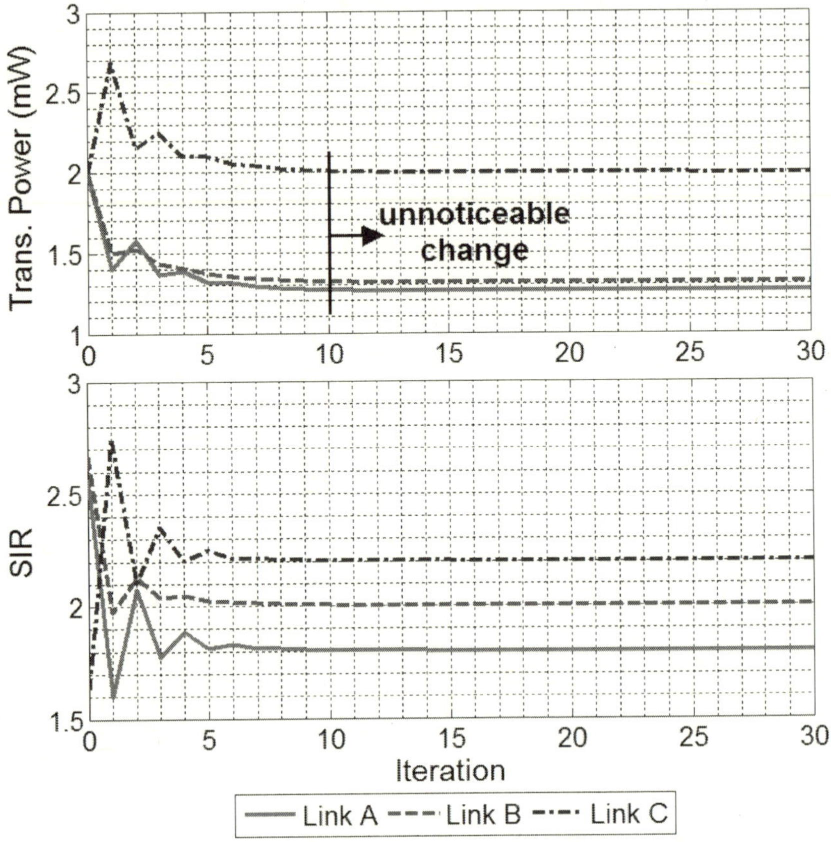

Figure 1.21 Graphs of the transmit powers (top) and the SIR levels (bottom) for 30 iterations of the algorithm. By Iteration 10, there is very little change in the graphs, indicating that the DPC algorithm has converged. We can verify that the SIRs converge to the correct values.

$$\frac{0.9 \times 1.99}{0.2 \times 1.26 + 0.2 \times 1.31 \; + 0.3} = 2.2.$$

Since the measured SIRs are the same as the target SIRs, we see from Equation (1.2) that "The Ratio" will be unity, so the power will not change. Hence, the algorithm has converged.

One benefit of the DPC algorithm is that each device really doesn't need any knowledge of how the other links are operating. To calculate its next power level, all it needs is its current transmit power, target SIR, and current measured SIR. These are all *its own* parameters, and

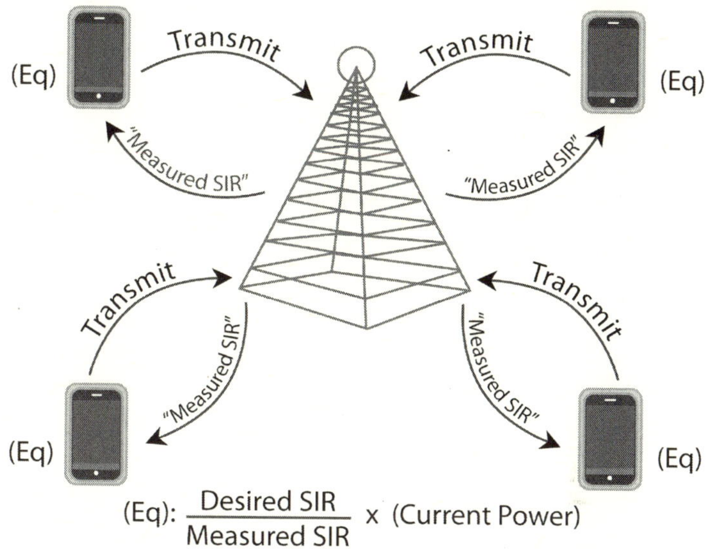

Figure 1.22 With distributed power control, each device performs its own computations, and there's no need for any information to be shared between them. As a result, we say that DPC is a fully distributed algorithm.

it makes its current decision *independently* (e.g., there's no need for it to know the SIRs of any of the other links). This allows each device to perform its computations internally, without the need to share information with the others. As a result, we say that DPC is a completely **distributed** algorithm (see Figure 1.22), as opposed to other more **centralized** ones we will encounter.

DPC illustrates a recurring theme in networking: allowing each user to make independent decisions driven by self-interest can aggregate into a fair and efficient state across all users.

The DPC algorithm is the solution Qualcomm proposed to deal with interference issues. But in a real cell, there can be thousands of phones, and as conversations are started and stopped and people move from location to location, you can imagine that the channel conditions and SIR values from link to link will change quite rapidly. As a result, it is required to have power control implemented up to 1500 times each second.

Soft handoffs

Besides interference management, there was another major issue with DPC that Qualcomm had to overcome: the way a code-based wireless

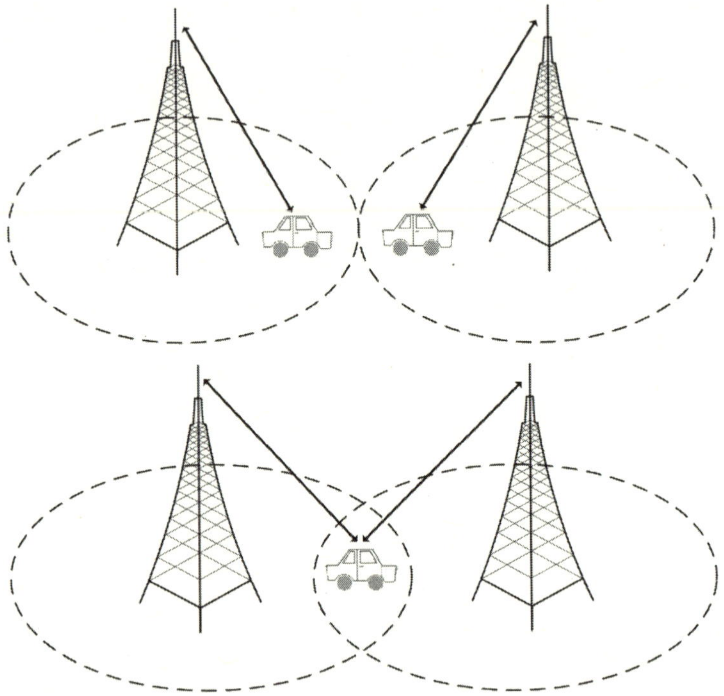

Figure 1.23 Difference between a hard handoff (top) and a soft handoff (bottom). In the case of a hard handoff, an MS is only connected to one cell tower at a time, and when it moves from one cell to another its connection is severed and picked up almost instantaneously. On the other hand, a soft handoff allows connection to be established with the new BS before the original is dropped, and hence there is coverage overlap between adjacent cells. In this diagram, a cell is drawn as an oval for visual simplicity.

network accomplishes **handoffs**, or transition of a call from one BS to another. This is necessary when a user has moved to a new location where his signal quality is better with a different BS than the one he is currently on. In analog and time-based wireless networks, when a user moves to a different cell, he switches frequency channels, and this process happens very quickly. This is referred to as a *hard handoff*. But hard handoffs don't work with code-based wireless for two reasons. First, different cells aren't assigned different frequencies with CDMA. Second, there's no way a power control algorithm such as DPC would be able to update fast enough the instant a user entered another cell. As a result, the user's call would be dropped, and whenever users entered and exited BSs rapidly, call quality for everyone would diminish.

Qualcomm's solution to this problem was the *soft handoff*, a more sophisticated technique where a user would make a connection with the next cell tower before breaking that with the original. We illustrate this in Figure 1.23. In fact, with soft handoffs, the network would constantly monitor the signal strength from each call to various towers, so that many stations were processing any given call. Hence, this technique requires coverage overlap between adjacent towers. Though it added complexity, the combination of CDMA with soft handoffs was shown to lead to even better transition quality than TDMA with hard handoffs.

Even after Qualcomm had developed the power control and soft handoff solutions, the rest of the journey to making CDMA a 2G standard was by no means easy. Qualcomm advertised and promoted CDMA as possessing various features that TDMA did not: a ten-fold improvement in capacity over analog, better security, better voice quality, and longer battery life (this last point is true because under code-based wireless, phones are never transmitting at their highest output power, whereas with TDMA they typically do in order to maximize voice quality). It took numerous large-scale demonstrations throughout the US under realistic network conditions before major network operators would endorse CDMA.

Finally, in 1993, CDMA was approved as a 2G cellular standard under IS-95, with the brand name cdmaOne. And three years later, the first large-scale commercial deployment of CDMA in the US was made by Sprint PCS. Though it has largely been upgraded to 3G standards, IS-95 and its immediate revisions (IS-95A and IS-95B) are still used in parts of the world and operate in the 800 MHz range.

1.1.5 Where we are today

The growth in the number of mobile subscriptions over the past few decades has been tremendous, as shown in Figure 1.24 for the US. The mobile penetration rate in the US has been larger than 100% since 2011.

Largely since the turn of the 21st century, 3G phones have gained momentum worldwide. The International Telecommunication Union's (ITU) 3G specifications released in 2000 essentially called for cell phones to function as handheld computers: in addition to phone calls and texting, these phones are capable of Internet access, video calls and mobile TV. The two main 3G standards are UMTS, used primarily in Europe, Japan, and China, and CDMA2000, used especially in the US and Korea. Both technologies are based on CDMA and are typically deployed in the frequency range of 1.9 - 2.1 GHz. According to the ITU, roughly

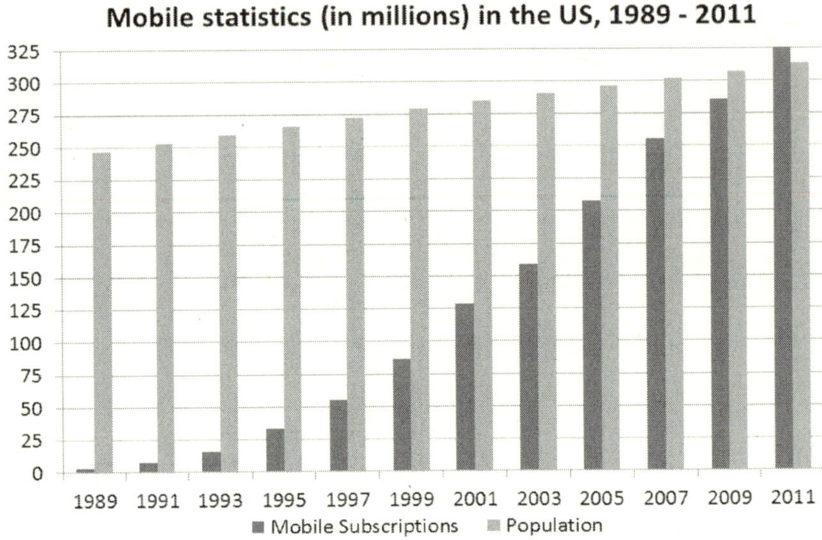

Figure 1.24 The annual number of mobile subscriptions in comparison with the total population in the US, from 1989 to 2011. This illustrates the tremendous growth of the number of cell phone users in only a few decades.

45% of the world's population was covered by at least one 3G network as of the beginning of 2012. And this number is expected to increase to 85% by 2017.

Adoption of the 3G standard and the entry of Apple into the wireless industry gave rise to the commercial **smartphone**. As of 2012, over 30% of the US's population owned one, the most popular brands being iPhone, Android, and Blackberry. We show statistics by year for the US in Figure 1.25, with predictions out to 2016.

Since 1G AMPS networks were commercialized in the 1980s, a new generation of cellular has emerged roughly every ten years. Keeping on this track, the performance requirements for 4G were released in 2008. They specified higher speed requirements and capacities than the previous 3G specifications. Since then, the major standard that has emerged is Long Term Evolution (LTE). Rather than using CDMA, LTE is based on a technology called **Orthogonal Frequency Division Multiplexing** (OFDM).

The first LTE smartphones in the US appeared in late 2011. Contrary to popular belief, LTE's current release version does not support the data rates specified in the original 4G requirements, though it is

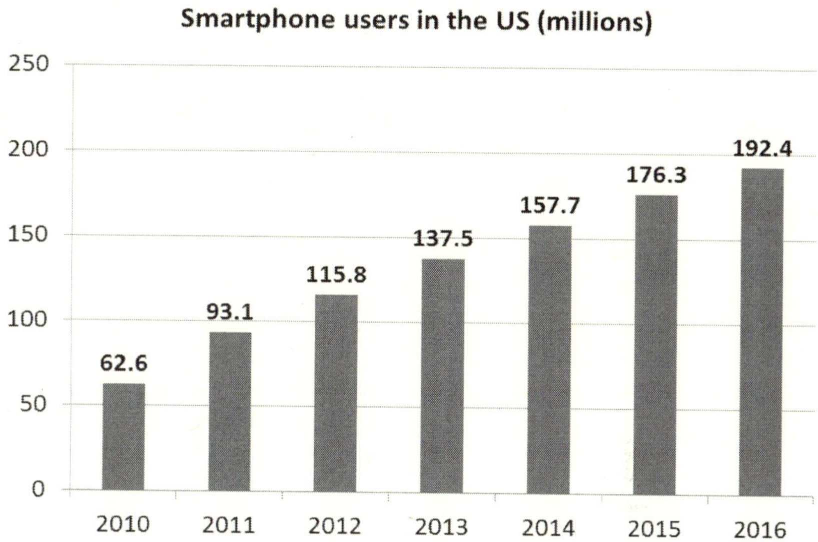

Figure 1.25 Number of smartphone users in the US with predictions out to 2016.

sometimes branded as such. However, the predicted performance improvements over 3G are expected to attract 1 billion users by 2017.

The story of cellular is a perfect example of how networks have struggled throughout the years to meet capacity demands of consumers. Different methods of sharing – frequency, time, and code-based wireless – have all been developed in accordance with our needs. Even if we don't realize it outright, real-time updating and management of the power at which our calls are operating is essential to cellular network's operation. All it takes is one phone that begins transmitting at too high of an output power to tip the balance, causing reduced sound quality for everyone else or an arms race.

In the next section, we will investigate WiFi, another type of wireless network. Rather than having stringent power control algorithms, it relies on "random access" methods to manage interference among users in the same location.

1.2 Random access in WiFi

By the mid-1980s, the second generation of cellular had gained momentum worldwide. The two competing brands – TDMA and CDMA

Figure 1.26 The trademark WiFi symbol used by the Wi-Fi Alliance.

– each revolutionized the wireless industry, bringing about much needed improvements in capacity.

At the same time, engineers began wondering whether they could think of a fundamentally *different* way to share the air. Efforts in this direction led to the invention of the WiFi technology that we all know and love today (Figure 1.26).

1.2.1 Unlicensed spectrum

Over the years, engineers have faced one major obstacle to operating cellular networks. You need a *license* to operate those frequencies. And governments around the world sell these licenses for tens of billions of dollars, which is typically only affordable by major carriers like Verizon, AT&T, Orange, and Vodafone.

But in 1985, the US FCC officially released some spectrum to the public that did *not* require a license. These bands, centered around 2.4

Figure 1.27 WiFi provides wireless connections to devices (i.e. laptops, phones, tablets) that are close to an access point. This contrasts with cellular, which operates over long distances.

Figure 1.28 Traffic scenarios tend to offer great analogies to communication networks. We can illustrate the main differences between cellular and WiFi by contrasting stoplights with stop signs. Stop signs (WiFi) are much more efficient when the traffic volume is small, but are not scalable as the number of cars (devices) increases. In these situations, we prefer stoplights (cellular) which regulate traffic flow by dedicating the intersection (the resource) to one side at a time.

and 5.8 GHz, are called the Industrial, Scientific, and Medical (ISM) frequency bands.

As the name suggests, the ISM bands were originally established for use by industry, science, and medicine. But engineers were quick to seize the opportunity. And not only for communications. In fact, the most commonly encountered ISM device today is the microwave oven, because these frequencies happen to excite water molecules well.

What if we could create small stations to provide wireless Internet connection to people who don't move too fast? This way, they wouldn't have to contend with cellular users in the first place. This idea forms the basis for the technology **WiFi**. Rather than use the ISM bands for cellular communications, they are used to provide wireless to devices that are *stationary* and *close* to the *access point*. We illustrate this in Figure 1.27.

Traffic light vs. stop sign

To start off, let's consider a simple analogy to car transportation. These systems tend to offer great comparisons to communication networks, as we will see repeatedly thoughout this book.

Suppose you're driving in your car, and you approach an intersection as in Figure 1.28. Perhaps the intersection is governed by a traffic light. We can draw a comparison between this and cellular networks: while the light is green, the intersection is *dedicated* to your side, just like time, frequency, and/or codes are allocated to callers for the duration of their sessions. The intersection is the resource, and that resource is dedicated while it's being used.

As long as there's traffic volume on the other side, the red light makes sense. But what happens when you're stuck at it and nobody is driving by the other way? That seems pretty wasteful (and quite frustrating). Why shouldn't you be able to drive through if nobody is coming? In such situations, *stop signs* are a more efficient way to regulate traffic. When we approach one, we stop, look both ways, and then move on through as long as nobody is coming the other way. We just need a form of *coordination* (stopping to look) to minimize the chance of an accident.

With the stop sign, rather than *dedicating* the intersection, we let everyone *share* it as long as they follow the rules. This is more efficient when there is a small amount of traffic, to prevent wasted periods of time where you're waiting for no reason.

Stop signs are quite similar to the way WiFi works: there's no dedicated resource. We just have devices "listen" (i.e., "look both ways") before transmitting to prevent colliding with someone else.

But this method becomes highly problematic as traffic volume increases. When lines build up, the nature of "stop" and "go" one at a time can cause extremely long waits. This is especially true when the other side doesn't have a stop sign. In these situations, the dedicated nature of stoplights may be preferred. As we will see, WiFi has the same problem: its performance degrades drastically as the number of devices increases.

WiFi standards

Now, some more on WiFi's background. Officially, it's called **IEEE 802.11**, because of the scheme that the IEEE uses to name their standards. The 802 part is for Local Area Networks (LANs), echoing the fact that it is short-range, and the .11 part is for wireless. The more attractive name Wi-Fi, short for "Wireless Fidelity," was coined and stuck.

Various groups began developing WiFi technology. While spurring innovation, this caused a problem because a lot of the devices designed by different people weren't *compatible* with one another.

As a result, they needed a way to ensure *interoperability* among products claiming to use WiFi. The WiFi Alliance was established in 1999

Standard	Year	Frequency (GHz)	Maximum Speed (Mbps)
-	1997	2.4	2
b	1999	2.4	11
a	1999	5	54
g	2003	2.4	54
n	2009	2.4 & 5	100

Figure 1.29 Illustration of the progression of WiFi standards over time, and some of their characteristics. The speed is assuming ideal channel conditions. There have been other variations that are not listed here.

to do just that. If they stamp their logo (Figure 1.26) on your device, it means it conforms to an IEEE 802.11 standard.

Just like cellular, WiFi was improved significantly over a small time period. Various upgrades have increased customers' connection speeds significantly. We measure transmission speed as a *rate*, meaning per second. The unit for this is the number of transmitted **bits per second**, or bps. WiFi speeds today are on the order of millions of bps, or Mbps.

The first WiFi standard was introduced in 1997, which offered a speed of 2 Mbps at 2.4 GHz. To make the naming convention even more confusing, each WiFi upgrade adds a letter to the end of 802.11, but not in chronological order:

- In 1999, 802.11b was released, which uses the 2.4 GHz band and can transmit up to 11 Mbps. In the same year, 802.11a was introduced, transmitting up to 54 Mbps at 5 GHz band.
- In 2003 we moved to 802.11g, which improved the 2.4 GHz speed to 54 Mbps.
- More recently, 802.11n released in 2009 has pushed the transmission rate to over 100 Mbps.

We summarize this progression in Figure 1.29.

With improving transmission speeds, the demand for WiFi services continues to increase dramatically. By 2011, there were already more than a billion WiFi devices around the world, with hundreds of millions being added each year. We show the approximate number of WiFi-

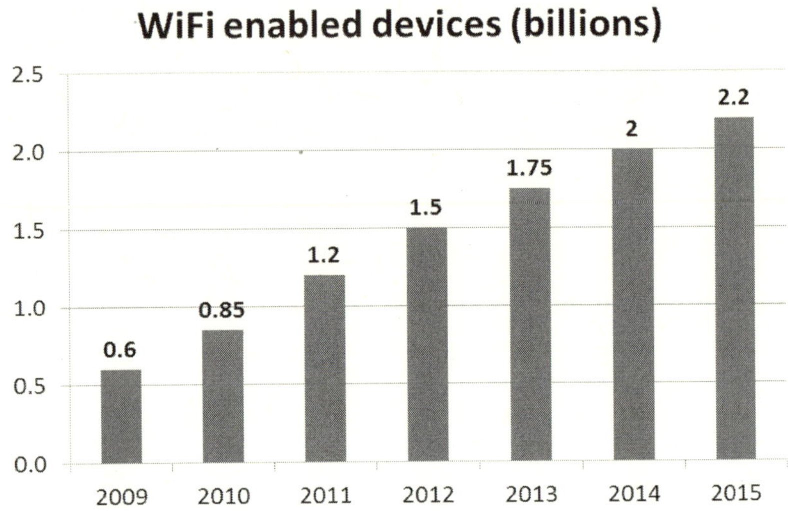

Figure 1.30 Approximate number of WiFi-enabled devices (billions) in the world, by year. It hit 1.5 billion in 2012, rising from roughly 900 million in 2010. Looking forward to 2015, it is expected to hit 2.2 billion.

enabled devices in the world, by year, in Figure 1.30. By 2015, this number is expected to reach 2.2 billion worldwide.

WiFi deployment

As we've said, WiFi's method of "sharing the air" is very different than cellular's. But the way it's deployed is quite similar.

- Users in WiFi networks lie in what's called a **Basic Service Set** (BSS), or **Extended Service Set** (ESS), as opposed to a cell.
- Within each BSS or ESS, users communicate directly with an **Access Point** (AP), similar to a base station.

We illustrate the WiFi network layout in Figure 1.31. When your device searches for WiFi connectivity, it sends messages to discover which APs are in its transmission range. This results in the list of names that you can choose from on your screen, which are the user-friendly names used to identify a service set. These names are more commonly known as **Service Set Identifiers** (SSIDs). You've probably been frustrated at one time or another, because often you see a lock next to some of the SSIDs with good signal strength. If it's password protected, your device can *associate with* (i.e., connect to) the AP only if you have the

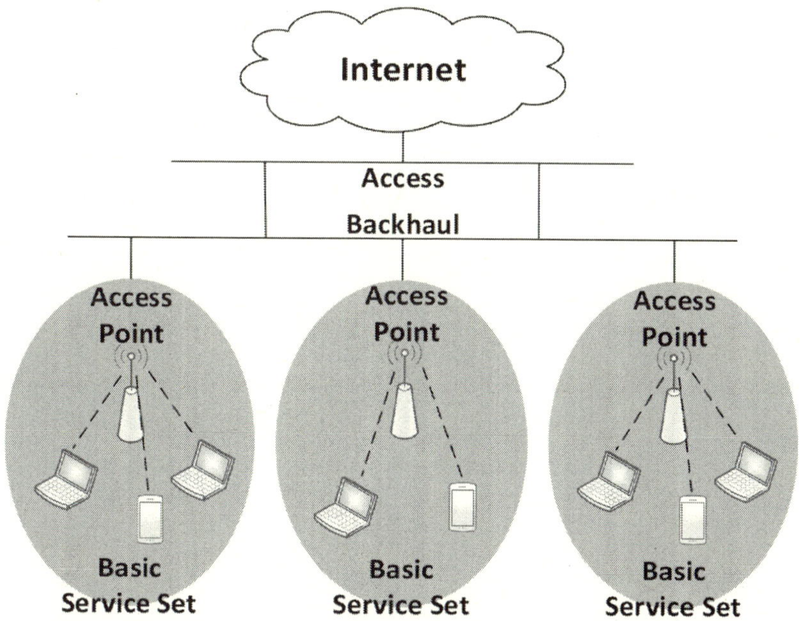

Figure 1.31 A WiFi deployment. User devices transmit over the air to a designated Access Point. Users and their AP constitute a Basic Service Set. Each of the BSSs are connected to a wireline backhaul, often Ethernet, which is connected to the rest of the access network. The access network links to the rest of the Internet.

password to authenticate your status, e.g., as a paying customer if the AP is part of the WiFi subscription service.

Now, what is the difference between a BSS and an ESS? A BSS would be in a location where there is a *single* AP, like a home (as we will see in a minute), small office, or a coffee shop. An ESS is very similar, but for larger deployments where multiple APs would be needed to provide adequate service to the users, such as a hotel or an office building. All of the APs in a single ESS will offer the same SSID.

Back to Figure 1.31, each AP is typically connected to what's called the backhaul. This is a wireline deployment, often **Ethernet**, which is another (and much older) IEEE 802 family member. Remember, "802" means local area network: Ethernet is a technology used for *wireline* LANs. From this, the Ethernet is connected to the access network, which can ultimately provide a connection between the WiFi deployment and the rest of the Internet. The backhaul is often provided by a

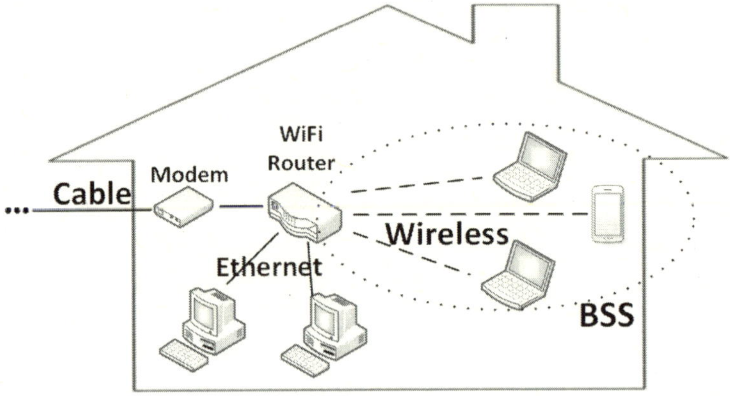

Figure 1.32 Typical home WiFi deployment. Your house probably has a wireless router, which is a router with an integrated access point (we will discuss routers in more detail in Chapter 6). Both wireless and wired devices connect to it. The wired devices, like your desktop, connect to it via Ethernet cables. The router is connected to a modem, in turn to the access network like DSL, cable, or fiber. Note that more recently, all three devices – the modem, router, and AP – may be found within the same device.

digital subscriber line (DSL), cable, or fiber service: all terms that we commonly hear.

To illustrate these terms more clearly, we give a specific example in Figure 1.32: a WiFi setup in a home. Your wireless devices connect directly to a wireless router (we will discuss routers in more detail in Chapter 6), which serves as the AP in your home's BSS. Other devices, like desktops, can connect to your router through wired connections. Your router is then connected to a modem, which goes to the access network (e.g., cable) extending outside your home.

And then there are those WiFi hotspots, which we find at public places that offer wireless Internet access. Today, we find them everywhere we go: coffee shops, hotels, airports, restaurants, and even around street corners. With growing demand, in June 2012, five leading cable providers in the USA announced that they would join their fifty thousand hotspots into a single WiFi service. One year earlier, the South Korean government announced a plan to cover the *entire city* of Seoul, including every corner of every street, with WiFi service by 2015.

If many WiFi users crowd around one street corner, how many hotspots are needed to take care of that demand? And, more importantly, how can this traffic be transported from the air to the rest of the Internet? These are all questions with answers that are constantly being updated by research and industry, to provide faster and more reliable service to an ever-growing customer base.

WiFi or cellular?

Your smartphone can offer two options to connect to the Internet: cellular or WiFi. Which to choose? Well, if there's no WiFi around, you're forced to stick with 3G (or 4G). You find this especially true when you're in transit. Cellular operators have done a good job to ensure that you can get reception pretty much anywhere you go. But when you're at a hotspot, you typically prefer WiFi. Of course, this depends on whether it is free and accessible or not. We all know that WiFi is often faster than 3G cellular, especially if your *channel conditions* are good.

- You need to have good *signal strength*. You want the number of bars that your WiFi connection shows to be as high as possible. Everything else constant, the closer you are to the AP, the better your signal strength will be.
- You need to experience little *interference*. You want to be sharing the AP with as few others as possible. In general, the fewer people using the AP, the less interference there will be.

If you're sitting next to the AP with not many people around, the WiFi data rate can be very impressive. If you're using 802.11n, it's definitely faster than 3G, and probably even faster than the access link that connects you to the rest of the Internet. But if you sit outside the limited range of the AP or move outside of your current BSS, you can easily get disconnected. And if you share the air with ten or so other WiFi devices, the speed can drop substantially. You've probably experienced this in a crowded public WiFi hotspot.

We illustrate differing channel conditions in Figure 1.33. Anna, who is right next to the AP, will have better signal quality than someone like Dana who is at the edge. And the devices cause interference for each other: think of any two trying to send to the AP at the same time. These transmissions will collide, just like a car crash resulting from someone running a stop sign.

A real life example of this happend when Steve Jobs, the former CEO of Apple, introduced the iPhone 4 at a conference in 2010. His demo phone had trouble connecting via WiFi since there were a lot of people in the autotorium. He had to ask all the attendants to get off the WiFi before he was able to get on.

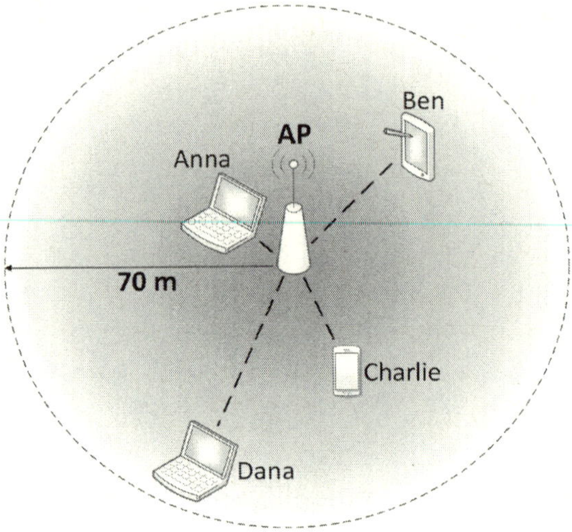

Figure 1.33 Illustration of a few people in the same BSS. The coverage area (for 802.11n) extends roughly 70 m indoors (more like 250 outside). A person like Anna sitting right next to the AP will have much better signal strength than someone like Dana at the edge of the BSS. Interference also affects the channel conditions: if any two people transmit at the same time, their signals will collide.

Is there some kind of limit as to how many users a WiFi hotspot can support? As witnessed above, the answer is yes. In the rest of this section, we will discuss some of the important concepts behind WiFi's operation. Specifically, we will try to get a sense for why our speed drops so much as more people use the same AP.

1.2.2 Selecting and connecting

At this point, we understand the purpose of WiFi: to provide *high speed* connections to *nearby* and *largely stationary* devices. Overall, we see that it's quite a different type of wireless networking than cellular, and as a result, it requires some different management approaches. We will investigate these in the remainder of this chapter.

Accessing WiFi

The first question is, how does your device *access* WiFi? To do that, it needs to select the correct access point, channel, and transmission rate.

- *AP selection*

  A WiFi device needs to make sure it has chosen the best access

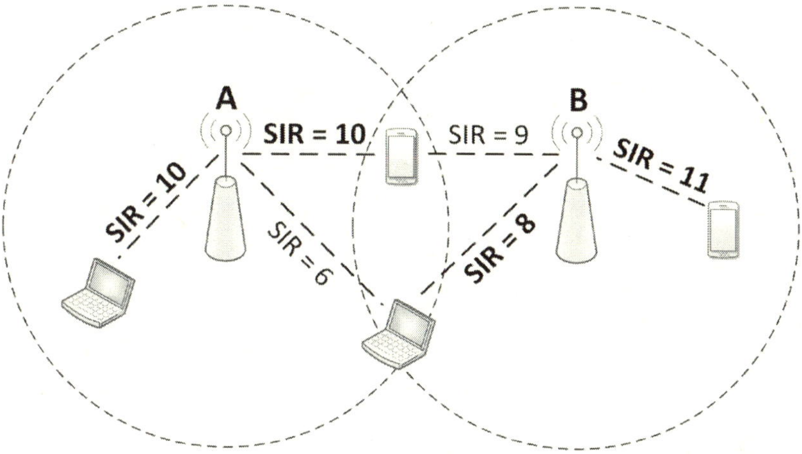

Figure 1.34 Our WiFi devices regularly scan the area to determine whether they are associating with the best AP or not. The best AP is the one to which the SIR is the highest. As a result, the phone and laptop in the middle would rather be with A and B, since 10 > 9 and 8 > 6, respectively. The laptop on the left has only one choice, A, and likewise, the phone on the right must use B.

point for itself. Which one is the "right" one? Well, among all the ones that it can access, it wants to make sure it has the highest signal-to-interference ratio possible (this is the same metric we discussed with cellular).

So, the device will regularly scan the air and determine whether its current AP offers the highest SIR. If not, it will switch. We call the AP it chooses the one it associates with. This would occur in an ESS if the client device is roaming; in this case, the session can be handed from AP to AP with no service interruption.

In Figure 1.34, we give an example of this. The laptop on the left and the phone on the right only have one choice: AP's A and B, respectively. For the devices in the middle, the cell phone will choose A (10 > 9), and the laptop will choose B (8 > 6).

- *Channel selection*

  Next, your device has to make sure it can communicate with its AP. To do this, it needs to select the proper frequency range to send/receive messages on.

  Each AP operates on a given frequency channel. The overall ISM band is divided into a number of such channels, each with a pre-determined width. Taking the example of 802.11b (in the US), there

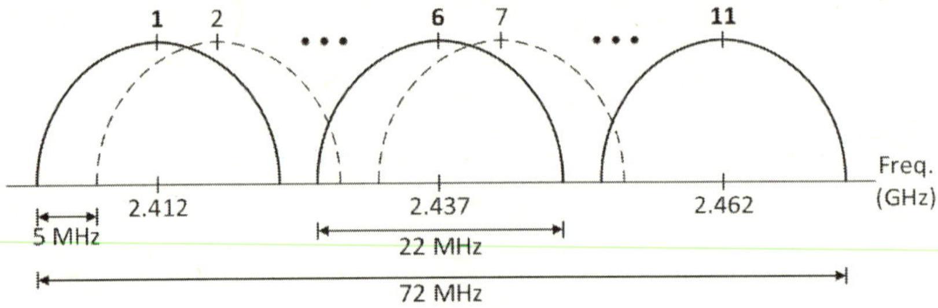

Figure 1.35 Illustration of the frequency range for WiFi's 802.11b, around 2.4 GHz. Each AP has a designated channel that they transmit and receive on, which is how they communicate with devices. There are 11 channels, separated by 0.005 GHz (5 MHz) on the frequency axis. Channel 1 is centered at 2.412 GHz, channel 2 at 2.417 GHz, and so on. As we can see, channels 1, 6, and 11 are non-overlapping.

are 11 located around 2.4 GHz, each with a width of 0.022 GHz (22 MHz). On each side of a channel's **center frequency**, it occupies half of this, or 0.011 GHz (11 MHz).

We show the allocation for 802.11b in Figure 1.35. Notice that the channels *overlap* with one another. Why is that? Well, we only have 72 MHz to start with. In order to have 11 non-overlapping at 22 MHz each, we would need 11 $\times$ 22 MHz = 242 MHz in total. To find the maximum number of non-overlapping channels possible with our range, we can divide the total width (72 MHz) by the width per channel (22 MHz) to get 3.27. Rounding this down, we see that there are at most 3 non-overlapping channels (1, 6, and 11).

In order for a device to communicate with its access point, it will select the channel that its AP is using. This means that all devices in the same basic service set are using the same channel. If two APs are close in proximity, then selecting the same channel causes unnecessary interference: the devices not only have to contend with others in their basic service set, but also those in the neighboring one. Ideally, we would want the access points to use different channels, as illustrated in Figure 1.36. Unfortunately, many WiFi deployments just use the default settings of their access points, meaning that many are on the same channel unnecessarily, hurting performance.

- *Rate selection*

  We saw that one of the distinguishing features between different versions of WiFi is the transmission rate. More precisely, this is the *maximum* transmission rate, and will only be achieved under very ideal conditions.

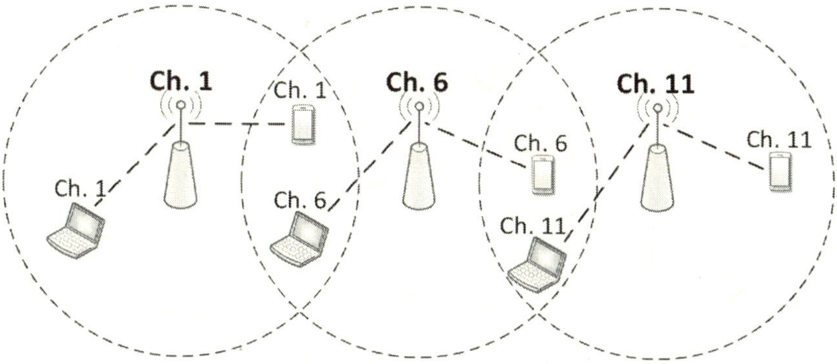

Figure 1.36 Each device that wants to associate with an AP has to use its corresponding frequency channel. Here, we show three APs with coverage areas that intersect. Ideally, they would each be on non-overlapping channels, which are 1, 6, and 11 with 802.11b. This allocation would cause users in the intersection regions not to interfere. For instance, the phone and laptop between Ch. 1 and 6 are close to each other, but since they are on separate channels, they don't interfere. Likewise with those in-between Ch. 6 and 11.

In many WiFi hotspots, the channel condition fluctuates as the interferers come and go. You may have 25 Mbps now, but once someone else turns their laptop on, you are down to maybe 15 Mbps. The maximum rate is rarely achieved, and the AP will tell the devices to *back off* to a lower rate if there is too much interference to handle the current one. The rate could be as low as 1 Mbps in 802.11b.

Now, suppose your WiFi device has gotten the above three things right. It has the right AP, the right channel for that AP, and the right transmission rate. We need to take care of *interference* now.

Interference

When two transmitters are within interference range of each other, and they send at similar times, they will collide. More specifically, we speak of *frames* colliding, where a frame is a piece of digital data transmission (more on that in Chapter 6).

There are three possible outcomes of a **collision**:

1. *Both frames are lost*: This means that the receivers of each frame could not correctly decipher them.
2. *The stronger frame is received, but the weaker frame is lost*: Here, "strength" refers to SIR. So, the frame with the higher SIR is suc-

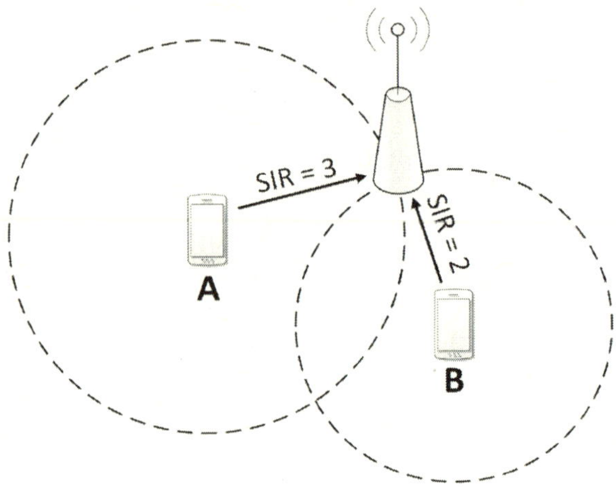

Figure 1.37 When two WiFi mobile stations close to each other transmit a frame at around the same time, the frames will collide. Here, we show the frames' energy propagating through the air, which intersect on their way to the AP. There are various situations that could occur as a result. If only one frame makes it, it will be that of A, since it has a higher SIR than B. But perhaps both will make it, or maybe neither will make it. It's generally very difficult to model these situations mathematically. We show the intersection in Figure 1.38.

cessful, while the one with the lower SIR isn't. This is called **capture**.

3. *Both frames are properly received*: This is called **double capture**.

So, which outcome will prevail? Well, that depends, very strongly, on a few factors, as illustrated in Figures 1.37 and 1.38.

- *How long the frames overlap*: The longer they clash, the more chance we have of disrupting the communication. Remember, each frame is nothing but noise for the other, so this just prolongs the effect.
- *How big the SIR difference between the frames is*: If one frame has a much higher SIR than the other, then it has a much better chance of surviving some interference than the other one does. On the other hand, the one with a low SIR can't tolerate very much noise.
- *How large of an SIR is required at the receiver*: Receivers with less internal noise can tolerate more interference than ones with more.

The above three events are particularly difficult to model. So let's take a more conservative approach, by assuming the worst case: whenever a collision happpens, both frames are *always lost*.

1.2.3 Random access protocols

How do we deal with this interference? We saw previously how 3G cellular does it: give everyone different codes and use power control to adjust their volumes. Maybe we can just use that here? Not quite: WiFi has some properties which prevent power control from being a good solution.

1. *WiFi is all unlicensed.* There's a lot more interference that we have no control over, since the spectrum is free for anyone to use.
2. *WiFi has a much smaller cell size* (a BSS vs. a cell). Adding another person to a basic service set can make a big difference in interference conditions, because there are only a few to start with. With 3G, there are a lot more people, making the power control conditions more stable.
3. *WiFi has a much smaller maximum transmit power.* Since the ISM band is unlicensed, we are not allowed to crank up the power too high. The dynamic range of powers required with cellular may not be allowed with WiFi.
4. *WiFi is typically indoors.* There tends to be a lot more objects and walls that can obstruct transmissions when indoors. Interference conditions due to these factors can quickly change.

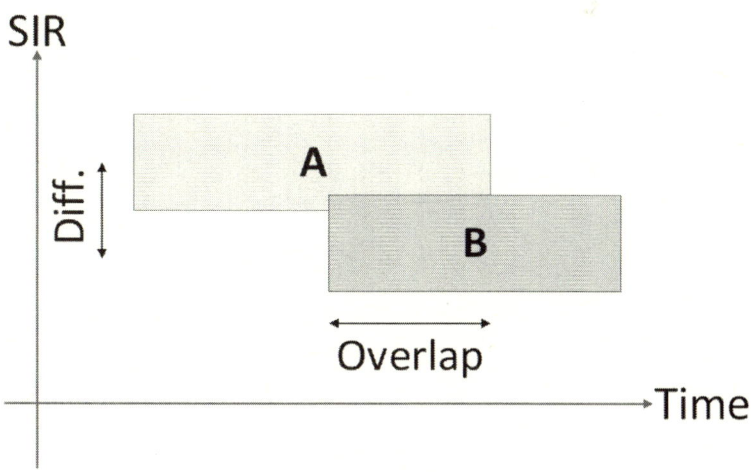

Figure 1.38 Here, we show the overlap of the two frames in time from Figure 1.37. The longer this overlap is, the higher the chance that neither frame will be properly decoded at the receiver. At the same time, since A has a higher SIR than B, it has a higher chance of making it through the collision.

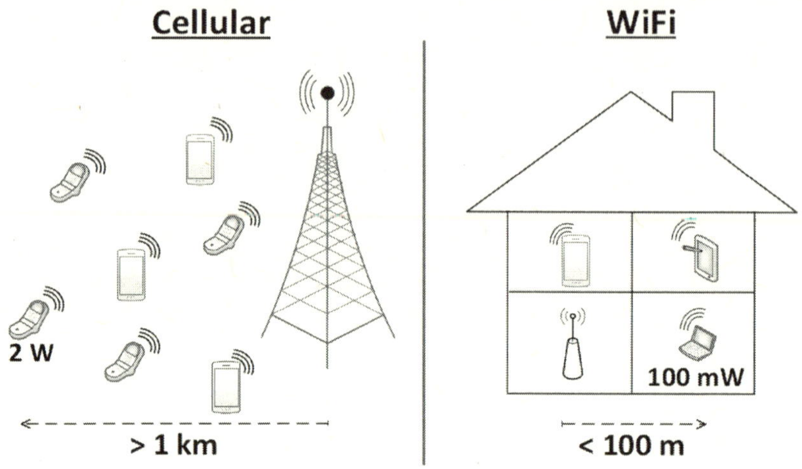

Figure 1.39 Illustration of the main differences between a cell (on the left) and a basic service set (on the right). Cell service typically extends over 1 km, and their devices may transmit up to around 2 W, depending on power control. On the other hand, WiFi's range is often less than 100 m, and devices are limited to around 100 mW (depending on the standard). These differences make it hard to use power control in WiFi networks.

We illustrate these differences in Figure 1.39.

Coordination is key

If we can't use power control, then how do we deal with interference? Well, WiFi takes an entirely different approach: it tries to avoid collisions from happening in the *first place.*

How can we go about this? With cellular, we got really used to power control and CDMA. We have to revert back before that, to the basic ideas behind *TDMA*, where everyone has their own timeslot in which they can transmit. Except in WiFi, we don't dedicate these timeslots.

With TDMA, users take turns. If we have three transmitters A – C, it would go something like A, B, C, A, B, C, and so on. We are *guaranteeing* that each user can have the medium completely to herself for her timeslot each round. This is just like intersections run by stop lights, where we only allow one side to come through at a time.

With WiFi, we don't do this. Instead, we have people transmit when they need to, but require them to remain cognizant of what other people are doing to prevent collisions. This is like when you come to a stop sign, and you can proceed as long as nobody is coming the other way.

Both of these techniques are methods of **medium access control**

Figure 1.40 Revisiting the cocktail party analogy. With enough interference, we need to enforce some sort of coordination. Power control is like asking each guest to adjust their volumes to mutually compatible levels. TDMA is like assigning everyone different times to talk. And random access is like allowing people to speak as long as they wouldn't be interrupting anyone else.

(MAC), e.g., defining how people share the communication medium (in this case, the air). But while TDMA is a form of *controlled* access, WiFi is governed by **random access**.

We can think again of the cocktail party analogy (see Figure 1.40), where we have guests' voices overlapping in the air. With enough interference, you can't understand what your friend is trying to say.

- Power control in CDMA is like asking the guests to adjust their volumes to reasonable levels.
- TDMA is like arranging for the guests to talk at different times, by assigning them different times to speak.
- WiFi separates the guests in time, but doesn't assign slots. Rather, at a given point in time, a guest is allowed to talk as long as nobody else is.

With WiFi, we ask each guest to obey a certain procedure for deciding locally when to talk, and how long to talk. The timeslots are *shared* by all the devices, since they can try to use the medium when they need it, if it's not currently occupied.

In other words, WiFi requires everyone to have some courtesy. We have a name for this: **Carrier Sensing Multiple Access** (CSMA), since each device tries to "sense" the presence of the others. CSMA is fully distributed, just like distributed power control: each device executes it completely on its own without some central coordinator.

CSMA is a random access protocol, and is one way to go about

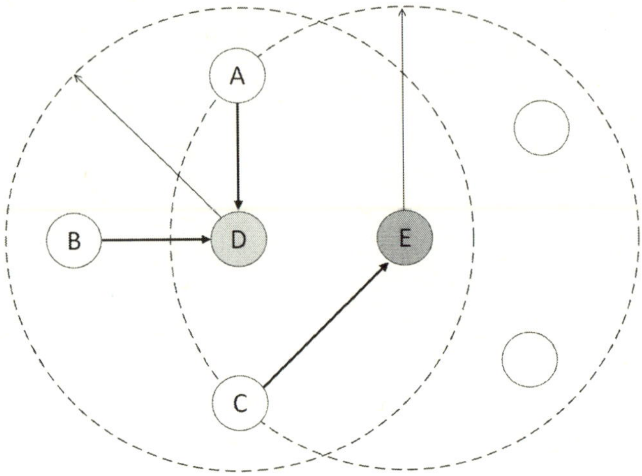

Figure 1.41 Here we represent WiFi stations (either a device or an AP) as circles. There are two receiving stations, D and E, and we have denoted their ranges accordingly. A and B want to send to D, while C wants to send to E. In order to avoid collisions, we must have *at most* one of these three transmitting at any given time. Even though A and B are sending to a different station than C, their frames will still collide.

WiFi. But it's pretty complicated mathematically, and those specifics are beyond the scope of this book. So we will take a conceptual approach to explaining CSMA. But before we do that, let's consider a simpler random access protocol that we can go into more detail with.

ALOHA

The protocol we will be looking at first was invented in the early 1970s, at the University of Hawaii. It was coined with a fitting name to the location: **ALOHA**. The ALOHA System was the first demonstration of a wireless data network that could be used on the Internet.

Suppose we are a WiFi device and we have some message we want to transmit over the air. If we were the only device that could possibly be using the interface, then the solution would be simple: just send. But unfortunately, there will most likely be other devices that want to be transmitting close in proximity to us. And if two frames meet in the air at any time, we have a collision, and nothing gets through. This is illustrated in Figure 1.41: we need *at most* one of A through E to be transmitting at any time to avoid a collision (note that the access points, D and E, can be transmitting too).

Our objective here is to get the most data through as possible. If

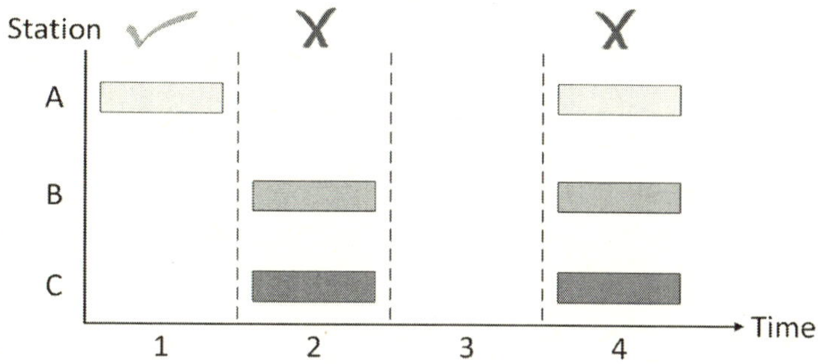

Figure 1.42 A timeslot diagram for three WiFi stations to illustrate collisions (see Figure 1.41). At the beginning of each timeslot, a station needs to decide whether or not to transmit. In the first slot, only A transmits, so it's successful. In the second, B and C both do, so they collide. In the third, nobody transmits, so we have a "wasted opportunity." In the fourth, everyone transmits, again resulting in a collision.

everyone's constantly transmitting, we will have a high sending rate, but the problem is that we will have a lot of collisions. At the same time, we want to make sure we have the air occupied as much as possible, since otherwise it's just wasted space. We can quantify performance by looking at **throughput**, which is the rate of *successful* message delivery. We can speak of per-user throughput, which is how much each individual person is getting, or aggregate throughput, which is how much everyone is getting in total. We will refer to devices and APs as WiFi **stations**, so we can speak of per-station throughput, etc.

We take a simple approach here. Divide time into slots, and let a single timeslot be the amount of time it takes any station to transmit exactly one frame over the air. We assume that there's a lot of demand for the channel, so that at any given time, a station has something to send.

At the beginning of each timeslot, each device is faced with the following question: should I send a frame, or not?

We illustrate potential results of this "decision-making" process in Figure 1.42. If all the stations always answered "yes" to this question, then we would always have collisions, and the throughput would be zero. That's clearly an undesirable state to be in. A given station has to refrain from sending sometimes, to allow others to get their data through. In turn, it should expect that the others will refrain so that it can get its data through soon as well.

We could make whether or not a station will transmit dependent

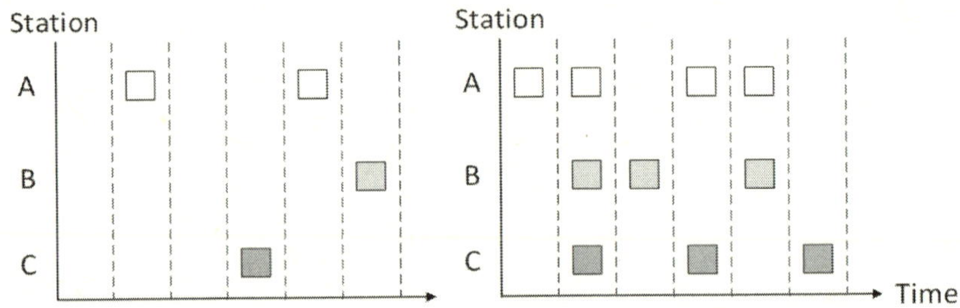

Figure 1.43 Conceptual understanding of what happens as we increase `ProbTrans`, from left to right. The diagram gets "denser," with more frames being sent. On the left side, we show a low probability of transmission: we have no collisions, but we pay the price of a lot of timeslots not being filled. On the right side, we show a higher probability, which results in collisions.

upon a lot of different things: whether it has had a collision recently, whether its current message is important, etc. But for the time being, let's just assign a *fixed probability* to it transmitting, which is simply the *chance* that it will choose to send at each timeslot. If it's 50%, then a station will send half the time, and hold back the other half. If it's 10%, then it will send one in every ten times on average, and hold back the other nine. Let's call this chance `ProbTrans`, which is always between 0 (0%) and 1 (100%). Note that in math equations, we always write probabilities in their decimal forms.

Figure 1.43 gives a conceptual understanding of what happens when we increase `ProbTrans`. The higher it is, the more frames are sent, and the higher the chance of a collision. But if it's too low, we will have a lot of wasted opportunities, which is also undesirable. The question is, what is the *best* choice of `ProbTrans` to make the throughput as high as possible? Let's try to figure that out now.

In each timeslot, there are three possible outcomes (see Figure 1.42).

1. *Successful transmission*: this will occur if exactly *one* station decides to send.
2. *Collision*: this will occur if *more than one* stations decide to send.
3. *No transmission*: this will occur of *no* station decides to send.

The first case is of particular interest here. We want to find the chance of having a successful transmission in each timeslot. For this to happen, we need two things to occur:

1. *Someone transmits*

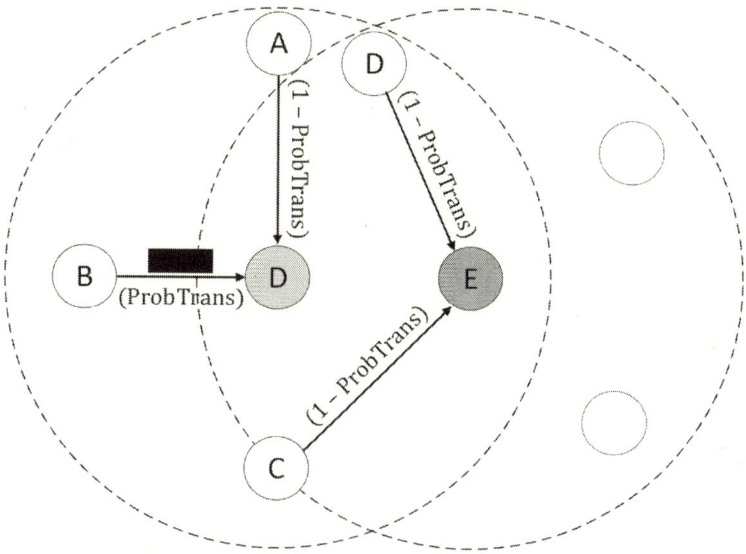

Figure 1.44 In order for a station, say B, to have a successful transmission in a timeslot, we need two things to happen. First, it needs to transmit, which has probability ProbTrans. Second, everyone else needs to *not* transmit. This has probability $(1 - \texttt{ProbTrans}) \times (1 - \texttt{ProbTrans}) \times (1 - \texttt{ProbTrans})$, since there are three others.

One station has to transmit a frame. For any given station, the chance of this happening is the probability of transmitting, `ProbTrans`.

2. *Nobody else transmits*

At the same time, the rest of the stations have to *not* be transmitting. The chance of a given station not transmitting is simply one minus the chance that it does, or $1 - \texttt{ProbTrans}$. For instance, if there was a 40% chance of you transmitting, there would be a 60% chance of you not transmitting: $1 - 0.4 = 0.6$.

But we need *all* of the others not transmitting, except the one that is. For instance, if we have three stations A, B, and C, and A is transmitting, we need B *and* C not to transmit. For our purposes here, *and* means we *multiply*. So, we multiply the probabilities: $(1 - \texttt{ProbTrans}) \times (1 - \texttt{ProbTrans})$. If $\texttt{ProbTrans} = 0.4$, then the chance of one station not transmitting is 60%, so the chance of two of them refraining is 36%: $0.6 \times 0.6 = 0.36$.

We can extend this to any number of stations. Call that `NumStat` for short. We choose one station to transmit, and we need the remaining $\texttt{NumStat} - 1$ of them not to transmit. So we multiply (1 −

ProbTrans) "NumStat - 1" times: (1−ProbTrans)× (1−ProbTrans) × ⋯. With ProbTrans at 40% and four stations, we would want three not to transmit, so we'd multiply three times to get 21.6%: $0.6 \times 0.6 \times 0.6 = 0.216$.

We illustrate this process in Figure 1.44.

So, what's the chance that a given station has a successful transmission? That's the probability that it transmits *and* nobody else transmits. So, we multiply the two cases. This gives an indication of the *per-station* throughput of the system:

$$\texttt{ProbTrans} \times (1 - \texttt{ProbTrans}) \times \cdots \times (1 - \texttt{ProbTrans}), \tag{1.4}$$

where we do the right multiplication NumStat − 1 times. This is the chance that *she* will decide to send and the remaining decide *not* to send.

Normally, we speak of throughput in bits per second, which is not the same as Equation (1.4). We'd have to multiply it by the number of bits per frame, and the number of frames per second. But for comparative purposes, this will suffice, because it gives us how well we are doing relative to an *ideal* case. The ideal case is how much throughput we would be getting if the station always transmitted, and it was always successful. So if Equation (1.4) was 0.5, the station would be getting half of what is physically possible with no interference.

Now, how do we find the *total* throughput? Well, Equation (1.4) gives us the probability that a *given* station has a successful transmission. We have to add up these probabilities for each station, to get the total:

$$\text{Probability A is Successful} + \text{Probability B is Successful} + \cdots .$$

This is the same as multiplying Equation (1.4) by the number of stations. So, the *total* throughput of the system is

$$\texttt{NumStat} \times (\text{Per-station throughput}). \tag{1.5}$$

Equation (1.5) tells us that we have two variables that affect the total throughput:

- ProbTrans, or the probability that a station transmits in a timeslot.
- NumStat, or the number of stations.

While it's hard to control when people come and go from our WiFi area (NumStat), we can enforce the transmission probability (ProbTrans)

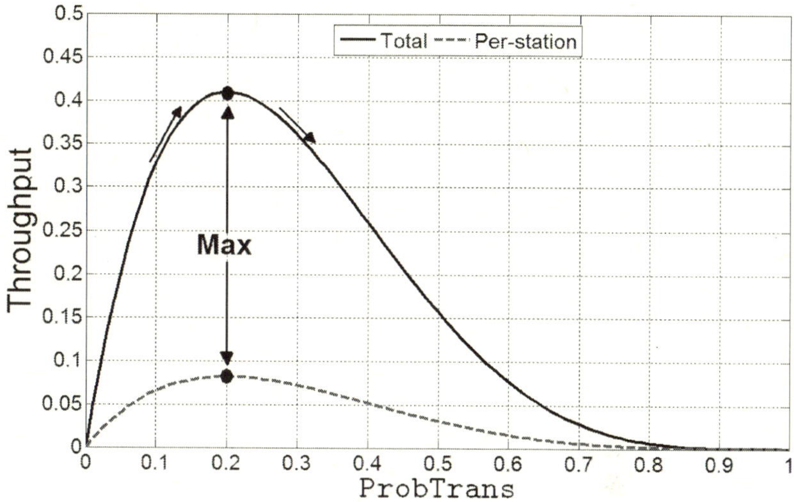

Figure 1.45 Plot of the total and per-station throughput for the ALOHA protocol, for 5 stations. We used Equations (1.4) and (1.5) here. The trend is that as ProbTrans increases from zero, the throughput initially increases, until `ProbTrans` = 0.2, when it's the highest. Then, it decreases down to zero as ProbTrans approaches one.

as part of our "courtesy procedure." So, what value of `ProbTrans` should we choose?

First, let's fix the number of stations and see what happens when we change the probability of transmitting. In Figure 1.45, we plot the total and per-station throughputs for `NumStat` = 5, using Equations (1.4) and (1.5). Starting from the left side of the graph, we see that increasing `ProbTrans` initially makes the performance go up quickly. With a small `ProbTrans`, there are little to no collisions, so we are just filling up the wasted timeslots and helping the throughput.

But once `ProbTrans` = 0.2, the figure hits its maximum, and then the throughput starts to drop: when `ProbTrans` > 0.2, transmitting more is causing more collisions than it's helping. In fact, it decreases all the way back down to zero, which should make sense because having everyone send in every timeslot is just as bad as having nobody send at all. It's worth mentioning also that the only difference between the two curves in Figure 1.45 is that the total throughput is five times larger than the per-station throughput.

Clearly then, to make the throughput as high as possible, we should have `ProbTrans` = 0.2. So each station should transmit 20% of the time. Are we done?

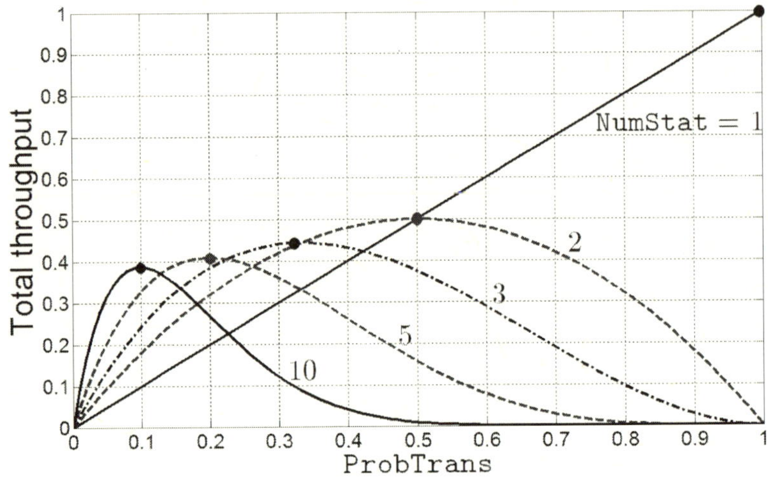

Figure 1.46 Here, we plot ALOHA's throughput as the transmission probability changes, for different numbers of stations (1, 2, 3, 5, and 10 as indicated). It's important to note that the best choice of `ProbTrans` changes for different `NumStat`. For one station, we want to be transmitting with 100% probability. For two, we want each at 50%. For three, we want each at 33.33%, and so on. Note that the maximum total throughput that we can achieve gets lower as the number of stations increases.

Not quite. We have to look at how the optimal transmission probability changes with the number of WiFi stations sharing the air. We only looked at the case of five so far.

To this end, in Figure 1.46, we plot the total throughput for various values of `NumStat`. Each curve corresponds to a different one: 1, 2, 3, 5, and 10, as indicated on the plot. We can make a few remarks here.

- For one station, the total throughput increases *linearly* with the transmission probability. This makes sense, because there's no chance of it colliding with anyone else. So it would want to be transmitting in each timeslot, with `ProbTrans` = 1. Already, we can see that it's not in our best interest to always have `ProbTrans` = 0.2.
- For two stations, we start to see the effect of collisions once `ProbTrans` is larger than 0.5. Therefore, to maximize throughput, we should have each of the two stations transmitting 50% of the time. Notice that the *total throughput* at 50%, of 0.5, is only *half* of the maximum, of 1, that we saw in the previous case.
- As the number of stations increases, we see collisions take over for lower and lower `ProbTrans`. For 3 stations, the maximum is reached

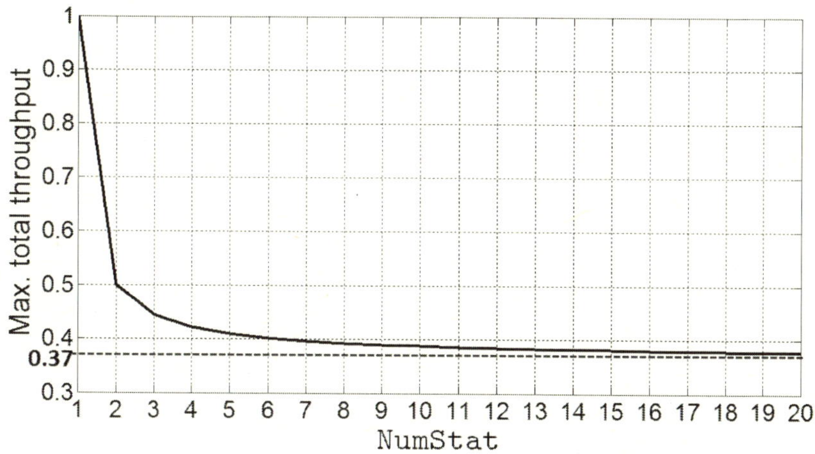

Figure 1.47 Plot of the maximum total throughput for ALOHA as we increase the number of stations. With two stations, it has already cut in half, and by six, it's down to 0.4. As we increase `NumStat` higher and higher, we approach a limit of roughly 0.37, which is not very high.

at 33.33%. For 5, the maximum is reached at 20%, and so on. This maximum is decreasing as the number of stations increases.

In Figure 1.47, we plot the maximum total throughput as the number of stations is varied. Clearly, the throughput always decreases as we add more stations in. Once we hit two, the throughput has already reduced to 0.5, and by the time we hit six, its at 0.4. This means that if we added up the number of successful frames that two stations get, it would be *half* of what a single person could get. And with six people, they would collectively get only 40% of what one person would get.

As we increase the number of stations higher and higher, we approach a *limit* as to what the total throughput will drop to. We can spot this in the figure after about 20 stations, when the throughput is reducing only a tiny bit with each additional station. As `NumStat` goes to infinity, the throughput will go to 0.37, meaning that we are successfully transmitting only 37% of the time.

Overall, we see that ALOHA doesn't scale well to a large number of users. This is a general theme for WiFi, and the price we pay for simple protocols.

CSMA

So, ALOHA has many problems. Have they been fixed? Somewhat: the current WiFi protocols use CSMA, and it gives us better throughput.

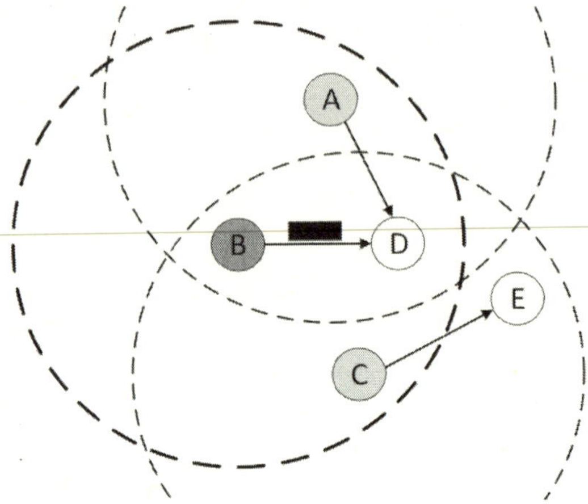

Figure 1.48 With CSMA, we have all stations perform carrier sensing. This way, they can see what's going on in their near vicinity. The circles around the stations indicate how far they can hear. So with B currently transmitting, both A and C can hear it, since B's transmission lies within their circles. Hence, they will refrain from sending anything until B has finished.

With ALOHA, we make no attempt to *coordinate* the transmissions of the stations. Sure, we reduce the number of collisions by having people transmit less, and randomly at that. But what happens if I have a lot of data to send right now, and a timeslot opens up when nobody is sending anything? If my "randomness" tells me not to send, then I won't. And that's a wasted opportunity. Similarly, if the channel is currently occupied, I should be able to tell that, and refrain from sending.

The problem is that ALOHA is purely random, and there's no *sensing*. In our stop sign analogy, that would be like putting a blindfold over your eyes, and randomly choosing when to go without looking to see if anyone was coming. Clearly, that would be extremely dangerous, risking the chance of a collision. (Keep in mind, however, that two frames of data colliding is less likely to be damaging than two cars colliding.)

So, we need to add some *sensing* into the mix. As a transmitter, I need to be able to sense when the channel is and isn't occupied, so I can determine whether or not to send, just like you need to look both ways before proceding past a stop sign.

Now, suppose you are a transmitter. Before you send any frame, you will regularly listen to the air. This is called **carrier sensing**, as shown in Figure 1.48: A and C can hear that the channel is currently

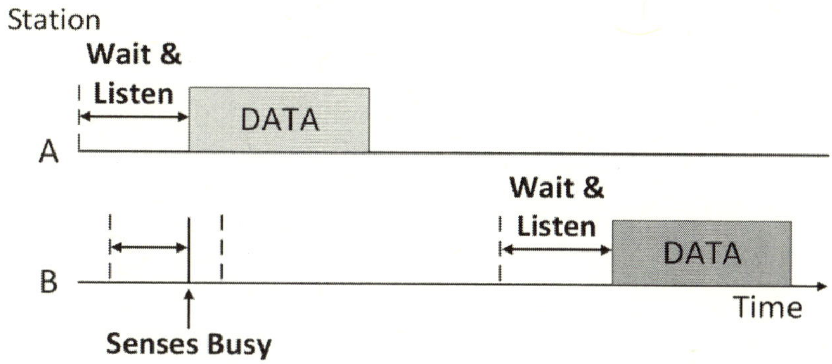

Figure 1.49 Before sending, a station has to observe a wait and listen period, throughout which they don't hear any transmissions. First, A initiates his period, and a little after, B starts his. A's ends, and as such, it begins transmitting. B hears this before the end of his own period, and so it has to wait until the channel is free before it can start listening again. Eventually, it gets through the wait and listen period and sends his data as well.

occupied by B, and hence they will refrain from sending anything until it's finished.

And once you get a first glimpse of a quiet channel, you can't just immediately begin sending. Rather, you have to first observe a wait-and-listen period, as shown in Figure 1.49. If you sense the channel is busy at any time during the wait-and-listen period, then you just stay silent. The purpose of this is analogous to what happens in typical conversations. When you think someone has finished talking, it's common courtesy to give them a few seconds before you respond, just in case they weren't done.

Once you make it to the end of a wait-and-listen period without hearing anything else, you can assume the channel is idle, and start sending. Note that even though you waited, it's still possible that your frame will collide with someone else's. Perhaps they began their waiting period at the exact same time as you, and decided to begin transmitting when you did (there are other possibilities as well). In any event, your receiver will not get your frame.

How do you know your destination did not get it? By using feedback. CSMA requires that upon receipt, an *acknowledgement* (ACK) frame is sent back to the source (we will study acknowledgements in detail in Chapter 7). This will notify the sender that everything went OK. So, if your frame does not get there, you will not get an acknowledgement back. This is how you can tell whether or not a collision occured.

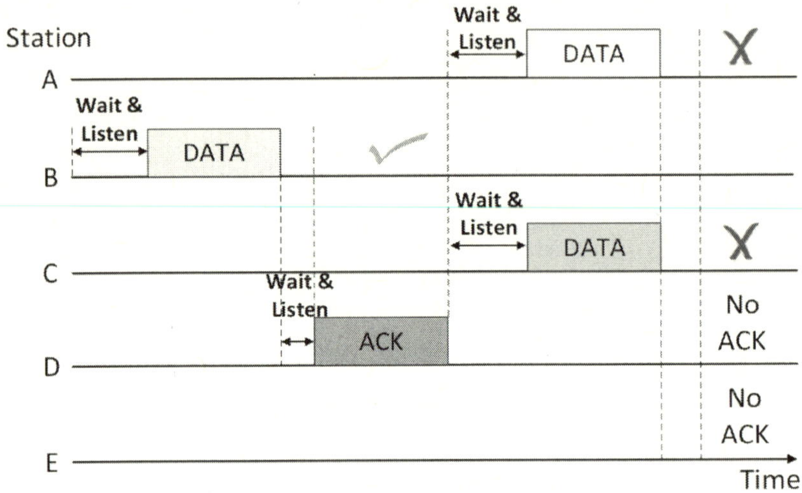

Figure 1.50 Illustration of acknowledgements. First, B needs to send. Sensing that the air is free, it goes through the wait-and-listen period, and at the end, begins the transmission to station D. Then, D sends an acknowledgement message back to B. B receives this and knows everything went smoothly. After this, A and C don't hear anything, so they go through the wait-and-listen period, and begin transmitting at the same time. Collision occurs, and they realize this when they don't get acknowledgements back.

Note that the recipient also goes through a wait and listen period before sending the acknowledgement message. We illustrate these ideas in Figure 1.50, whose senders and receivers correspond to those in Figure 1.48.

So then, what do we do upon a collision? Well, we need each station to back off to a later time, at which point they will try to send again. The question is, how do we choose what time to back off to?

Well, clearly, we don't want the stations to choose the *same* time: this would just cause another collision. Instead, we have them each choose some *random* time in the future to re-transmit.

This "point in the future" is determined by the current **contention window** size. If the current contention window was 3, then we'd choose some random number between 0 and 3. The number we get tells us how many timeslots to wait before re-transmitting: if we came up with 2, we would wait 2 timeslots. If we came up with 0, we would transmit immediately.

We illustrate this process in Figure 1.51, where we show two stations

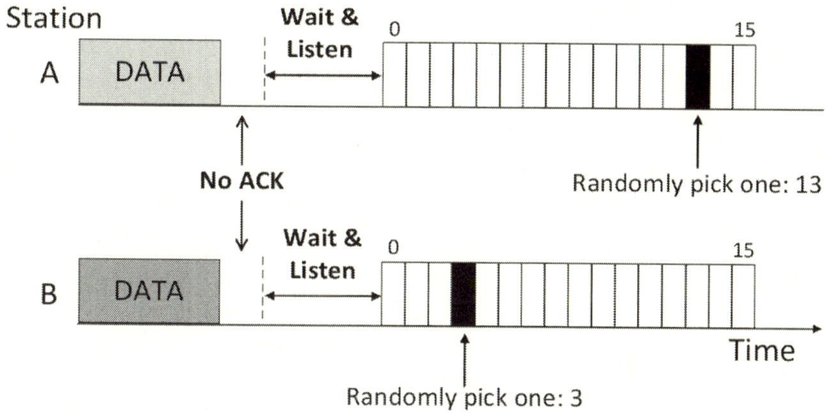

Figure 1.51 Here we show A and B interfering with one another. Once they realize they have collided (no acknowledgement), they go through another wait and listen period. If they reach the end, they have to determine how long to back off. So they draw a random number between 0 (meaning transmit immediately) and their current window size, 15. This number tells them how many slots to wait before sending again. B sends after 3 slots, and A sends after 13, hence avoiding a collision.

having a collision. First, they go through the wait and listen period a second time. Then, they choose random numbers between 0 and the window size, 15, to determine when to transmit again.

The point of randomizing this is to make the chance that these stations collide again low. This is why they both choose random numbers separately. Of course, it's possible that the stations will choose the same one, and collide again. Or maybe they will just collide with different transmitters altogether. If frames keep colliding, you know the interference condition in the air must be pretty bad. So you, as well as all those stations experiencing persistent collisions of their frames, should start backing off more aggressively. This is done by increasing your contention window size.

To help understand this, we can turn again to the cocktail party analogy. Suppose you're in a room having a chat with a group of people. You have something you want to say. Someone is currently talking, and then they stop, which begins your internal "wait-and-listen" period. You give them maybe two seconds to keep talking if they want to, or for someone else to respond abruptly.

After that two seconds is up, you start talking. But what happens? Someone else tries to begin talking, and you collide. You're both caught

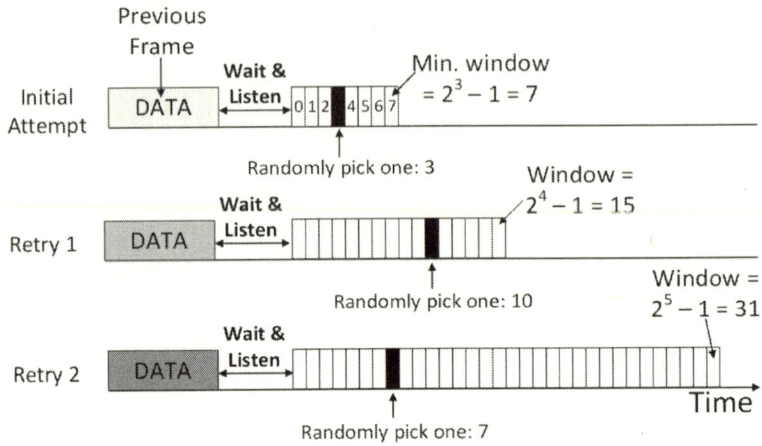

Figure 1.52 Process of repeated backoffs in WiFi's CSMA. Once the previous data frame finishes, the sender goes through the wait and listen period. The number of backoff slots is initially 8, making the window size 7. So, it randomly chooses a number between 0 and 7, coming up with 3. It then waits 3 timeslots before transmitting, but suffers a collision, causing the number of backoff slots to double, to 16. It then goes through wait and listen again. Once that finishes, it chooses a random number between 0 and 15, coming up with 10. But it suffers another collision, so the number of slots goes up to 32. The next random number is 7. Finally, the data gets through, and the window size resets.

off-guard, and pause. You back off, wait two seconds, and then try to "re-transmit." But she waits the same two seconds, and you collide. This time you wait longer, and back off farther, say four seconds. And so does she, causing another collision. Then, being polite, you decide to wait eight seconds.

Finally, she senses the lack of your voice so she takes advantage of it, and begins talking herself. Maybe this was after two seconds. So she gets her words in, and then in eight seconds, you get yours (assuming she needs less than six seconds to talk).

So what was the problem here? Why didn't "backing off" work the previous tries? Because you both waited the same amount of time, as luck would have it. Two seconds, then four seconds. Finally, in the third retry, you each get your message through because you wait eight seconds while your friend waits two. This is why randomization is important: otherwise, we would keep choosing the same times to re-transimt, and would keep colliding.

The progression 2, 4, 8, ... is how the window size is increased with

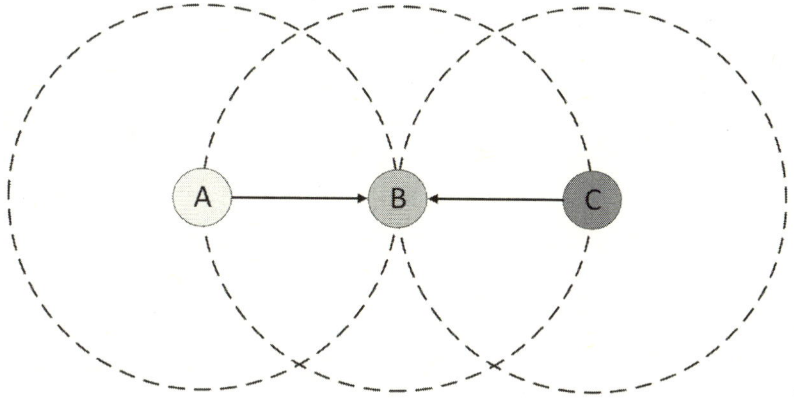

Figure 1.53 Transmissions by A reach no farther than B, and likewise for C. By just listening to the air, A and C have no way of figuring out that each other exist, since their transmissions are not in each other's circles. If they both want to transmit to B, they will interfere with each other at B. If A is currently occupying B, C will not know this. Hence, he may decide to transmit, causing a collision in the middle of A's session, and vice versa.

CSMA: by factors of two. Linearly increasing the window size (2, 3, 4, ...) is one option, but people thought that would not be aggressive enough. So instead, CSMA mandates *multiplicatively* backing off. And since the multiplicative factor is 2, we call it **binary exponential backoff**. In our analogy with only two people talking, multiplicative backoff might not make sense to you. But as more and more people want to get a word in, you can imagine the interference conditions would get very bad. That's why we want everyone who's colliding to increase the window size rapidly, to reduce the chance of it happening again.

We illustrate the process of binary exponential backoff in Figure 1.52. To find the current window size, we multiply the previous number of slots by 2, and subtract 1 from the result. We subtract 1 simply because we want to include zero: it's possible that we won't backoff at all, and transmit immediately. Initially, we have 8 potential backoff slots, 0 - 7, and so we say the window size is 7. Next, we have 16 potential backoff slots, 0 - 15, and so we say it is 15, and so on.

There are many more intricacies of CSMA that we won't have time to get into. For instance, it's possible for you to collide with someone you can't sense. We illustrate this so-called **hidden node problem** in Figure 1.53: A and C can't hear each other, but their transmissions can interfere at B. For instance, suppose A is currently transmitting to B. C, not knowing this, may decide to send, thereby interrupting A's transmission. This becomes a problem unless other precautionary measures are taken.

CSMA vs. ALOHA

CSMA gives us performance improvements over what ALOHA can do, when the parameters are tuned right. Let's take a look at some of those benefits here.

- *Per-station throughput*: The per-station throughput decreases as we add more stations with both CSMA and ALOHA. This should make sense, because we have more people sharing the air. But with CSMA, it decreases *less rapidly* than it does with ALOHA.
- *Aggregate throughput*: The aggregate throughput actually increases as we add the first few stations in CSMA. This is contrary to what we saw with ALOHA in Figure 1.47, where it was always decreasing.
- *Distributed implementation*: Each station can implement CSMA completely on its own, without the help of some central coordinator. All it needs to do is sense its immediate surroundings. With ALOHA, we would need a way of constantly updating all the stations with what probabilities they should set, based on the number of stations surrounding them.

But CSMA doesn't solve it all. In fact, after just a few stations, the aggregate throughput will begin to go down in CSMA, too. Though it's an improvement over ALOHA, the rate of decrease is still substantial.

Overall, neither CSMA nor ALOHA is scalable to a large number of WiFi devices, which is why we see such poor performance at hotspots when there are a lot of people around. That's the price we pay for such simple protocols.

Summary

In this first chapter, on the principle that "sharing is hard," we discussed two types of wireless networks: cellular and WiFi. We use both of these daily, as we work and play, but probably don't take much time to think about how so many people can share the air without ruining each other's conversations. We saw how the long-range, regulated cellular networks are privy to techniques such as distributed power control through CDMA, where everyone adjusts their levels without causing an arms race. On the other hand, we saw how the short-range, unregulated WiFi networks are more feasibly dealt with using random access techniques, where everyone tries to avoid collisions through sensing. Different as they may be, WiFi and cellular are both prime examples of the constant struggle we face to fit more and more users into the same communication medium, while relying on distributed algorithms to keep the coordination methods as simple as possible.

2 Consensus Is Hard

In the last chapter, we saw how challenging it can be to efficiently divide channel resources among many users sharing the same communication medium. Now, we will turn our attention to another networking principle, which has been investigated in a wide variety of flavors - finding *consensus* from a set of data. In doing so, we will look at two famous websites, Google and Wikipedia. We will describe how Google uses bids to allocate ad spaces to buyers, and also how they use hyperlink connectivity among webpages to determine a rank-ordered list for their search results. We will also survey the methods Wikipedia uses to aggregate votes from editors to form a consensus on which article edits should be kept. Overall, we will see how each technique works well, even though the problem of finding a consensus usually lacks a "right" answer.

2.1 Match making on Google

If you've been on the internet in the past decade, you've used Google search. We go so far as to 'google' everything from the closest coffee shops to the detailed answers of our college coursework, and have, as such, made the company's name a commonly used verb in everyday speech. In fact, "google" was added to the Oxford English Dictionary in 2006, and refers to using the Google search engine to obtain information from the Internet. A large portion of readers were perhaps referred to this book by Google, too.

We see Google's logo (Figure 2.1) on the Internet everywhere, and may have wondered at one time or another how the search engine works so well. Indeed,

Figure 2.1 Google's ubiquitous, trademarked logo.

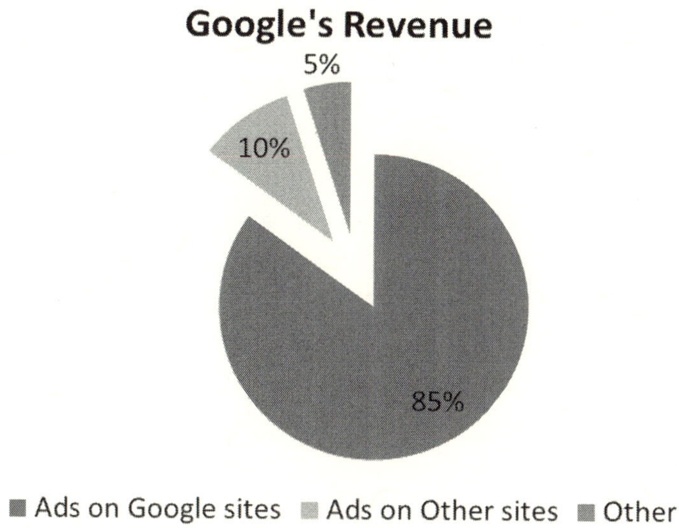

Figure 2.2 Google's (approximate) net revenue by source from the third quarter of 2011. About 85% came from advertisements on Google's own sites, and 10% from those on partner sites. Only 5% came from other sources, such as Google Apps and search appliances.

- How does Google *rank webpages*?

Even before this, you have probably considered something more fundamental. With close to 55,000 employees, it generated a revenue of over $14 billion in just the first quarter of 2013. So,

- How does Google *make money*?

In this section, we will look into how Google generates its revenue stream, and in the next we will turn to understanding the mechanisms of PageRank.

2.1.1 Online ads

Besides its convenience, Google search has another nice feature: it's free. How is this possible?

The main reason that all the search engines we know (besides Google, there's also Yahoo!, Ask, etc.) don't charge for their service is due to online advertisements. Websites, ranging from those of small businesses to large corporations, pay huge amounts of money to have their information posted in search query results. It is estimated that the online

Figure 2.3 A simple example of a search results page. The query here is "Food stores," so the links that show up will be related to food shopping. The ads are arranged by how much the advertisers bid for the first slot. Ben's Bakery offered the most, at \$5, followed by Matt's Market (\$3) and Gabby's Grocery (\$1). As a result, the stores will pay \$5, \$3, and \$1 whenever someone clicks on their links.

ad industry worldwide reached over \$90 billion in 2012. In fact, Google generated over 90% of its revenue from ads alone in 2011, as shown in Figure 2.2.

Pay per what?

Online advertising started in the early days of the Internet (mid-90s). It was originally sold to advertisers as banners on a **Pay Per Thousand Impressions** (PPTI) basis. This means that each time a banner aggregates 1000 new views, the advertiser has to pay the seller some predetermined amount of money. But is this really the best way to charge advertisers? Naturally, *seeing* an ad does not guarantee that a person will *click* on it, let alone *buy* something from it. So it has limited indication as to how much the advertiser will gain.

In 1998, a search engine company called GoTo began offering advertisers an alternative. They could bid on how much they were willing to pay to appear at the top of the results page, in response to specific search queries. This was on a **Pay Per Click** (PPC) basis: the advertiser would pay their bid to GoTo every time a person clicked on the

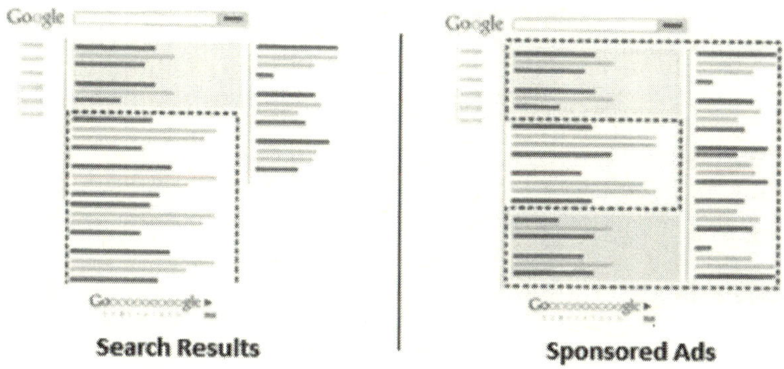

Figure 2.4 Illustration of the Google search layout. The standard search results (left) are shown in the middle of the page. On the other hand, the sponsored ads (right) are shown along the edge of the page. Results in the middle of the page (including ads) are over a white background, while those at the top and bottom are over a colored one.

link to their website. The list of search results provided by this engine was in descending order of the bid amounts. This is illustrated in Figure 2.3, which gives a simple example of three advertisers. GoTo was later renamed Overture Services in 2001, and was eventually acquired by Yahoo! for their own search marketing services in 2003.

Google was founded in 1998, and its AdWords division was put in charge of advertising by the turn of the century. They were offering keyword advertising with pay per click pricing as a feature of their search engine. But the creation of the so-called *sponsor search* model, also known as **search ads**, was widely credited to GoTo. So, Overture filed a lawsuit against Google in 2002. It ended with a settlement in which Google gave stock to GoTo in return for a PPC keyword search license. Since then, Google has become the largest search ad provider, possessing almost 45% of the market in 2010.

The ad giant

AdWords allows advertisers to create ads, mark them with keywords, and send them to Google's database. When someone searches for a keyword, Google returns a standard list of search results, and a list of ads that have keywords matching the query. The Google search layout is depicted in Figure 2.4. The standard search results are shown in the middle of the page, while the sponsored ads are shown along the edge of the page - on the top, right panel, or bottom.

As an example, we typed "Online Education" into Google search in

Figure 2.5 The click-through rate of an ad space, measuring the expected number of visits in a given amount of time, is an important consideration for an advertiser when submiting a bid. The higher the click-through rate, the higher the expected revenue, and the more valuable the ad space.

May 2013. Aside from the ten standard search results that have links to certain online programs and rankings, there are ads at the top and on the right panel that Google returned based on the keywords. The first one at the top is for a site that will list a number of online programs that match your credentials (`www.educationconnection.com/online-degree3`). This space, as well as the one at the top of the right panel, probably receive the most clicks.

So, where will your ad appear on the list? This answer depends on how much you are willing to pay for sponsorship, and is allocated through an auction. So the highest bidder gets the best spot, the second highest the second best, and so on. We will look at different auction mechanisms in this section.

Also, when do the advertisers pay Google? With PPC, Google gets paid every time an ad is clicked. The average number of times an ad space is clicked by a viewer, say over one hour, is called the **click-through rate** (`ctr`). We can say that the payment by advertisers to Google is *proportional* to the click-through rates, since this is an indication of the benefit Google is providing them. The higher the click-through rate, the more valuable the ad space is (see Figure 2.5).

Finally, what's in it for the advertisers? In addition to the received number of clicks per hour, say `ctr`, each advertiser has an average revenue in dollars, say `rev`, generated from each click. So the product, `ctr` × `rev`, gives the expected revenue of a buyer for a particular ad space. This is the **valuation** of the ad space to the buyer, and each advertiser will have a different valuation for each space. For instance, suppose an ad space receives 20 clicks each hour, and for each click, there's a 50% chance that the viewer will purchase an item. Assume the average cost of an item purchased is \$70. The expected revenue is then

$$20 \frac{\text{clicks}}{\text{hour}} \times 0.5 \frac{\text{purchases}}{\text{click}} \times 70 \frac{\text{dollars}}{\text{purchase}} = 700 \frac{\text{dollars}}{\text{hour}}.$$

We have to make a few assumptions moving forward. As a side note, whenever you see the phrase "we assume," the remaining portion of the sentence is never true in general. We assume them to be true for the sake of further analysis or design. Once in a while, practical implications drawn from such assumptions are empirically verified to be valid.

- We assume that the observed number of clicks per hour is the same as the estimated click-through rate.
- We also assume that the click-through rate is independent of the actual content of the ad placed in the space. In reality, some ads are naturally more attractive than others, and as such will obtain more clicks in the same space.
- Finally, we assume that the average revenue generated from each click is independent of the specific ad space that is chosen. So whether I have Google's first ad space or their last, I expect to make the same money *each time I get a click* (obviously, my click-through rate will be much lower, and that's why my revenue plummets). This is a much more reasonable assumption than the previous two.

2.1.2 Auctions

The process by which Google sells ad spaces to buyers is an ad **auction**. In general, an auction has a number of bidders, items, and sellers, but we will focus on the case of a single seller.

Though auctions gained much of their popularity with the rise of the Internet, they were used as a means to negotiate the exchange of commodities for many centuries before the days of Google. In fact, they are recorded as early as 500 B.C. Throughout history, items ranging from horses and livestock to real estate and entire empires have been placed on the auction block. Here are two interesting examples:

- In 193 A.D., the Roman guards assassinated their emperor and put the entire empire up for auction. The highest bidder, Julianus, became successor to the throne.
- During the American Civil War (1861 - 1865), goods seized by armies

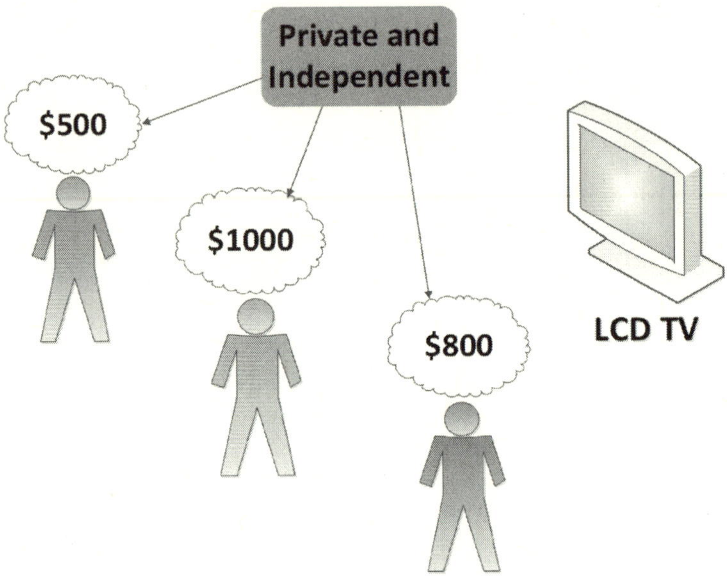

Figure 2.6 Three bidders' valuations for an LCD TV are shown here: $500, $1000, and $800. For our purposes, we assume that each person knows his own valuation, but that they are private and independent.

were sold via auction by the colonel. This is why some auctioneers in the US today are called "colonels."

We will begin our auction analysis focusing on the case of a number of bidders for a single item, before focusing on the more general case of multiple items. But before moving on, let's point out a few more assumptions we will be making.

We assume that each bidder has an *intrinsic* valuation of the item, which they all know. In reality, bidders may not know precisely how to quantify their own valuation. And further, we assume that each bidder's valulation is *private* and *independent*, as illustrated in Figure 2.6.

- A **private valulation** means that nobody knows it except the bidder himself. This is not true in general: sometimes people make their valuations public so everyone can get an idea of how much an item is worth.
- An **independent valulation** means that the bidder's valulation will not change depending on what other peoples' are. This is also not true in general: if you are thinking of buying a house, your real-estate

Figure 2.7 Depiction of the classic, open, public auction room.

agent's valuation of the property will likely be able to influence your own.

2.1.3 Open auctions

When you picture an auction, you likely think of a public venue in which there is a single auctioneer and multiple bidders, as depicted in Figure 2.7. The auctioneer stands in front of the room next to the item being auctioned, mediating the process, while the bidders stand in the crowd.

This type of auction, in which everyone's bids are publicly announced, is called an **open auction**. There are two main types: ascending and descending price.

- **Ascending price auction**
  The auctioneer first announces a base price, and then any of the bidders can raise their hands to bid to a higher price. He may say "Bidding starts at ten dollars! Any takers for ten dollars?" Then someone raises their hand: "We have ten dollars, anyone going for more than ten dollars?" Then someone offers twenty dollars, and so on, until nobody will increase more: "One hundred going once, going twice, gone!" This last bidder wins the item and pays his most recent bid. This is the type of auction we are most used to.
- **Descending price auction**
  The auctioneer announces a high price first, so high that nobody will

Figure 2.8 An eBay auction sometime in 2008 for a Blackberry. The "Current bid" is the ask price, and the "End time" indicates the hard close time of the bid.

accept it. Then he gradually lowers it down until there is a bidder that shouts "OK!" This bidder wins the item, and pays the price at which he stopped the bidding.

Open auctions are used quite frequently, but not by Google and other search ad sponsors. The alternative to a public venue is private disclosure of bids, called **sealed envelope** auctions, which are much more practical in many settings.

eBay auction

Before delving into sealed envelope auctions, let's go through an example that will illustrate some key points. eBay is a multi-billion-dollar corporation that runs an online auction and shopping website that you may have visited, or even purchased from, at one time or another. Founded in 1995, it currently has more than 40 million active users. Figure 2.8 shows a standard eBay auction, for a Blackberry in 2008. Here's how it works:

- *Start price*
  A seller announces a start price, and can also choose to set a secret *reserve price*: If the end result of the auction is less than this, he can choose not to sell the item. In the rest of this section, we will assume the user waives the right to the reserve price.
- *Duration*
  eBay has a publicly announced time horizon, after which the auction

ends. In Figure 2.8, the "End time" gives the hard close time of the Blackberry auction.

- *Minimum increment*
  There is a minimum increment, say `inc`, between successive bids, so as to prevent inefficiencies in the auction.
- *Final sale*
  Let `Bid 1` be the highest bid, `Bid 2` be the second highest bid, and so on. At the conclusion of the auction, the winner pays the smaller of the highest bid and second highest bid plus the increment, i.e., the minimum of `Bid 1` and `Bid 2` + `inc`.
- *Ask price*
  Some information about the current bid is continuously released to the public, but not necessarily the current highest bid. eBay displays an ask price, which is the lowest value that will be accepted for the next bid. To determine this, eBay takes the price that the highest bidder would have to pay if the auction ended now (i.e., the minimum of `Bid 1` and `Bid 2` + `inc`), and adds `inc` to that. Bidders get a sense of where the current bids lie from the ask price, but cannot be certain whether it is derived from `Bid 1` or `Bid 2`. In Figure 2.8, the "Current bid" represents the ask price.

Suppose there are three bidders: Anna, Ben, and Charlie, participating in an eBay auction, as shown in Figure 2.9.

- *Start*
  Steve, the seller, lists a used iPhone on eBay, with a start price of \$50 and a duration of 4 days. The minimum increment, `inc`, is \$5.
  Highest bid: n/a; Ask price: \$50.
- *Day 1*
  Anna enters the eBay auction, with her first bid at the ask price of \$50. As such, eBay changes the ask price to \$50 + \$5 = \$55.
  Highest bid: Anna, \$50; Ask price: \$55
- *Day 2*

  Ben decides he wants an iPhone but doesn't have the time to sit and watch the auction. As such, he uses a "proxy agent," which is automated bidding offered by eBay. This allows a user to set a *maximum bid*: If necessary, the proxy will keep bidding the ask price until it is too high.

  Ben sets the proxy at \$80, and as such, it bids the ask price of \$55. The highest bid is now \$55, and the second highest is Anna's original \$50. Since `Bid 1` = \$55 and `Bid 2` + `inc` = \$50 + \$5 = \$55

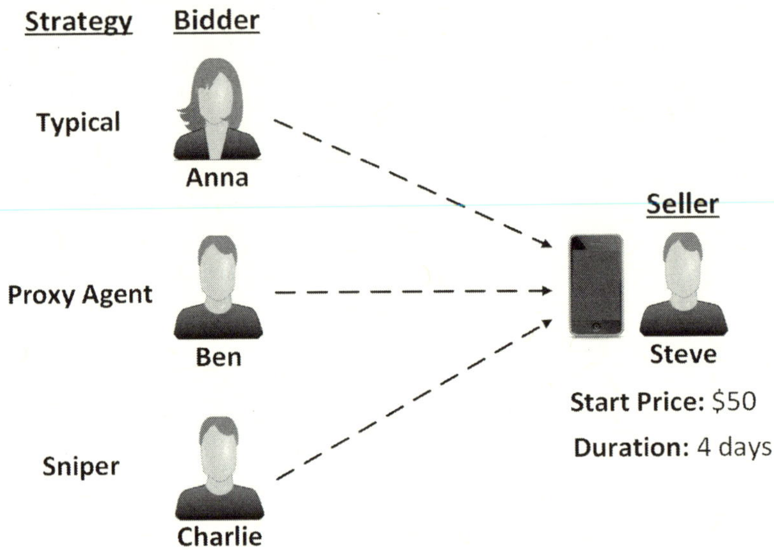

Figure 2.9 Scenario we use for our eBay auction example. There are three bidders, each of whom uses a different strategy.

are the same, the new ask price is \$55 + \$5 = \$60.
Highest bid: Ben, \$55; Ask price: \$60

- *Day 3*

  Anna sees that she's no longer the highest bidder. Rather than the ask price, she bids ten dollars higher, \$70, making this the high bid and \$55 the second highest. Since \$55 + \$5 = \$60 is less than \$70, the new ask price is \$60 + \$5 = \$65.

  Ben's proxy agent immediately bids \$65, since it is lower than his maximum. But Anna is still the highest, at \$70. Since \$65 + \$5 is the same as \$70, the new ask price is \$70 + \$5 = \$75.

  The proxy sees that Ben is *still* not the highest bidder, so it takes the \$75. Now, Ben has the highest bid at \$75, and Anna's \$70 is the second highest. Hence, the ask price becomes \$75 + \$5 = \$80.
  Highest bid: Ben, \$75; Ask price: \$80

- *Day 4*
  Anna is frustrated, and as such, enters \$95 as her bid. Ben's proxy agent now knows he can no longer win, since Anna's bid is higher than his maximum. Now \$95 is the highest and \$75 is second. Since \$75 + \$5 = \$80 is less than \$95, the new ask price is \$80 + \$5 = \$85. We illustrate this computation in Figure 2.10.
  Highest bid: Anna, \$95; Ask price: \$85

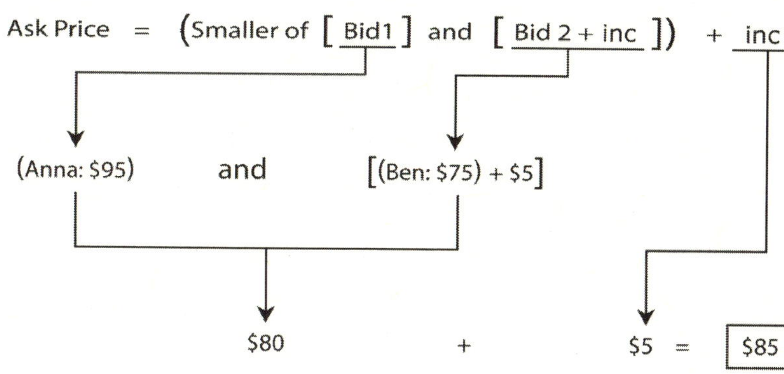

Figure 2.10 Illustration of how the ask price is computed on Day 4.

- *Day 5*
  Anna sees she is the highest bidder and gets confident. But on the last day of the auction, Charlie, who has been watching all along and really wants an iPhone, bids \$110. Now, \$110 is the highest and \$95 is second. Since \$110 is larger than \$95 + \$5 = \$100, the ask price becomes \$100 + \$5 = \$105.
  Highest bid: Charlie, \$110; Ask price: \$105
- *End*
  Anna didn't have time to react to Charlie's bid and loses the auction. This type of behavior on Charlie's part is known as *sniper bidding.* Charlie wins the iPhone for whichever of his bid, `Bid 1`, or the second highest plus an increment, `Bid 2 + inc`, is lower. Since \$110 is more than \$95 + \$5 = \$100, he pays \$100 for the phone.

The evolution of bids in this auction is summarized in Table 2.1.

	Day 1	Day 2	Day 3	Day 4	Day 5
Anna	**\$50**	\$50	\$70	**\$95**	–
Ben	–	**\$55**	**\$75**	–	–
Charlie	–	–	–	–	**\$110**
Ask price	\$55	\$60	\$80	\$85	\$105

Table 2.1 The bidding history and ask price evolution in our eBay auction example. The highest price at each stage is bolded.

eBay auctioning is neither strictly sealed nor open. It progresses through a series of stages in which the bidders have *some* idea of what the current highest bid is. This feedback is given through the ask price.

2.1.4 Sealed envelope auctions

Let's now turn to sealed envelope auctions. Before considering the more general case with multiple items (e.g., ad spaces), we again focus on just one.

As stated, each bidder submits his bid privately, and they are all revealed simultaneously to the auctioneer. Figure 2.11 illustrates this. The auctioneer then decides the **matching** (i.e., allocation) and how much to charge for it. The matching part is easy: the highest bidder gets the item. But the amount charged will vary, depending on what type of sealed envelope auction it is.

- In a **first price auction**, the winner pays the highest bid, which is his own.
- In a **second price auction**, the winner pays the second highest bid.

We illustrate this difference in Figure 2.12.

Intuitively, a second price auction sounds wrong. If I want an item, and I know the winner will pay the *second highest* bid, why not bid an extremely high amount, way above my own valuation? Well, it turns out that this intuition is wrong, for a simple reason: if everyone engaged

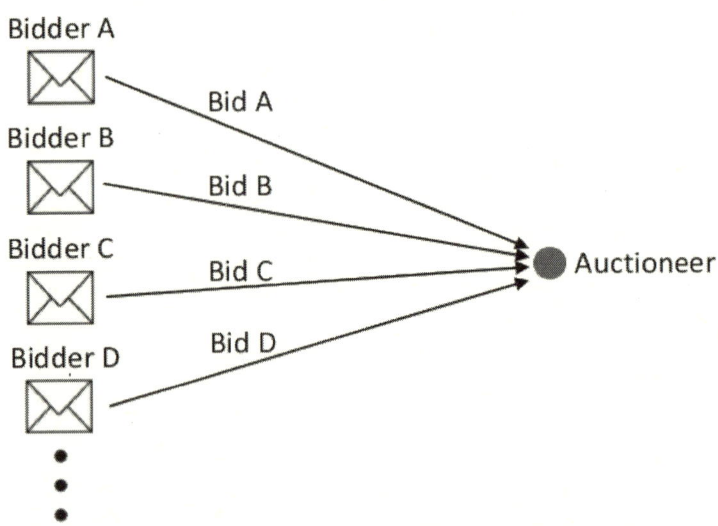

Figure 2.11 Illustration of a sealed envelope auction. Each of the bidders (A, B, C, ...) submits their bids privately to an auctioneer. The outcome of the auction is clear: for a single item, the highest bidder will win it. Determining what to *charge* is a different story.

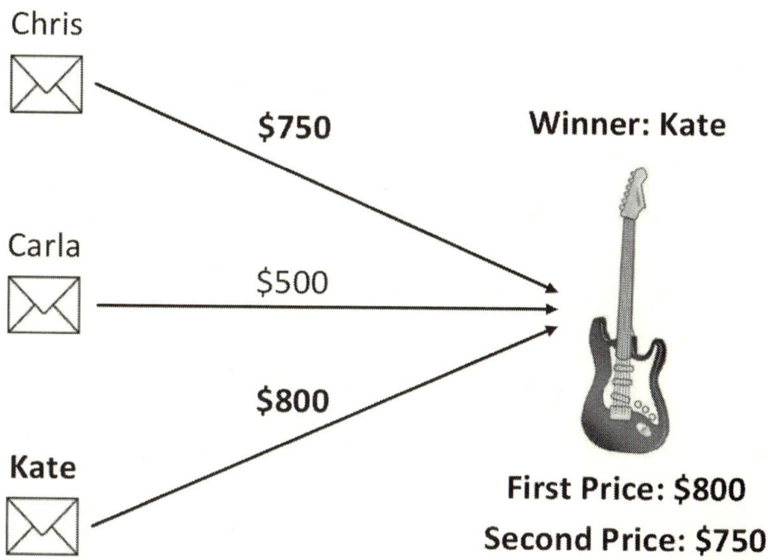

Figure 2.12 In this sealed envelope auction, Kate is the winner of the guitar, with the highest bid of $800. Under first price, she would pay her bid of $800, and under second price, she would pay the second highest bid (Chris') of $750.

in that same strategic thinking, then this "second highest bid" would be extremely high. So, the winner would be paying much more for the item than she thinks it's worth.

A bidder's goal in an auction is to maximize her **payoff**: the net benefit she receives at the end of the auction. If a bidder wins an item, her payoff is the difference between her valuation and the price she pays:

$$\text{Payoff} = \text{Valuation} - \text{Price paid}. \tag{2.1}$$

On the other hand, if she loses, her payoff is zero.

The second price is right

In a first price, sealed envelope auction, the price the winner pays is her own bid, so her payoff is her valuation of the item less what she had bid. In general, it's difficult to determine a person's *optimum* bid to maximize her payoff, because whether or not she wins depends on everyone else's strategy. But one thing is clear: the bid should *always* be less than the valuation, since otherwise winning the item can only

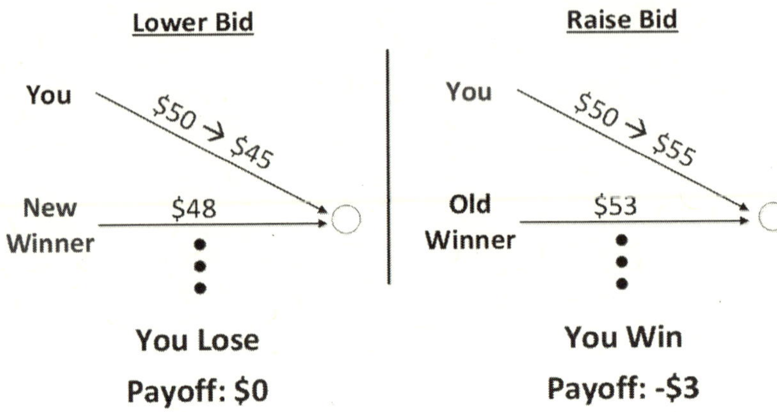

Figure 2.13 With sealed envelope, second price auctioning, it is in your best interest to bid truthfully. If you lower your bid, you run the risk of losing the item when you would have won with positive payoff. And if you raise your bid, you could end up winning and paying a price higher than your valuation, resulting in negative payoff.

result in a *negative* payoff. In this case, the bidder would lose more than she gains.

In a second price auction, the winner pays the second highest bid, so her payoff is her valuation less this number. It turns out that this type of auction encourages **truthful bidding**, meaning that the best strategy of each player is to submit her true valuation as her bid. To see this, suppose you are bidding on an item, and you initially decide to submit your true valuation, say $50. We will walk through two scenarios: if you try to increase your bid, and if you decide to decrease your bid.

- *Lower your bid*
  Suppose you decide to lower your bid, to say $45. The only way this will change the outcome is if you were originally going to *win*, and you lower it enough that the *second highest* bidder now wins. Maybe this person had submitted $48. Originally, you had a positive payoff: you were going to pay the second highest bid of $48, and your valuation was $50 – $48 = $2. But now that you lose the auction, your payoff is zero! Clearly, you should have bid your true valuation: positive payoff is better than zero payoff.
- *Raise your bid*
  On the other hand, suppose you raised your bid, to say $55. The only way this would change the outcome is if you were originally going to *lose* the auction, and you raise it enough that you now win and the

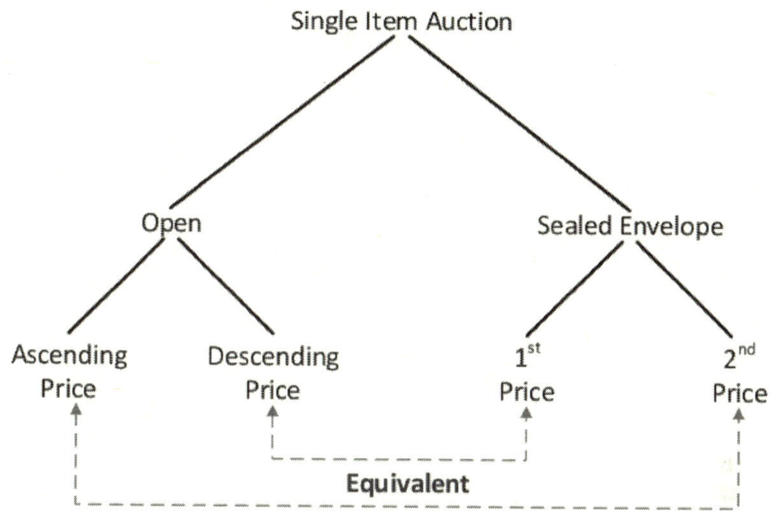

Figure 2.14 A taxonomy of the single-item auctions we have discussed. It turns out that an open, ascending price auction is essentially equivalent to a sealed envelope, second price auction, as is descending price to first price.

previously highest bidder loses. Maybe this person had submitted \$53. Originally, you had zero payoff. Now that you win the item, you pay \$53, meaning your payoff is $\$50 - \$53 = -\$3$. This value is *negative*, because your valuation is *lower* than the previous highest bid. Clearly, you should have bid your true valuation here, too: zero payoff is better than negative payoff.

We summarize these conclusions in Figure 2.13. In either case, you should have stuck with your original bid. So for a second price auction, it's in your best interest to bid truthfully.

This argument seals the math, but what is the intuition? For example, why not a third price auction? It turns out that a second price auction is the only one that makes each user capture her own **negative externality**, or the damage that is caused to everyone else by her very presence. If we took the winner out of the auction, the second highest bidder would win and gain her associated payoff. In this way, the highest bidder is paying for the fact that this person did not win.

Comparison between auctions

We can draw striking similarities between the auctions we have discussed so far. First, consider an ascending price, open auction. While the price is increasing, all the bidders have their valuations in mind, and each will stay in the game until the current bid is higher than their valuation. The current price will keep escalating until the bidder with the *second highest* valuation opts out. So, unless the winner increased the current price drastically, he pays the second highest bidder's valuation (plus some small difference). In this way, *ascending price* is equivalent to *second price* auctioning.

Now, consider a descending price, open auction. The auctioneer will keep decreasing the price until it reaches the highest bidder's level, at which point the highest bidder will stop the auction. And given that he doesn't wait for it to drop further (*i.e.*, he is cautious), he will pay his own valuation, which is the *highest* price. In this way, *descending price* is equivalent to *first price* auctioning.

A taxonomy of the single item auctions we discussed, as well as these equivalence relations, is given in Figure 2.14.

Generalized second price auctions

So far, we have been examining auctions with only a single item to sell. But AdWords, as well as other search ad companies, have multiple ad spaces to sell in each auction. As a result, there are multiple items facing each bidder.

Let's assume there are 3 advertisers (bidders) and 3 ad spaces, as shown in Figure 2.15. Each of Bidders A, B, and C have different expected revenues per click, denoted `rev A`, `rev B`, and `rev C`. At the

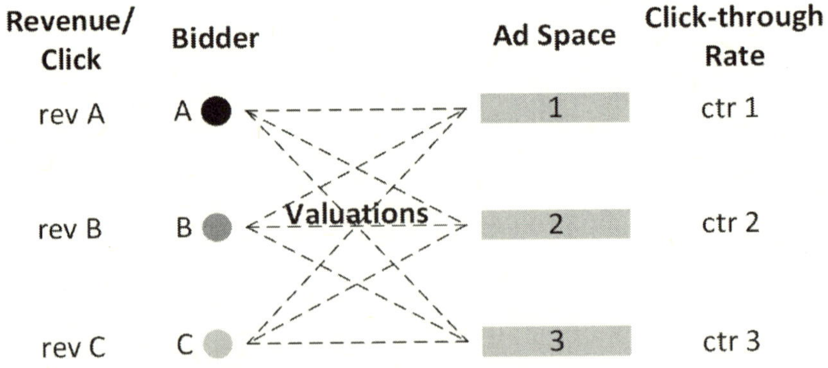

Figure 2.15 An example of a Google ad auction with 3 bidders (A, B, and C) and 3 spaces (1, 2, and 3). An AdWords GSP auction will allocate one space to each buyer, where the winner of a space has to pay the price the second highest bidder would have paid for that space.

same time, each of Ad Spaces 1, 2, and 3 have different click-through rates, denoted `ctr 1`, `ctr 2`, and `ctr 3`. As a result, there are 9 different valuations, three for each bidder. For instance, Bidder B's expected revenue (per hour) for Ad Space 3 is `rev B` × `ctr 3`, and Bidder C's for Ad Space 1 is `rev C` × `ctr 1`.

The job of a multiple-item auction is to allocate a separate item to each advertiser. In general, there can be more buyers than spaces, or vice versa, but this is a simple extension that just leaves advertisers with no space, or some empty spaces, respectively.

Rather than submitting a bid for each space, Google only wants a single number from each advertiser: the amount he will pay for *one click*. Google assumes the bid for each ad space to be proportional to this number. This is actually a significant simplification: why not have each person submit a separate bid for each ad space? Maybe I value the first ad space at $100 per click, while I only value the second at $10 per click. For simplification, the sponsor search industry feels that entering a single number suffices.

For ad space allocation and charging, Google AdWords uses a **generalized second price** (GSP) auction, as follows.

- *Bid*: Each buyer submits a bid, which is the price they are willing to pay Google per click.
- *Matching*: The ad spaces are allocated to buyers in descending order of their bids. So, the highest bidder gets the first ad space, the second highest the second space, and so on.
- *Payment*: The price each buyer pays follows the second price for a single item methodology. The highest bidder pays the price the second highest bidder would pay for the first ad space, the second highest bidder pays the price the third highest bidder would pay for the second ad space, and so on. If there are at least as many ad spaces as buyers, the last buyer likely pays Google a standard minimum bid.

Suppose that Anna, Ben, and Charlie are bidding on three ad spaces. Their revenues are 20, 40, and 10 dollars per click, while the click-through rates of the spaces are 10, 5, and 1 click per hour, respectively. Assume that each person bids his or her true valuation ($20, $40, and $10), and that Google's minimum bid is $3 per click. Let's take a look at the GSP auction.

- *Bidding*: Anna, Ben, and Charlie will submit $20, $40, and $10 per click, respectively, as their bids.

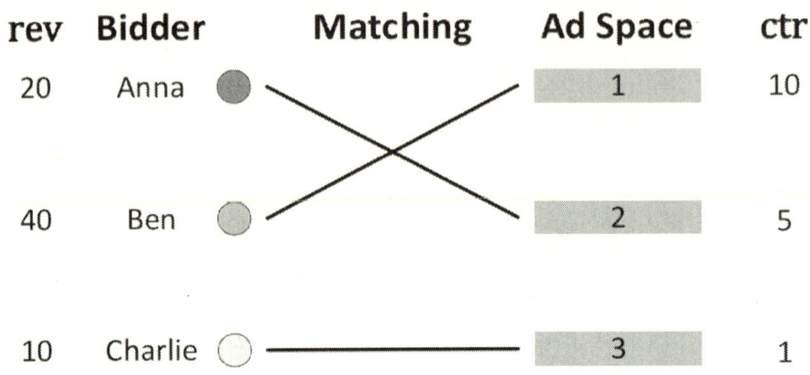

Figure 2.16 GSP allocation of ad spaces to advertisers assuming truthful bidding.

- *Matching*: By the GSP mechanism, the highest bidder gets the most valuable ad space, the second highest the second most valuable, and so on. Therefore, Ben gets the first space, Anna the second, and Charlie the third. This allocation is illustrated in Figure 2.16.
- *Pricing*: By GSP, a buyer pays the price the next highest bidder would have paid for her space. So, Ben is charged \$20 per click (Anna's bid), which amounts to `rev A` $\times$ `ctr 1` $= \$20 \times 10 = \$200$ per hour. Similarly, Anna is charged \$10 per click (Charlie's bid), meaning `rev C` $\times$ `ctr 2` $= \$10 \times 5 = \$50$. Finally, Charlie is charged the minimum of \$3 per click, amounting to \$3$\times$`ctr 3` $= \$3 \times \$1 = \$3$.
- *Seller's revenue*: Google has collected $\$200 + \$50 + \$3 = \$253$ per hour from the auction.
- *Buyers' payoffs*: We compute each buyer's payoff by taking the difference between the valuation and the price. Ben's valuation of the first ad space is `rev B` $\times$ `ctr 1` $= \$40 \times 10 = \$400$, so his payoff is $\$400 - \$200 = \$200$ per hour. Anna's valuation of the second ad space is `rev A` $\times$ `ctr 2` $= \$20 \times 5 = \$100$, so her payoff is $\$100 - \$50 = \$50$ per hour. Finally, Charlie's valuation is `rev C` $\times$ `ctr 3` $= \$10$, so his payoff is $\$10 - \$3 = \$7$ per hour. In summary, the total payoff is $\$200 + \$50 + \$7 = \$257$. We illustrate this computation step by step in Figure 2.17.

Table 2.2 summarizes the results of the auction from the buyers' perspective.

GSP is precisely Google's solution to finding consensus among ad space bidders, determining their ordering in the search results, while

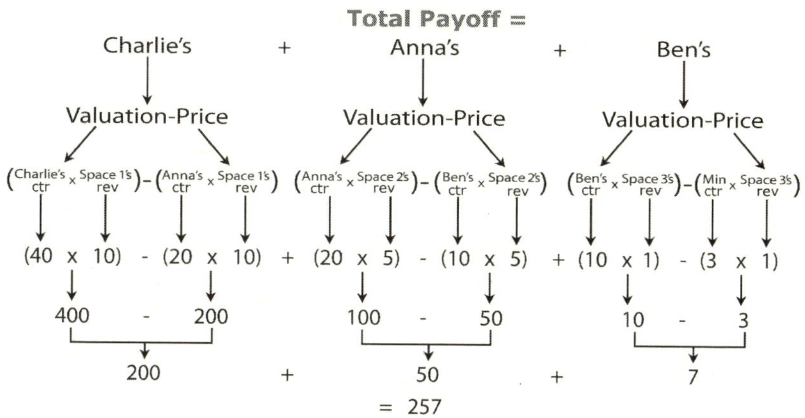

Figure 2.17 Illustration of how the total payoff is computed for our example.

Buyer	Revenue ($)	Ad	Valuation ($)	Price ($)	Payoff ($)
1	20	2	100	50	50
2	40	1	400	200	200
3	10	3	10	3	7

Table 2.2 Summary of the GSP ad auction example from the point of view of the buyers, assuming truthful bidding. The revenue is per click, while the valuation, price charged, and payoff are per hour.

maximizing profit. Note that GSP is not the only mechanism that can be employed to determine a matching. Others, which we won't have time to discuss, may yield a different result. In fact, GSP for *multiple* item auctions does *not* encourage truthful bidding, even though it does for single item auctions. Another method, called **Vickrey–Clarke–Groves** (VCG), will encourage it in either case. However, there are disadvantages to VCG as well. This is why the question of what the "right" consensus is can have more than one "right" answer.

We now understand how Google makes over 90% of its profits. Let's now explore how they ranks their standard webpage results to make searching as efficient and high quality as possible.

2.2 PageRank by Google

There's no doubt about it: you and most everyone you know use Google search. Type a phrase into the search query, hit enter, and out pops

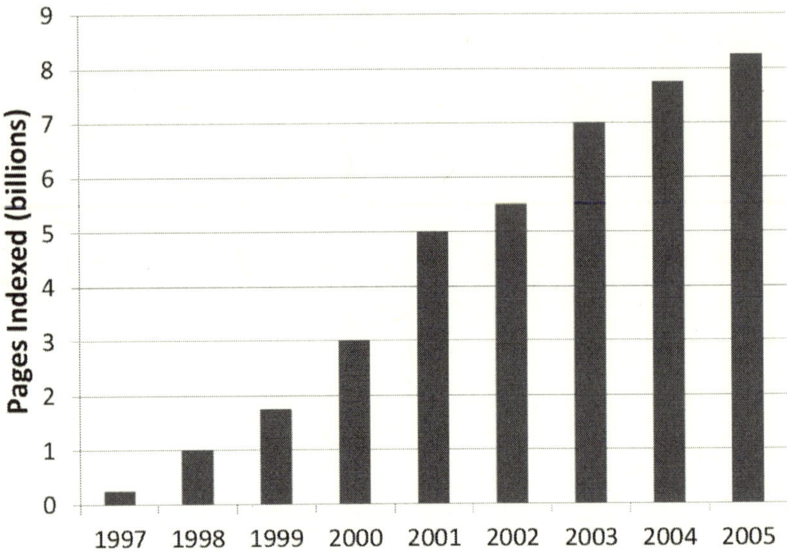

Figure 2.18 The number of webpages indexed by Google by year from the search engine's origins in 1997 until it stopped posting the number on its home page in 2005. In 2012, it was estimated that Google has 40 billion webpages in its database, a five-fold increase over the 8 billion in 2005.

perhaps hundreds of millions of relevant links. And it's likely that you'll find exactly what you're looking for within the first five or so results.

The amount of information on the Internet has been growing at a rapid rate since the introduction of the web in 1989. It's hard to estimate exactly how many webpages exist today, but current statistics place the number at possibly 60 billion. Google is constantly *crawling the web* to update the pages in its database, as well as to add newly published ones. The estimated size of the web that is *indexed* by Google, *i.e.* the number of webpages Google "knows of," is currently 40 billion. This number is up from 8 billion in 2005, and 24 million back when Google was a prototype at Stanford in 1997. We show the number of indexed pages over time in Figure 2.18. The graph only shows data up to 2005 because that is the year Google stopped publicizing the database size on its home page.

If Google indexes this many pages, how is it that when you enter a query, you can usually find more than enough of what you need within the first few results? Clearly, Google doesn't display the pages it finds matching your search one-by-one. In fact, it has a *ranking* algorithm that solves huge systems of equations each time you search, to return

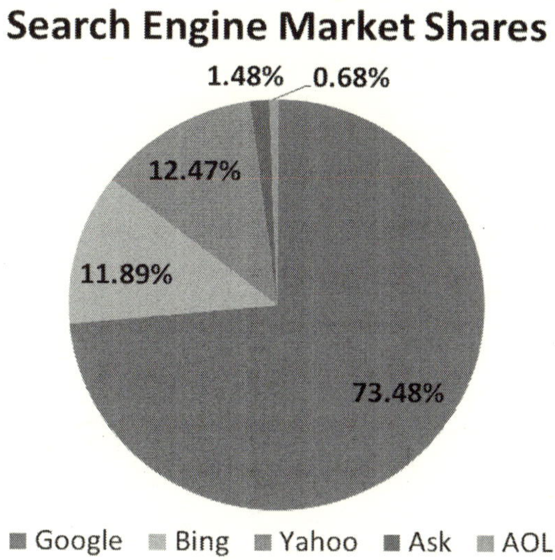

Figure 2.19 Market share of the five most popular search engines in May 2012. Bing's and Yahoo's portions are close, as are Ask's and AOL's. Google search possesses almost three quarters of the total market, roughly six times that of Yahoo search.

your results in order of *importance*. We will look how a page's "importance" is determined in this section.

2.2.1 Relevance vs. Importance

The idea of ranking webpages dates back to the early 90s, when the initial search engines were created. The first well-known search engine to provide **full text search**, where the words in the query are matched to the full content of the documents, was WebCrawler in 1994. Soon after, between 1995 and 1996 some prominent engines appeared, including AltaVista, Yahoo, and Ask Jeeves. All of these engines ranked their results based upon how often the search terms occurred in the pages, or how strongly associated the terms were with the pages. In this way, they would display the most *relevant* webpages first.

Then along came the two Google founders, Sergey Brin and Larry Page, in 1997 with a novel approach for designing a search engine. To them, the best way to rank web pages effectively in a short amount of time was to take two main factors into consideration:

- A **relevance score**: How relevant the content on a web page is to the search query.

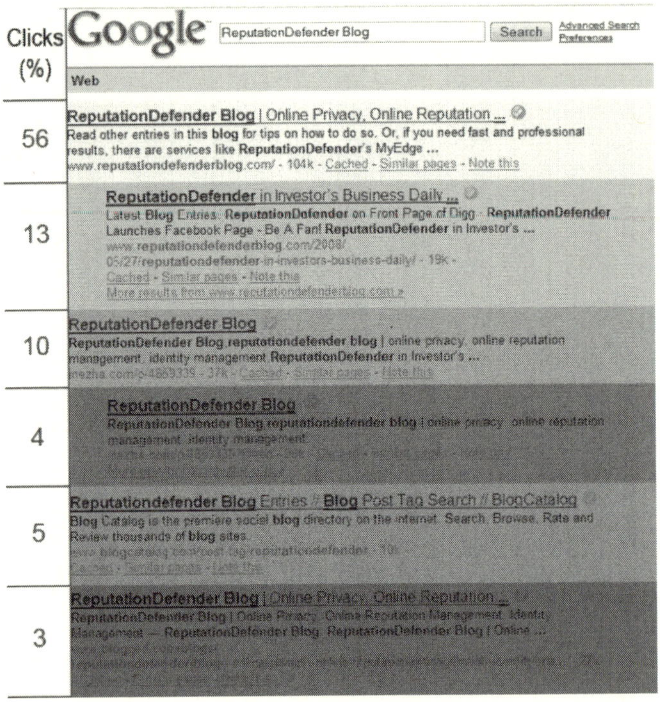

Figure 2.20 A study at Cornell circa 2008 found that the first three search results on Google receive, on average, roughly 80% of clicks. Further, the page of the first ten results receives over 98% of all clicks.

- An **importance score**: How *important* a webpage is, irrespective of the content on the page.

Computation of the relevance score did set Google's vision apart from previous search engines, because it takes factors such as capitalization, font, and position of content into account. But it's the notion of *importance* that made PageRank, Google's ranking algorithm (a clever pun on Larry Page's name), much more successful than previous approaches. It has been a driving force in bringing Google to fame over the past decade. And today, Google possesses almost three quarters of the total search engine market, as shown in Figure 2.19.

In a Google search result, the rank of each webpage is really a composite of the relevance and importance scores. But for our purposes, we will focus on the latter, since that usually determines the order of the top few webpages in any reasonably popular search. And naturally, these first few pages are by far the most visited. Typically, the first three links receive roughly 80% of all clicks, and the entire first page receives over 98%. This is illustrated in Figure 2.20.

2.2.2 Webgraphs

Webpages are connected to one another through **hyperlinks**, i.e., references to external data from within a page that a user can follow. The idea of embedding hyperlinks in text was integral to the creation of the World Wide Web, because hyperlinks are the means by which webpages connect to one another. That is, we say that one page *references* another if it has a direct hyperlink that will take the user to it.

Connections between webpages can be represented succinctly using a **graph**. We could spend an entire course on the mathematics of graph theory, but simple terminology here will suffice: a graph consists of a set of **nodes** (or vertices) which are interconnected by **links** (or edges). In our context, we view nodes as webpages, and links as an indication of whether one page references another or not. As such, we are constructing a **webgraph**. A webgraph is **directed**, meaning its links are asymmetric: just because Page A references Page B doesn't mean Page B will reference Page A.

Webgraphs are essential to understanding the computation of impor-

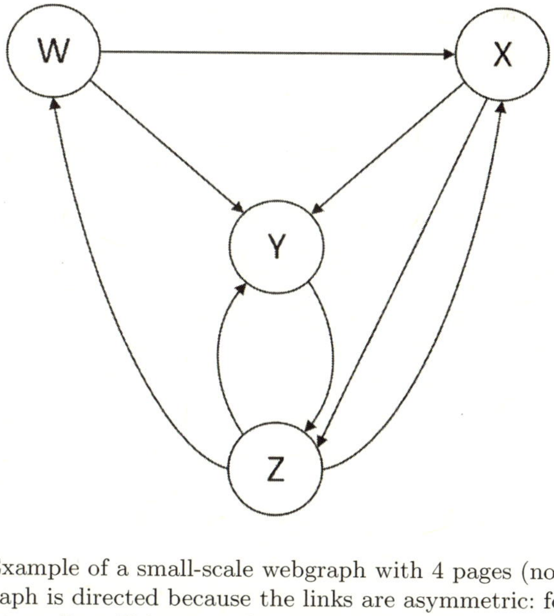

Figure 2.21 Example of a small-scale webgraph with 4 pages (nodes) and 8 links. The graph is directed because the links are asymmetric: for example, Page W points to Page X but not the other way around.

tance scores because they encapsulate relationships among webpages. But a webgraph of the Internet would take quite a long time to write down: it would have somewhere around 60 billion nodes. Additionally, it is extremely **sparse**, meaning most webpages are not connected. For instance, a Wikipedia page may have a few hundred links, depending on its size, which is a very small subset of the billions in total.

To fit the size of a page, we will use the very small graph shown in Figure 2.21 to illustrate the key steps in computing importance scores. Here, we have four pages (W, X, Y, and Z), and eight hyperlinks. We assume that all of these pages are relevant to the given search query.

So, what makes a page important? Perhaps the number of others that point to it? This is called the **in-degree** of the webpage, and measures how many **incoming links** the node has.

- From Figure 2.21, we see that Page Y has incoming links from Pages W, X, and Z, giving it an in-degree of three.
- On the other hand, Page X has an in-degree of two, with incoming links from W and Z.
- Page Z also has an in-degree of two, with links from X and Y.
- Finally, Page W has an in-degree of 1, with a link from Z.

By this logic, Google would return Y, X, Z, and then W (X and Z are interchangeable) as its results, in that order. But does this tell the whole story? Google thinks not.

2.2.3 The "Random Surfer"

Google explains the PageRank concept through an analogy to someone surfing the web at random. This surfer is given a web page, and keeps clicking on links randomly. During this process, she may eventually get bored and enter some other address into the browser. We illustrate this idea in Figure 2.22: the person goes from A to D and then E based on hyperlinks, then randomly enters F into her browser, then clicks on G, and so on. According to PageRank, the probability that a given page is visited in this process is the page's *importance*.

We will break the random surfing process down step-by-step. Let's first ignore the idea of "getting bored," and assume the surfer follows the hyperlinks. From a given page, the probability that he will select a specific hyperlink is *inversely proportional* to the **out-degree** of the webpage. The out-degree is simply the number of **outgoing links** the node has.

Back to our example in Figure 2.21. The out-degree of Page W is

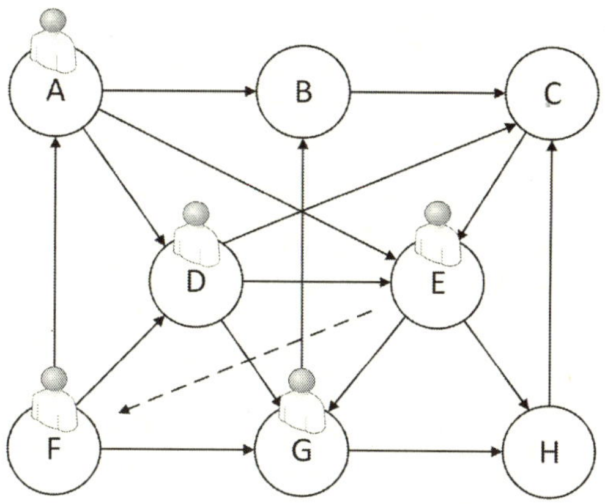

Figure 2.22 Simple illustration of the random surfer philosophy. A person starts on webpage A, and randomly clicks on one of the hyperlinks, selecting D. From D she chooses a reference again, ending up at E. Once at E, the person decides to enter a random URL into the browser, ending up at F. From F, she clicks on a link to G, and so on.

2. Therefore, there's a 50% chance a random surfer on Page W will navigate to Page X, and a 50% chance he will choose Page Y. He can't go directly to Page Z because there's no such link.

But what's the chance that the user will be on a given page in the first place? This will surely depend on the in-degree of the page, as we mentioned previously. But it depends on something more: the *importance* of the links that point to the page. For instance, while Z only has two incoming links, one of those links is from Y, which we previously said might be the most important. If Y is that likely to be visited, surely Z will be *at least as likely*, because once on Y, the only option the surfer has is to click on Page Z!

We can represent these concepts of in-degree, out-degree, and page importance conveniently with the webgraph.

- *Links*
  Each link has an *origin* and a *destination*. A link indicates the probability that a random surfer will click to go to the destination page from the origin.
- *Nodes*
  Each node has an importance score. Call them w, x, y, and z, for Pages W through Z, respectively. In general, we can equate the importance of a page to the chance that it will be selected in the random surfing process. For instance, Page X could be selected if

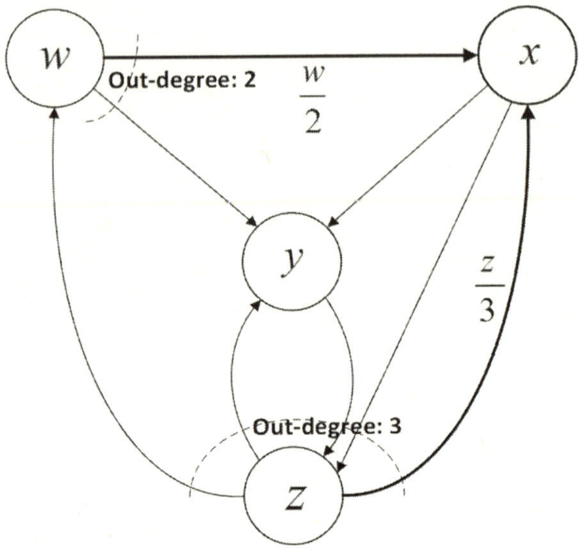

Figure 2.23 The importance score for Page X is dependent upon those of W and Z. Since W and Z have out degrees of two and three, these these terms become $w/2$ and $z/3$.

the surfer is on either of Pages W or Z. If on Page W, the chance of transitioning to Page X is 50%, and if on Page Z, the chance is 33.33% (three possibilities, so 1/3). And the chance that the surfer is on either of these two pages in the first place is in turn dependent upon their *own* importance. From this, we see that $x = w/2 + z/3$. We illustrate this in Figure 2.23.

Similarly, take Page Y. It can be selected from any of the other pages in the graph, since they all point there. While at W, there's a 50% chance of picking Y; from X, there's also a 50% chance; and from Z, there is a 33.33% chance. Therefore, $y = w/2 + x/2 + z/3$. We illustrate this in Figure 2.24.

Applying this to each page in Figure 2.21, we have

$$\begin{aligned} w &= \frac{z}{3}, \\ x &= \frac{w}{2} + \frac{z}{3}, \\ y &= \frac{w}{2} + \frac{x}{2} + \frac{z}{3}, \\ z &= \frac{x}{2} + y. \end{aligned} \tag{2.2}$$

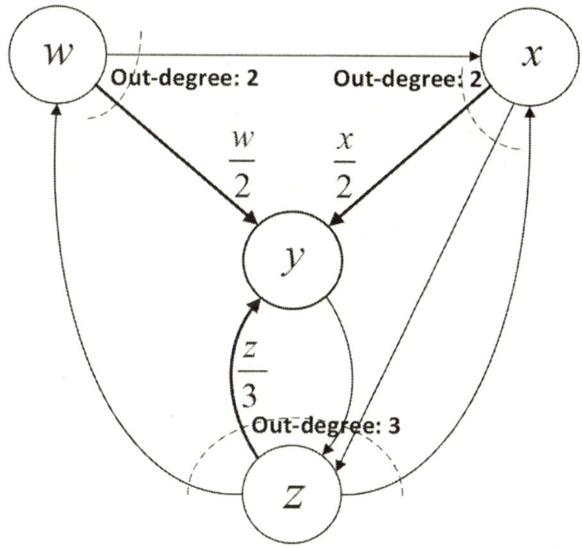

Figure 2.24 The importance score for Page Y is dependent upon those of W, X, and Z. Since W and X both have out degrees of two, these terms become $w/2$ and $x/2$. Since Z has an out degree of three, this term becomes $z/3$.

In general, a webpage's importance score is dependent upon other webpages', which is in turn dependent upon the original's. This type of *seemingly* circular logic is called **recursion**, and requires us to solve a system of *simultaneous equations*. The webgraph is an easy way to visualize the equations, by following the procedure that we illustrated in Figures 2.23 and 2.24:

1. Label each link as the ratio of the importance of the origin node to its out-degree.
2. At each node, equate its importance score to the sum of all the values on incoming links.

The number of equations you have will be the number of nodes. We show this procedure for the entire webgraph in Figure 2.25.

What we have now are four equations and four unknowns (w, x, y, and z). Is there a solution? Let's find out in the least tedious way possible.

The calculation

To simplify everything, let's consider what each of the importance scores represent. They are the *probabilities* of each node being visited

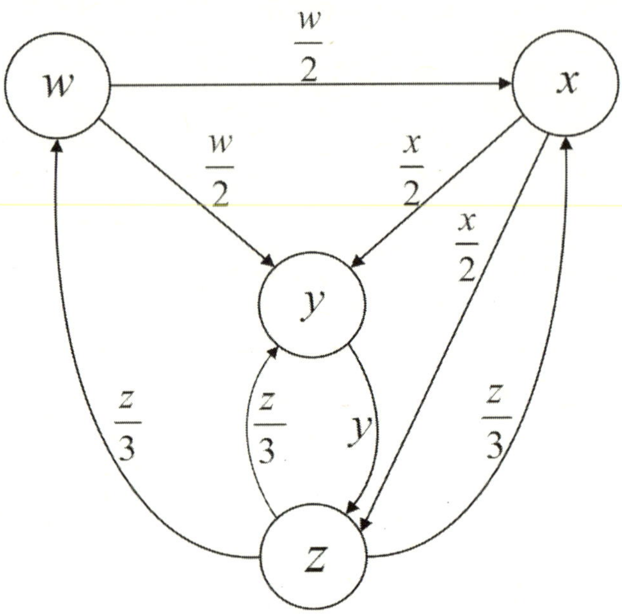

Figure 2.25 The webgraph provides a convenient way of visualizing the equations for importance score.

during the random surfing process. And these probabilities must sum to 1, since at least one page must be visited:

$$w + x + y + z = 1. \tag{2.3}$$

Equation (2.3) *normalizes* the sum of the importance scores, and is particularly convenient to work with. In fact, the easiest way to solve the system is to choose one importance score as a reference, get all other scores in terms of this one, and then plug the expressions into Equation (2.3). We will do that now.

From the first of our equations in (2.2), we have w in terms of z already. As a result, it will be convenient to get everything in terms of z. We can use the second and fourth equations (the simpler ones) to do this.

From the first, we have that

$$w = \frac{z}{3}.$$

We can use this to substitute for w in the second equation:

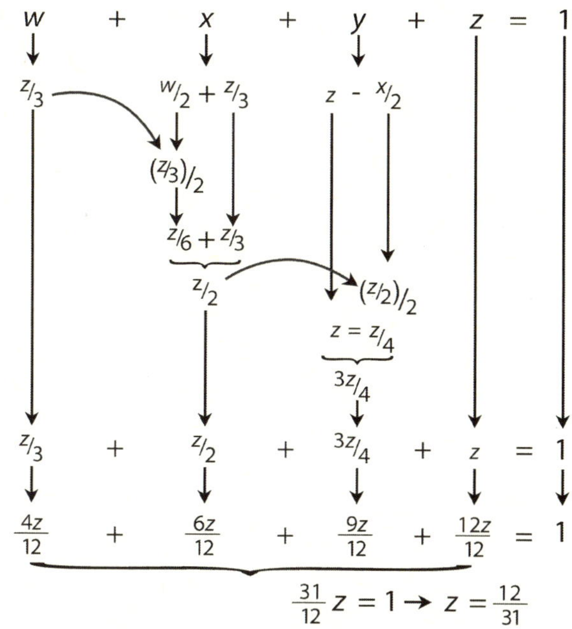

Figure 2.26 Illustration of the steps involved in computing the importance score of Page Z in our example.

$$\begin{aligned} x &= \frac{w}{2} + \frac{z}{3} \\ &= \frac{z}{6} + \frac{z}{3} \\ &= \frac{z}{2}. \end{aligned}$$

Now, we have x in terms of z. We can use the fourth equation to do the same for y. First, we re-arrange to get y on the left:

$$y = z - \frac{x}{2}$$

Then, we can substitute for x, using our previous result that $x = z/2$:

$$\begin{aligned} y &= z - \frac{z}{4} \\ &= \frac{3z}{4}. \end{aligned}$$

With all the importance scores now in terms of z, we can plug into Equation (2.3):

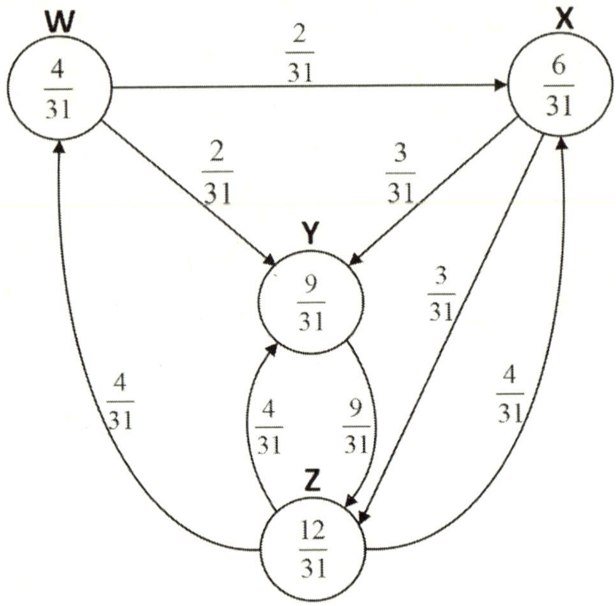

Figure 2.27 Importance scores and link weights for our example webgraph. You can see that each page's importance is the same as the sum of the incoming as well as the outgoing importance of the page. For instance, at Page X, $6/31 = 2/31 + 4/31 = 3/31 + 3/31$.

$$\begin{aligned} 1 &= w + x + y + z \\ &= \frac{z}{3} + \frac{z}{2} + \frac{3z}{4} + z \\ &= \frac{31z}{12}. \end{aligned}$$

With this, it follows that $z = 12/31 = 0.387$.

For clarity, we illustrate the steps in computing z in Figure 2.26. Now, we can backtrack and solve for the rest of the importance scores:

$$\begin{aligned} w &= \frac{z}{3} = \frac{4}{31} = 0.129, \\ x &= \frac{z}{2} = \frac{6}{31} = 0.258. \\ y &= \frac{3z}{4} = \frac{9}{31} = 0.290. \end{aligned}$$

You are encouraged to plug these values back into the equations in (2.2)

Page	In-degree		PageRank	
	Importance	Rank	Importance	Rank
W	0.125	4th	0.129	4th
X	0.250	2nd (or 3rd)	0.258	3rd
Y	0.375	1st	0.290	2nd
Z	0.250	3rd (or 2nd)	0.387	1st

Figure 2.28 Importance scores and rankings computed for the webgraph in Figure 2.21. The results for both Google's PageRank algorithm and the simple in-degree metric are shown. PageRank leverages all the connectivity in the webgraph, whereas in-degree doesn't.

to verify that all are true. In particular, it is important to note that the third one is satisfied, which we did not make use of explicitly. The solution is summarized in Figure 2.27, where you can see that adding the values on the incoming or outgoing links to each page gives the importance of the page.

Importance summary

What do we notice about these importance scores? The webpage rankings from most to least important are Z, Y, X, W. With in-degree, we had them as Y, X, Z, W. There are two main ways that these are different.

1. *With PageRank, Y is not the most important.* This is because Page Y aggregates its importance from half of Pages W and X, the two less-important nodes, and only one-third from Page Z.
2. *With PageRank, Z is the most important.* This is because Page Z aggregates its importance from the *entirety* of Page Y, which points to no other nodes, *and* a portion from Page X as well.

We summarize the importance scores and rankings for each method in Figure 2.28. Note that the in-degree importance is obtained by dividing each page's in-degree by the sum of the in-degrees (8), to normalize the result. So for Page X with an in-degree of 2, we get $2/8 = 0.25$.

2.2.4 Robust ranking

What we've just discovered is a procedure that allows us to find a set of importance scores that is consistent across various equations, thereby achieving consensus. However, given a general webgraph, are we *always* guaranteed a *unique* consensus?

The answer is: not quite yet. In general, we have to make two modifications to our procedure to guarantee a single, unique solution. Otherwise, we could have either numerous or no sets of importance scores that result in a valid consensus on what the ranking should be.

Dangling Nodes

First, *there will be no solution* to the problem if the webgraph has one or more **dangling nodes**. A dangling node is one that does not point to any other webpages. For instance, if we augment Figure 2.21 with the fifth node V as illustrated in Figure 2.29, then V is a dangling node.

Let's think about what implications this has on V's importance score (call it v). As we said, the PageRank procedure has each node spread a portion of its importance score to all of its outgoing neighbors. With a single outgoing link, that link gets all the score, with two, each gets half, and so on. As such, the sum of the importance of the outgoing links is equal to the node's importance. But V has *no* outgoing links: as a result, v must be 0 for the equations to work out.

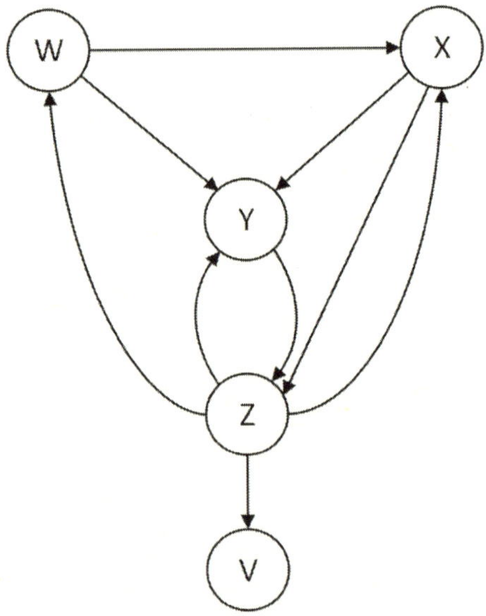

Figure 2.29 Here, we have modified our original example to include a dangling Node V. It has an incoming link from Node Z, but no outgoing links, which requires its importance score to be zero. The solution to this is to assume that V has a link to every node, rather than no nodes.

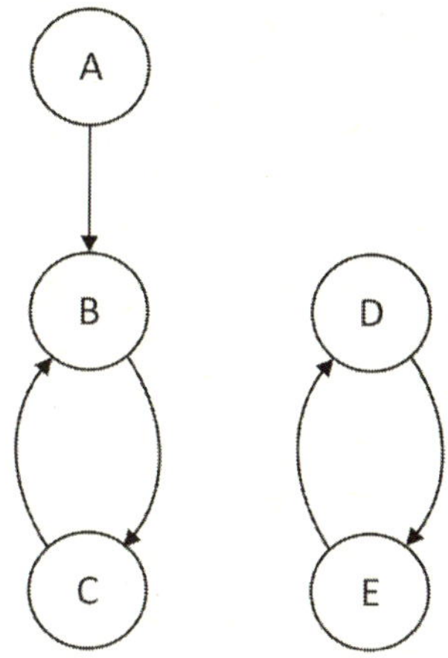

Figure 2.30 Unconnected graphs are problematic, because they allow for an infinite number of solutions in the PageRank algorithm. This is dealt with by adding some randomization into the procedure, so that a surfer can get between the different components of the graph.

Now, what does $v = 0$ imply for the rest of the graph? Well, Page Z spreads one-fourth of its importance score to V, and as a result $v = z/4$. So if v is zero, then z must be zero too! This logic cascades through the rest of the graph, requiring the scores for the rest of the nodes to be zero as well.

PageRank's solution to this problem is to assume that each dangling node has an outgoing link to *every* node (including itself), rather than none. This is intuitive: if a random surfer lands on such a page, he would enter some other link into the browser and keep going.

Disconnected Graph

But even if we apply this fix, *there will be many solutions* if the webgraph has more than one **connected component**. A connected component is simply a group of nodes in which any two can reach each other (directly or indirectly), but none can reach any outside of the group. For instance, in Figure 2.30, Pages A - C form one connected component, and Pages D and E form another. The problem here is that a random surfer can't get from one component to the other: depending on which he starts in, he would be stuck in that **subgraph** forever. We have no way of relating the importance of nodes in one component to the other.

The solution to this is precisely what we neglected initially: the surfer might "get bored" during the process, and enter some random website into the browser. So for a certain portion of the time, say 15%, he will follow the webgraph in deciding where to go next, and for the remaining 85% of the time, he will randomly choose from all the available webpages.

If we apply these two fixes to our procedure - one for dangling nodes, and one for unconnected components - then PageRank will always have a unique answer.

Between this section and the last, we have explored two of the fundamental notions behind Google's operation. We have seen how they uses large webgraphs to rank webpages, and how they leverage bids to mediate auctions involving large numbers of customers. Both of these are prime examples of how finding consensus in networks – whether among a set of equations or a set of bids – can be difficult.

2.3 Reaching agreement on Wikipedia

We will turn in this section to consensus formation from a potentially large body of conflicting opinions. While the general notion of finding a common ground among a set of data is the same, here we are ranking *preferences*, rather than objects as in PageRank.

Our case study in this section will be Wikipedia, which has become the world's largest encyclopedia among those both online and off. It is likely that the freely-open nature of Wikipedia would not work as well as it does if there weren't methods of achieving "rough consensus" in cases where contributors and editors disagree.

2.3.1 From Britannica to Wikipedia

Prior to the commercialization of the Internet in the 1990s, encyclopedias were book-based. The oldest English-language printed encyclopedia still being produced is the *Encyclopædia Britannica*, which began all the way back in 1768 in Scotland. The first edition was merely three volumes compiled by a single editor. Over the years, it has changed ownership many times, and has released 15 separate editions. Today, it has almost 4,500 contributors and 100 full time editors who continuously update its content.

Britannica's last printed edition was in 2010, culminating its knowledge into a 32-volume set (see Figure 2.31). But in March 2012, it announced that it would no longer publish printed editions, and would

Figure 2.31 After more than 244 years in print, Britannica announced that it would only continue supporting its online portion as of 2012. This is due largely to the advent of online encyclopedias, most notably Wikipedia.

instead focus on its online version. Why was that? Because electronic and online encyclopedias like Microsoft Encarta and Wikipedia had made knowledge much cheaper and more convenient to access. Today, Encyclopedia Britannica Online has roughly 120,000 articles.

With the advent of the Internet, many people attempted to develop online encyclopedias throughout the 1990s, but most of them were unsuccessful. In 2000, Jimmy Wales founded Nupedia together with efforts from his dot-com company Bomis, and hired Larry Sanger to be its editor-in-chief. All of its content was free, but pages were required to go through an extensive seven-step peer review process before being posted, and authors were expected to be experts in their fields. The hope was that by establishing a stringent editorial process, the quality of articles would reflect that of printed encyclopedias. But the downside was that it slowed down the process severely: only 21 articles were approved in Nupedia's first year.

Around this time, the concept of the **wiki** was established by Ward Cunningham: a page or collection of pages which allowed anyone with access to create or edit its content on the fly. Wales and Singer became interested in the idea, as they were looking for a simple and effective way to have content generated much faster. To that end, crowdsourcing the page writing seemed like a great idea. They created a new project called Wikipedia based on the wiki concept, and launched it on its own domain separate from Nupedia in 2001.

Within the first year, Wikipedia already had 18,000 articles, three

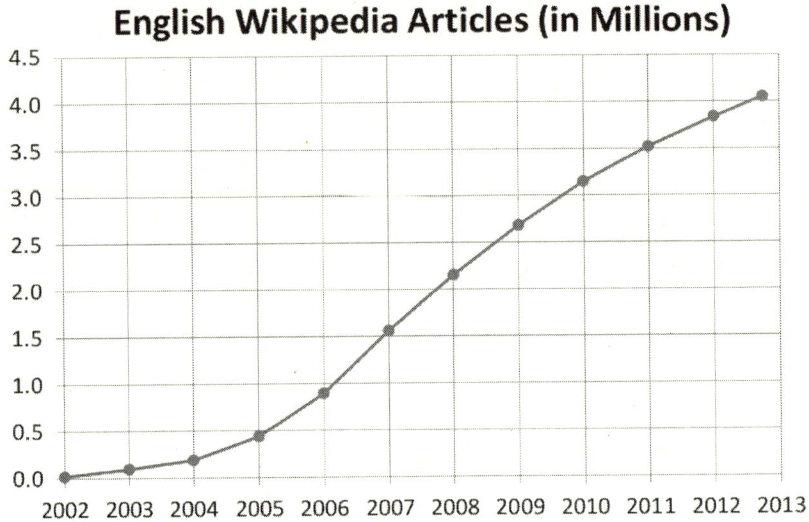

Figure 2.32 Number of Wikipedia articles in the English domain by year since 2002. In September 2012, it had over 4 million articles, which is about 17% of the roughly 23 million articles across all languages.

orders of magnitude higher than what Nupedia had accomplished. As a result, it quickly overtook Nupedia, which was discontinued in 2003.

According to some studies, by 2005 Wikipedia had become the most popular online reference site. Within a decade, Wikipedia generated over 4 million English articles and 23 million worldwide across 285 different language editions. Figure 2.32 shows the rapid growth in the number of English articles alone by year.

2.3.2 "Good faith" collaboration

Wikipedia has become one of the most popular sources of information online. As of September 2012, it was the sixth most popular website overall, receiving more than 2.7 billion visits each month in the US alone. Whether or not you cite it (or even look at it), a Wikipedia article will appear on Google search's results for most reasonable queries. In fact, one study showed that more than 99% of Google searches of nouns returned it on the first page of results. The exact percentages of times a Wikipedia page showed up in each rank are shown in Figure 2.33; there's over a 95% chance that it showed up in one of the first five positions. And in addition to being such a ubiquitous source of in-

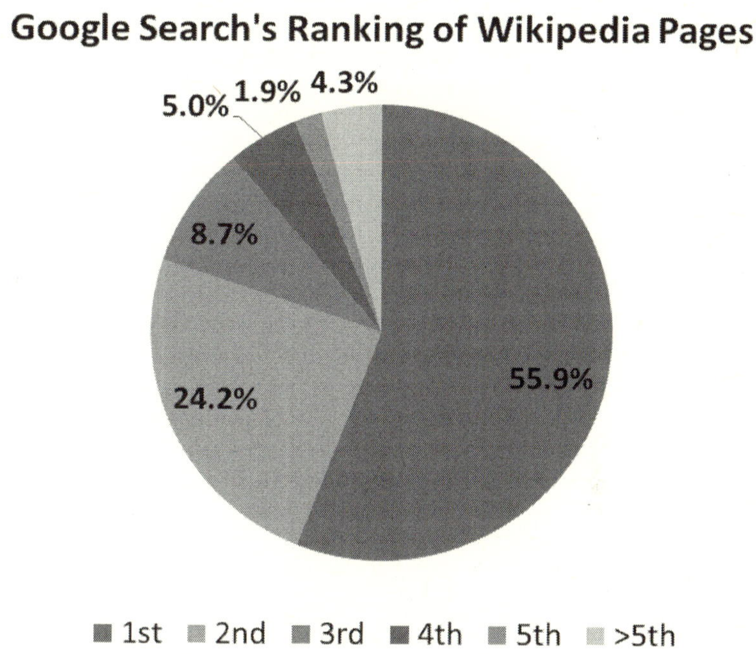

Figure 2.33 Chart of the percentage of times Wikipedia showed up in each place on the list of Google search results, when searching different nouns. So, if you search a typical noun on Google, this shows a 55.9% chance that a Wikipedia page will show up first in the results, a 24.2% chance it will show up second, and so on. There's over a 95% chance Wikipedia will show up on within the first five results.

formation, the quality of the articles on Wikipedia is consistently high, especially for certain fields like medicine.

Of course, there are also limitations to Wikipedia.

- *Misinformation*: Some information on Wikipedia is simply wrong. Even though anyone is allowed to edit articles, it's likely that errors contained in pages with a small audience will go unnoticed for long periods of time.
- *Mistakes*: Sometimes people make honest mistakes. These mistakes will remain as long as no contributor presents a stronger case otherwise.
- *Missing information*: It is impossible for an encyclopedia to be truly *complete* to everyone's liking. This is not just a limitation of Wikipedia, but of encyclopedias in general.

Due to these factors, there have been significant cases of people abusing

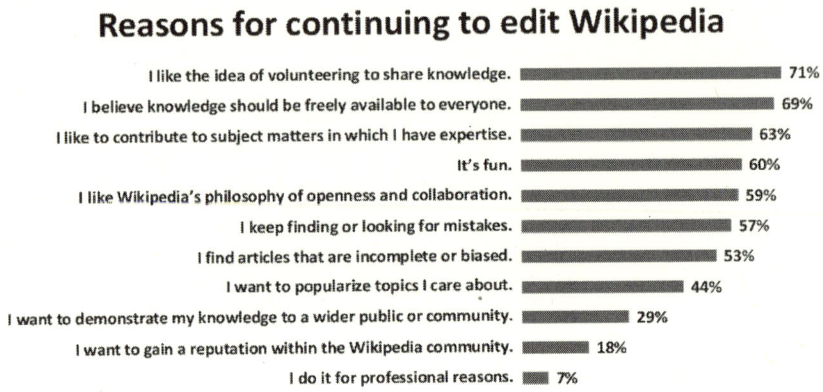

Figure 2.34 Results of a 2011 poll among Wikpedia contributors asking them what their reasons were for continuing to edit. Most are rather selfless: the ideas of contributing to knowledge, a free venue, and openness and collaboration are attractive in their own right.

Wikipedia. For instance, malicious individuals have vandalized articles, a typical case being when a person is falsely, but intentionally, reported as having deceased. For example, this happened to the musical artist Kanye West's page in 2009.

Network effect

Now, let's consider three features of Wikipedia.

1. *It's free.* If editors don't get paid, how can anyone be motivated to contribute? Well, it turns out that incentives don't have to be financial: the ability to influence others can be reward enough in its own right. It also gives people the ability to develop their reputation within the community. But according to polls taken among editors in 2011, this second reason is not as important: Figure 2.34 shows that the main motivations are selfless desires to contribute to a free knowledge base for everyone's greater good. Gaining a reputation and professional reasons are not priorities for most people.
2. *Anyone can write or add to an article.* This includes non-experts, anonymous writers, and people with conflicts of interest. The key is *checks and balances*: since anyone can contribute, Wikipedia has a large body of writers to check posts frequently through mechanisms for debating and updating. An example of this is the use of the talk and history pages that are provided through tags at the top of each

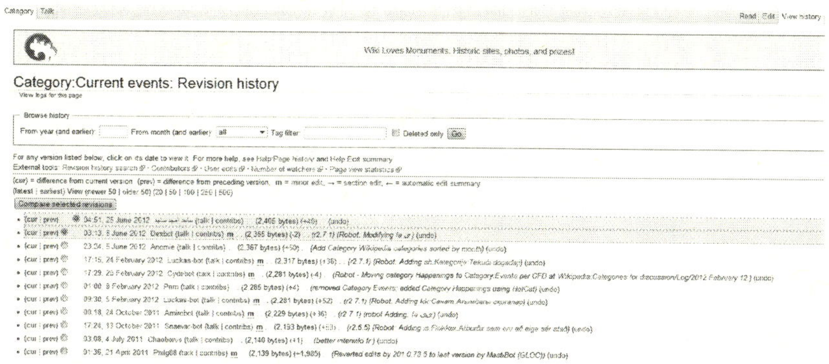

Figure 2.35 Illustration of a typical Wikipedia page. At the top, we see tabs for "Talk," "Edit," and "View history," among others, which contributors can use to update and debate on articles. On the current "View history" tab, we see a list of all the edits that have been made to the page, with the option to undo them.

article, as shown in Figure 2.35. All previous versions and all changes made to a page are recorded.

3. *Any subject can be contributed.* This includes controversial ones. Sometimes, certain articles can be "protected" from too-frequent edits to give time for the community of contributors to debate.

The first two items above provide Wikipedia with a strong, **positive network effect**. This emphasizes the *benefits* of having a larger network: a larger audience leads to more contributors, which in turn leads to more audience, in turn leading to more contributors, and so on. This success is often attributed to a "good-faith collaboration" environment within the Wikipedia contributor community.

On the other hand, the third feature highlights the **negative network effect** associated with Wikipedia, or the *cons* of having a larger network. Sure, as the audience and contributor numbers grow, we can expect more information. But this leads to more conflicts of opinion, which Wikipedia needs to deal with. We will see many more examples of positive and negative network effects throughout this book.

Consider the two graphs in Figure 2.36. Each represents a wiki, and the nodes (A – D) are people who are actively contributing to and reading it. All nodes are connected in each case because all edits are *public*, and as such, everyone can see and critique them. As we increase the number of people from three to four, we have a larger basis for information, and more people to monitor the pages for errors or spam.

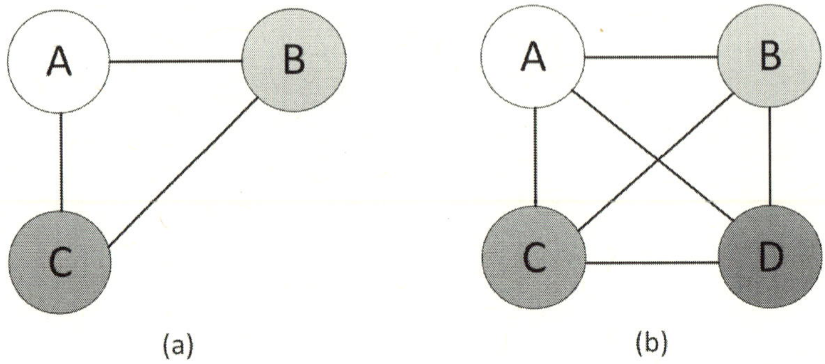

Figure 2.36 Here, each graph represents a Wikipedia article. A node is a person contributing to it, and all nodes are linked because everyone's contributions are visible to everyone else. In (a), we have a page where there are three people actively contributing, and in (b), we have four people. As the number of people increases, we have the potential of obtaining more information, but at the same time, there's more room for conflict of opinion.

But it also allows for more *clashing opinions*, and depending on the situation, more people who could write in false information.

So, how does Wikipedia enforce article quality? To start with, there are three core policies on Wikipedia to help ensure reliability and neutrality of the articles as much as possible.

1. *Verifiability*: Each key point and all data in an article must be externally verifiable, and the article must include a link to the primary source so that any reader can see where the information came from.
2. *No Original Research*: People cannot use Wikipedia as a publication venue of their new results. The purpose of an encyclopedia is to aggregate information, not to document uncited research.
3. *Neutral Point of View*: A reader should not be able to tell an author's bias when reading through a Wikipedia article. This is particularly important for controversial topics, like contentious political, social, and religious issues. For instance, in an article about abortion, a reader should not be able to tell if the author was pro-choice or pro-life. Unlike the above two policies, it is harder to enforce this one since "neutrality" is subjective.

There is also a reputation system for contributors: each article can be rated on a 1–6 scale. The higher the reputation of a person's articles, the more we will trust him or her to adhere to the above three principles. But perhaps the ultimate mechanism of enforcing quality and resolving

Figure 2.37 Depiction of a standard voting booth. This has become a popular method of submitting ballots privately.

disputes boils down to people *negotiating*. Depending on the stage of the article, experts and non-experts may join the discussion. There's a *hierarchy* of Wikipedia communities, and debates among contributors who cannot come to an agreement will be put forward to a group of the most experienced editors. This committee of editors acts like a jury, listening to the various sides of the debate, and then tries to decide by "rough consensus."

Now, how do we model the process of reaching a "rough consensus" through collaboration? Not easy. We will see in this chapter one underlying theory: voting. But we would need much translation from the standard voting theory to fully encompass what is actually practiced in the Wikipedia community.

2.3.3 Voting systems

Voting is essential in elections of those who will make and execute binding laws to be imposed on those casting the votes. Obvious examples are when political figures run for office: depending on the region, there are different systems in place to accumulate voters' preferences and determine the elected official. In our society today, voting in such elections is typically done *privately*, similar in concept to the sealed-envelope auctioning that we discussed with Google ad spaces. A typical way of accomplishing this is using voting booths, as shown in Figure 2.37.

But voting theory is useful in many other contexts where votes may or may not be private. These range from judging athletic competitions to decisions made by the jury in a court of law. It is a key component

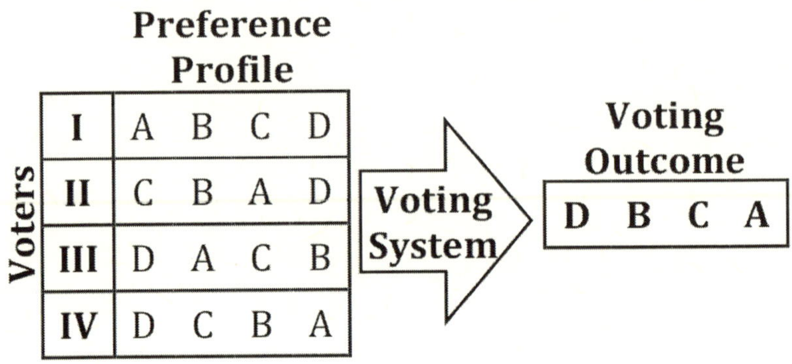

Figure 2.38 Illustration of the inputs to and output of a voting system. Each voter (I – IV) inputs a rank-ordered list of the candidates (A – D). The collection of these lists is called the preference profile. The specific output will depend on the voting system, but is always a single, rank-ordered list called the voting outcome.

of **social choice theory** that studies how individual preferences are collected and summarized.

The major concepts involved in a **voting system** are shown in Figure 2.38:

- We have a number of **voters** (I – IV), and a number of **candidates** (A – D) that are competing in the process.
- As the *input* to the system, each voter submits a single, rank-ordered list of the candidates. Here, Voter I ranks A first, B second, and so on. The collection of these rank-ordered lists from all voters (*i.e.*, the table on the left) is called the **preference profile**.
- The *output* of the voting system is a single rank-ordered list, called the **voting outcome**.

Note that not all situations call for each voter to submit a *complete* rank-ordered list that includes each candidate. Here are two examples:

1. To elect members of Wikipedia's arbitration committee, each voter simply divides the list of candidates into those she votes "for" and those she votes "against." The people with the most "for" votes will win.
2. In the US Presidential election, each person simply selects one candidate, out of however many possibilities, as his or her vote.

You can see an obvious tradeoff here. If we allow for the voting system to be *incomplete*, then the process becomes easier for the voters, but

Editor	Ranked List		
I	A	B	C
II	C	B	A
III	B	C	A
IV	C	B	A
V	A	B	C
VI	A	B	C
VII	C	B	A
VIII	A	B	C
IX	B	C	A

Figure 2.39 Voting list that will be used to illustrate three different voting systems. Here, we have nine voters/editors (I – IX), and three candidates (A – C).

we will *lose information*. There's growing research interest in quantifying the effects of such a tradeoff on the voting outcome, but for our purposes, we will assume all lists are complete.

Suppose nine Wikipedia editors need to vote on a contentious line in a page on abortion. Three contributors, A, B and C, have each attempted to make the line as least biased as possible. Each editor has ranked the contributions in what they think is least biased to most biased, as summarized in Figure 2.39. Let's look at different voting systems to see what outcomes we get.

Plurality voting

Perhaps the most popular voting system is **plurality voting**, where simply count the number of times each candidate appears first on voters' lists. The candidate with the highest number of first place votes wins and is put in the first place of the outcome's list. Then, the candidate with the second highest number of first place votes is placed

Freq.	Ranked List		
4	A	B	C
3	C	B	A
2	B	C	A

Figure 2.40 Summary of the preferences given in Figure 2.39. We indicate the frequency (e.g., the number of times) of each unique ordered list in the left column.

second, and so on. To simplify the discussion, we assume there are no ties.

First, let's simplify Figure 2.39 by putting one unique list in each row. This will make the analysis easier. We see that the ordering `A B C` appears 4 times, `C B A` appears 3 times, and `B C A` appears twice; No others appear. We illustrate this in Figure 2.40, where the frequency of each list is shown in the first column.

Now, how many first place votes does each candidate get? Looking at the first column of the ranked list, we see that A receives 4, C gets 3, and B gets 2. As a result, the outcome list by plurality voting is `A C B`. A's edit would win the line in the article.

Borda count

We can generalize plurality voting to **positional voting**. Looking at a voter's rank-ordered list, we assign some numbers to each candidate on the basis of its position in the list. It turns out that there's only one "right" way to do this, while still avoiding fundamental dilemmas in voting: **Borda count**. It is named after the French mathematician Jean-Charles de Borda, who initiated the scientific study of voting systems in the eighteenth century (though the origin of the Borda count has been recently traced all the way back to Ramon Llull, a thirteenth century philosopher).

Borda count looks at the preference profile and assigns points by "counting backwards" from the total number of candidates. The first-position in each list gets one less than the total. The second position gets two less, and so on, until the last candidate, who gets no points.

Then, we rank each candidate according to their total points across all the voters.

In the preference profile summarized in Figure 2.40, there are three candidates. We consider each list separately:

- For `A B C`, we reward 2 points to A, 1 to B, and 0 to C. There are four of these altogether: in total, A gets $4 \times 2 = 8$, B gets $4 \times 1 = 4$, and C gets 0 points.
- For `C B A`, we reward 2 points to C, 1 to B, and 0 to A. There are 3 of these: C gets $3 \times 2 = 6$, B gets $3 \times 1 = 3$, and C gets 0 points total.
- Finally, for `B C A`, we reward 2 points to B, 1 to C, and 0 to A. There are 2 of these: B gets $2 \times 2 = 4$, C gets $2 \times 1 = 2$, and A gets 0 points overall.

Now we add up the number of points for each candidate across the lists. The results are as follows.

$$\mathbf{A}: \ 8 + 0 + 0 = 8.$$
$$\mathbf{B}: \ 4 + 3 + 4 = 11.$$
$$\mathbf{C}: \ 0 + 6 + 2 = 8.$$

So with Borda count, B wins, and A and C tie for second place. Now B's edit would win the line on the page, not A's as we saw with plurality voting.

You are probably wondering: why do we start with one less than the number of candidates (2) for first place, rather than the number of candidates (3)? Well, that would require us to reward two to the second, leaving 1 to the last. But we really can't justify rewarding the last candidate any points, because the list is *required to be complete.* Last place carries no information: it only tells us that the voter prefers all the candidates to the last. As a result, we must start with 0 here and work our way up to 2 for the first.

Condorcet voting

Yet another method is **Condordet voting**, named after another French mathematician, Marquis de Condorcet, who rivaled Borda in the eighteenth century (again, surprisingly, the origin of Condorcet voting has been recently traced back to the thirteenth century). It is an aggregation of results from *pairwise* comparisons, which are the easiest to quantify. If we look at two candidates, say X and Y, with five voters preferring X to Y and four Y to X, then we can unambiguously say

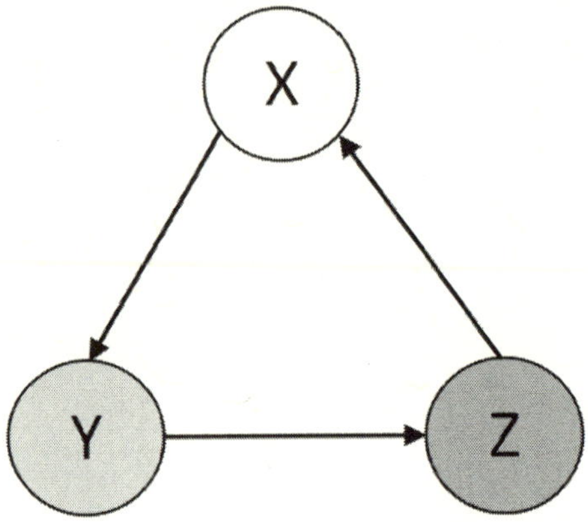

Figure 2.41 Illustration of a cyclic, logical inconsistency that can result from Condorcet voting. An edge implies that the origin beats the destination.

that X is the winner of the pair. There's not much you can say to argue against that.

Condorcet voting performs pairwise comparisons for all candidates. First place in the output list is awarded to the one who beats all others, second place beats everyone below him, and so on. Sounds intuitive, but there's a problem: we can have *logical inconsistencies.* For instance, suppose we have three candidates (X, Y, and Z), and we are given the following three lists: `X Y Z`, `Y Z X`, and `Z X Y`.

- Between X and Y, X beats Y twice, and Y beats X once. Hence, X beats Y.
- Between Y and Z, Y beats Z twice, and Z beats Y once. So, Y beats Z.
- Between X and Z, X beats Z once, and Z beats X twice. As a result, Z beats X.

From the first two, we have that X beats Y, and Y beats Z. So X should beat Z (indirectly through Y). But wait a minute: the third point tells us that Z beats X! This is logically inconsistent. It forms a **cycle**, as depicted in Figure 2.41: X beats Y, who beats Z, who beats X, who beats Y, and so on.

Let's return now to our example in Figure 2.40:

- **A and B**: A wins 4 times, and B wins $3 + 2 = 5$ times. Hence, B beats A.
- **B and C**: B wins $4 + 2 = 6$ times, and C wins 3 times. Hence, B beats C.

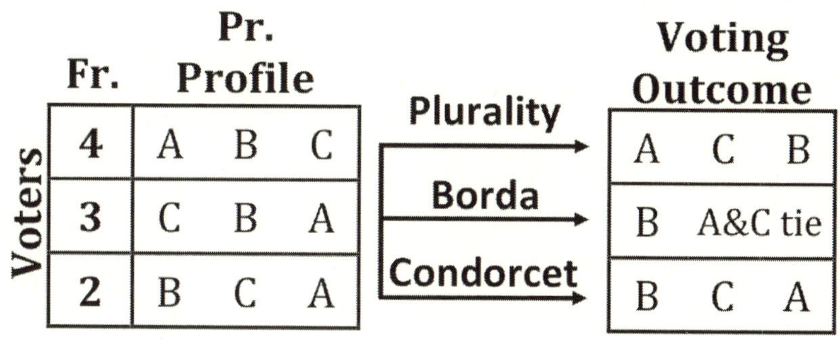

Figure 2.42 Voting outcomes for the three systems analyzed in our example. With Plurality voting, we get `A C B`. With Borda count, B wins, and A and C tie for second. Finally, with Condorcet voting, we have `B C A`.

- **A and C**: A wins 4 times, and C wins $3+2=5$ times. Hence, C beats A.

From the last two, we have that B beats C and C beats A. So, B beats A. Luckily, the first result (B beats A) is logically consistent with this. Hence, Condorcet voting works in this case, resulting in the output `B C A`. B's edit would be displayed on the article.

Different results

Each of the voting methods gave a different output list, as summarized in Figure 2.42. Strange, huh? You may object: "these preference profiles were made up specifically to have this happen. Real-world ones would not generate such inconsistencies between voting systems."

Well, you'd be right about one thing: we made this example up purposely to illustrate a point. Each of the three systems make sense in their own right, yet generate different outputs. Therefore, our intuition of what constitutes an accurate voting system can be incomplete at best.

How do you define "real-world cases" anyway? Well, we can start with some statements that realistically make sense to us, and take them to be true at the beginning of our analysis. These are called **axioms**. We can then try to reverse-engineer a voting system that will satisfy these fundamental properties.

2.3.4 Arrow's Impossibility Theorem

Let's construct a few statements that we can take as axioms for any voting system. Starting with the preference profile, what properties do we expect? Well, for one thing, we already stated that each voter must submit a complete list. Furthermore, in any real situation, each input list will be **transitive**: if there are three candidates A, B, and C, and someone prefers A to B and B to C, then they must also prefer A to C. This avoids logically inconsistent cycles at the input. We expect similar properties in the output list. To summarize:

1. Each input list (in the preference profile) is complete and transitive.
2. The output list (in the voting outcome) is complete and transitive.

Now that we have made some axioms about the input and output lists themselves, let's dig deeper into some of the properties of the voting system. For one, we want each input list to be able to influence the output in at least *some* way. Hence, there should be no "dictator," or some voter who will determine the output irrespective of the other voters' preferences. In other words,

3. The output list cannot just be the same as one input list no matter what the other input lists are.

Now, let's think about pairwise comparisons. After all, that's what got us in trouble with Condorcet voting in the first place. Obviously, if *all* of the input lists prefer one candidate to another in a pairwise comparison, then the output list should too. This is known as the *Pareto Principle*:

4. If all input lists prefer candidate A to candidate B, then the output list should do so too.

We can say more. When we compare just two candidates A and B, we really ignore all others. All that matters is whether A is before B, or vice versa. It doesn't matter whether A is third and B is fourth or A is tenth and B is twentieth: in both cases, we would say A beats B for this voter. So, we might say that whatever happens outside of the pairwise A, B ranking simply *does not matter*. As long as each voter kept his preference regarding A and B the same, the output's preference regarding A and B would never change. This is called the **Independence of Irrelevant Alternatives** (IIA), illustrated in Figure 2.43:

5. If between a given pair of candidates A and B, each input list's preference does not change, then even if their preferences involving other candidates change, the output list's preference regarding A and B does not change.

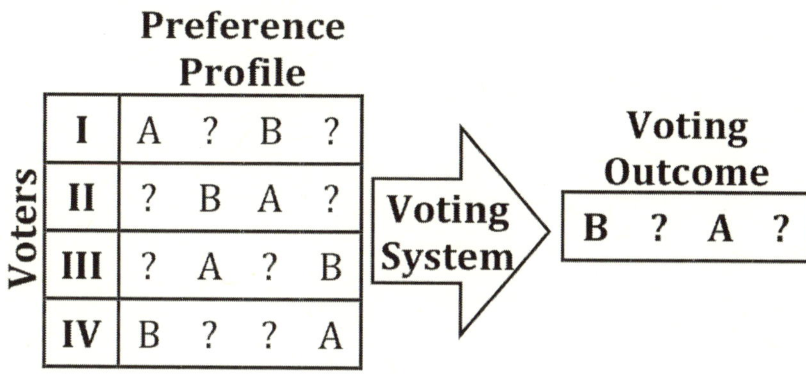

Figure 2.43 Illustration of the IIA criterion. As long as each voter keeps his preference the same regarding A and B, the preference in the output will not change, regardless of what candidates are placed before, in-between, or after the pair.

These statements sound intuitive. There should be a lot of voting systems that can satisfy all of them, right?

Actually, as soon as there are more than two candidates, it is *impossible* for any voting system to satisfy all five. This was proved by the economist Kenneth Arrow in his 1950 PhD thesis, and is known as **Arrow's Impossibility Theorem**.

Fallacy of IIA

How could that be? Something must be wrong with our axioms. But which one? The first two are about logical consistency, so we will have to keep them. And without the third one, there would be no point in voting, so we will have to keep that one too. So it must be either the Pareto Principle or IIA.

IIA looks suspicious. In fact, we can see pretty easily why this is, going back to our cyclic example in Condorcet voting.

Consider the set of three lists at the top of Figure 2.44. Under Condorcet voting, the output will be A>B>C (i.e., A beats B, who beats C). It's pretty easy to see that our first four axioms are satisfied: the inputs and output are complete and transitive, there is no dictator, and the fact that everyone prefers A and B to C is reflected in the output.

If IIA is to hold, we should be able to change C's location without affecting the positions of A and B in the output. So, consider the preference profile at the bottom of Figure 2.44, where we have moved C

Figure 2.44 Example to illustrate the fallacy of IIA. IIA and transitivity cannot be satisfied simultaneously.

around. Now, A>B and B>C, but C>A too! IIA has blocked transitivity, as the output list is now cyclic.

In summary, IIA opens up the possibility that transitive inputs can lead to cyclic outputs. When a voter compares A and B, there may or may not be a C in between, and that in turn determines whether or not the outcome is transitive. Arrow's Impossibility Theorem states that when it comes to ranking three or more candidates, pairwise comparisons are inadequate.

The next question is, what additional information do we need? It turns out that a numerical scale, rather than just relative order, will lead to a "possibility result." For example, by replacing IIA with the **Intensity form of IIA** (IIIA), there are voting systems that can satisfy all the axioms. In particular, Borda count will work.

So, what is IIIA? When we write A>B, we now also have to write down the number of other candidates that sit in between A and B; this is the *intensity*. If there are none, the intensity is zero. IIIA then states that, in the outcome list, the ranking of a pair of candidates depends on the pairwise comparison *and* the intensity. Intensity is needed to internalize the negative externality in a voting system, similar to the second price in an auction or the received signal to interference ratio in a cellular network.

Summary

Throughout this chapter, we have seen various ways in which certain amounts of "raw data" – whether that be bids, a webgraph, or a list

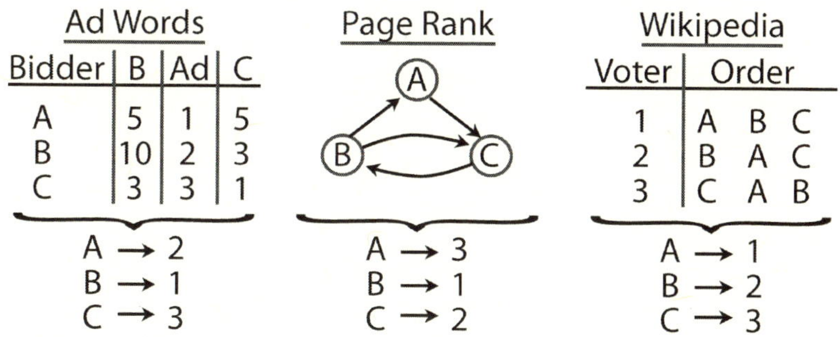

Figure 2.45 Summary of the three different case studies of consensus we've looked at in this chapter – Auctions in AdWords, PageRank, and voting in Wikipedia.

of votes – can be sifted through to find some consensus. Figure 2.45 summarizes the different case studies. In auctions, Google uses GSP to determine how to best match bidder to ad space, and in PageRank, they extract importance scores from enormously large webgraphs to determine which pages to display where in their search results. And when conflicts of opinion arise, we just saw how Wikipedia follows the notions of "rough consensus" to make decisions from rank-ordered voting lists. It is no wonder why finding a consensus is such a difficult, yet imperative, part of networked life.

3 Crowds Are Wise

Previously, we saw how two major companies – Google and Wikipedia – leverage raw data in various forms to find some consensus. In this chapter, we will study a similar topic: How online stores use product ratings to extract meaningful information about their items. First, we will explore some of the methods that Amazon uses when ranking products on their display page, and in doing so, we will introduce the "wisdom of crowds". Then, we will turn to a discussion on how Netflix recommends movies for customers to watch, based on users' rating histories and similarities in their tastes. Overall, we will gain an appreciation for how sophisticated rating and recommendation systems have become to work as well as they do.

3.1 Product ranking on Amazon

We will begin our discussion by focusing on Amazon, the dominant eCommerce company. Part of what makes their site so attractive is the fact that they aggregate customer reviews to give some sense of product "quality", and generally sort their search results accordingly. We will look at some of the ideas behind rating systems, and some of the conditions that are necessary to make review aggregation meaningful.

3.1.1 Amazon and eCommerce

Quite a significant portion of how we shop today, whether it be for shoes, DVDs, textbooks, or other everyday commodities, is done online. By 2012, the eCommerce industry in the US alone already had a net

Figure 3.1 Amazon's trademarked logo.

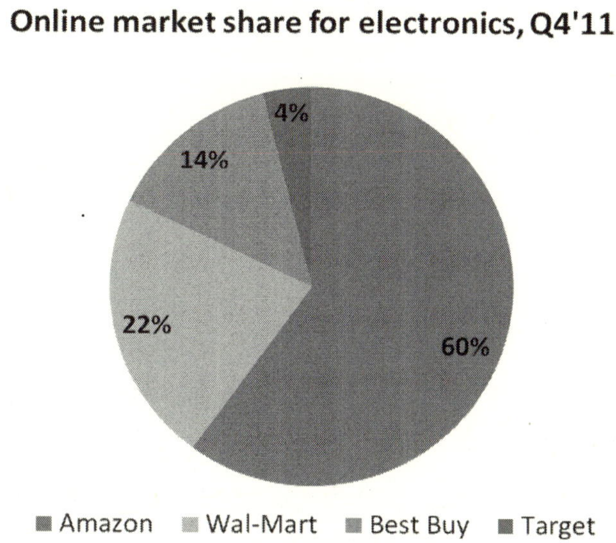

Figure 3.2 Fractions of consumer electronic units sold *online* by four of the largest such retailers in the fourth quarter of 2011. Amazon sold almost three times as many as Walmart, the second largest online retailer.

revenue of over $225 billion, and this number is increasing rapidly year by year.

Amazon (see Figure 3.1) has grown to be the world's largest online retailer. Founded in 1994 by Jeff Bezos, it initially served as an online bookstore. Though its initial growth was slow, it was one of the only eCommerce companies to survive when the "dot-com bubble" of Internet economic growth burst around 2000. Today, anything from clothing and shoes to software and electronic devices can be purchased from their site. You may have even purchased this book from there.

Besides eCommerce, Amazon has branched out to offer a wide variety of services. As of 2007, they moved into production, offering their own electronic book (e-book) reader, the Amazon Kindle. In fact, in July 2010, Amazon announced that e-book sales for its Kindle reader outnumbered the sale of hardcover books on their site. Amazon is also a pioneer in cloud services, but we'll save that topic for Chapter 8.

For many years, there was a clear distinction between street-side "brick and mortar" businesses that deal with customers face to face, and online companies. But today, many of the former have expanded to include eCommerce in an effort to stay competitive in the face of the web. A prime example of this is Walmart, which is by far the largest face to face retailer, and probably the second largest *online* retailer. But in the fourth quarter of 2011, Amazon sold almost three times as many consumer electronic units online as Walmart did, as shown in Figure 3.2. And in the second quarter of 2012, it was estimated that Amazon received 100 million unique website visitors, more than doubling that of Wal-Mart.

So, why has Amazon been so successful? Besides offering competitive prices, it allows customers to submit reviews for products, as do most other online retailers. And while these reviews are visible for other customers to see, Amazon leverages them to rank products in descending order of "quality." The focus of this chapter will be to explore some of the methods online retailers use to generate these rank ordered lists.

3.1.2 Average ratings

Suppose you are shopping on Amazon for a new HDTV. After browsing through the list of search results, you have finally narrowed your search down to two, on the basis of price: the ones in Figure 3.3.

Now, which to choose? At first glance, the one on the right seems like the better choice, because it has a higher rating: 4.5 stars as opposed to 4 stars.

But let's take a moment to think about how this rating was computed. Amazon allows its customers to enter reviews for products, each of which consists of three fields:

1. A rating, which some integer between 1 and 5 stars.

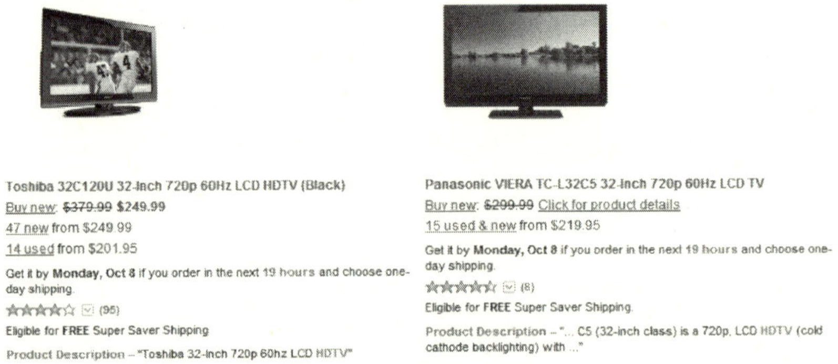

Figure 3.3 Two HDTV listings on Amazon. The one on the right has a higher average rating of 4.5 stars, compared to 4 on the left. But the one on the left has a higher review population of 95 people, compared to 8 on the right.

2. A textual review, which explains why the specific rating was given.
3. An indication of the number of people that found this review helpful.

The so-called "average customer review" of an item is obtained by finding the *mean* of all the ratings entered by customers. This is an attempt to *summarize* the opinions of the reviewers, and takes the form of a single number. It's highly desirable for most people, who don't have time to read through all the reviews. But shouldn't we be cautious of the people, both in total number and reviewer reputation, who have contributed to that "summary" rating?

Returning again to Figure 3.3, we see that 8 people have rated the HDTV on the right, while 95 have rated the one on the left. What does this mean? Intuitively, if more people have entered reviews for an item, then the *average* of their ratings is more trustworthy. It's less susceptible to those "harsh" and "lenient" reviewers that we saw with Netflix, and less prone to random reviewer *spam* entered by someone who has not even used the product before. To see this, consider a product which has received one 3-star review. The current average, then, is 3 stars. If someone comes along and decides to randomly enter a 5, the new average rating will be

$$\frac{3+5}{2} = 4,$$

which is an entire star over what it was previously. Now, suppose that instead of one 3-star review, there were 101. Then, if someone randomly entered a 5, the new average would be

$$\frac{101 \times 3 + 5}{102} = 3.02,$$

which is barely higher. In fact, rounding to one decimal place, the 3-star average would not change at all.

Obviously, not all 5-star reviews are too "lenient," and not all 1-star reviews are too "harsh." In fact, Amazon provides a comparison between the "most helpful" favorable and the "most helpful" critical reviews on each product for customers to see, as shown in Figure 3.4. While there can be wide variability between the number of stars of each, both are clearly well-received by other customers, given the number of people which found them "useful."

But we can say something for sure: the more reviews there are for a given product, more likely the average of these ratings can be trusted. So looking again at Figure 3.3, we would not be so quick to say that

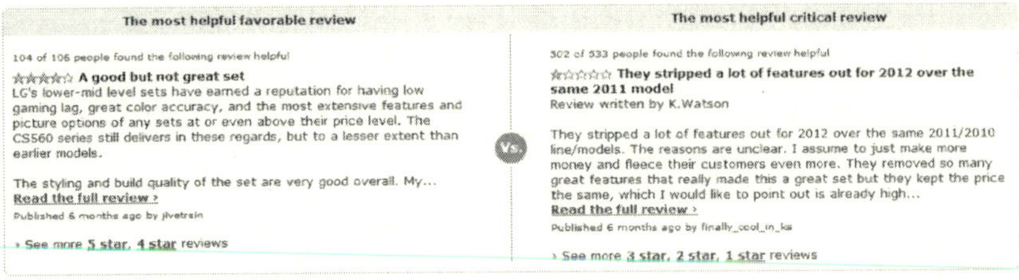

Figure 3.4 Illustration of the "most helpful favorable" and "most helpful critical" reviews for a product on Amazon. The mere fact that people can have such different views on a single product emphasizes the potential variability we face.

the 4.5-star average based on 8 ratings is better than the 4-star average based on 95. In fact, we are probably more inclined to take the HDTV on the left now.

This brings us to the core question we will explore in this chapter:

- *When can you trust an average rating?*

We can certainly see how reviews are potentially untrustworthy. Yet they are important in so many contexts of our lives: from recommendation letters written by teachers and employers to online purchases of every kind. So the first step is to see what approaches are needed to enhance the reliability of average reviews. The hope is that the following two approaches can help.

1. *Enforce quality*
   First, we need methods to ensure *accuracy*, by screening out really "bad" ratings. Unlimited and anonymous reviews have notoriously poor quality. Think about it: if you're promoting a competing product, it is in your best financial interest to enter a slew of poor reviews for the item you are battling with. But naturally, you also have to leave your name anonymous. On the other hand, in order to promote your own product, you may decide to enter many positive reviews to boost your average customer review. Additionally, as we said, review spammers can enter random reviews.

   So before anything else, we need to check the mechanism used to *enter* the reviews. Who can review? How strongly are customers encouraged to review? Do you need to buy the product before you can review it? Also, what is the range of numbers that can be entered for the review? For instance, it has been observed that a scale of 1-10 will often return an average review of 7, while a scale of 1-3 or 1 to

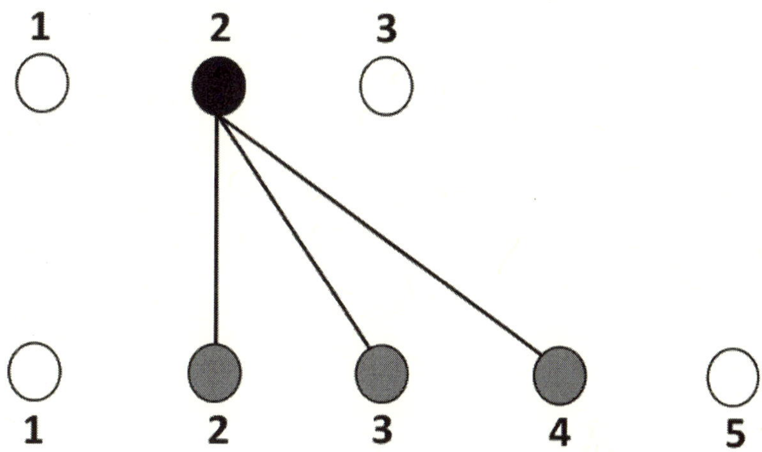

Figure 3.5 Illustration of one of the potential problems with rating scales. If you gave a '2' on a scale of 1-3, you will not necessarily choose the midpoint '3' on a scale of 1-5, since now you have more options inbetween: you could choose the below average '2', or the above average '4'. These are mixed psychological signals that can affect the rating outcome significantly.

5 will elicit very different psychological responses. This is illustrated in Figure 3.5, where we see that more options can naturally lead to confusion.

2. *Review population size*
   Second, the number of reviewers, or the review population size, needs to be large enough. But how large is large enough? And can we take advantage of the reviews we have to come up with a more useful number than the simple, *naive* average?

Both of these points raise tough questions that don't yet have good answers. Naturally, they depend on the type of product being reviewed. Movies (*e.g.*, on IMDb) are very subjective, whereas electronics (*e.g.*, on Amazon) are much less so. Hotels (*e.g.*, on tripadvisor) and restaurants (*e.g.*, on opentable) are somewhere inbetween. They also depend on the *quality* of the review. For instance, Amazon rewards people for submitting useful reviews by keeping a "top reviewer ranking." And if a reviewer is highly ranked throughout a year, they will be elevated to the "Hall of Fame." Still, reputation is difficult to put into numbers.

With these challenges, you may be thinking that opinion aggregation is unlikely to work well. But there have been notable exceptions recorded, and we will explore a famous one in the next section.

3.1.3 Two heads are better than one, but why?

Galton's experiment

Once upon a time, back in 1906 an intriguing contest occured on a farm in Plymouth, UK. At a livestock fair, an ox was placed on display, and the villagers were tasked with guessing the weight of the animal. Each of the 787 participants got a good look at the ox, and then wrote down his guess on a piece of paper without collaborating with anyone (see Figure 3.6).

Sir Francis Galton, a famous statistician around this time, happened to be present at the fair. Being the kind of man he was, he decided to run statistical tests on the results. Not one person guessed the true weight of 1,198 pounds: in fact, the numbers were all over the place. But remarkably, when he *averaged* the guesses, the result was 1,197 pounds, which was an error of less than 0.1 percent. Even the *median* (the middle number of the set) was 1,207 pounds, which is still still off by less than 0.8 percent.

How could this be? Everyone's guesses were so far off, still the simple *average* of their guesses was so close to the truth? There were actually several key factors at work here that made simple averaging work so well, which are by no means present in every "guessing game."

- The task was relatively easy. In particular, there was an *objective* answer with a clear numerical meaning. They were simply guessing the weight of an ox.

Figure 3.6 In 1906, an ox was put on display at a farm in the UK and roughly 800 villagers attempted to guess its weight. While they were all significantly off, the *average* of their guesses was only one pound away from the right answer.

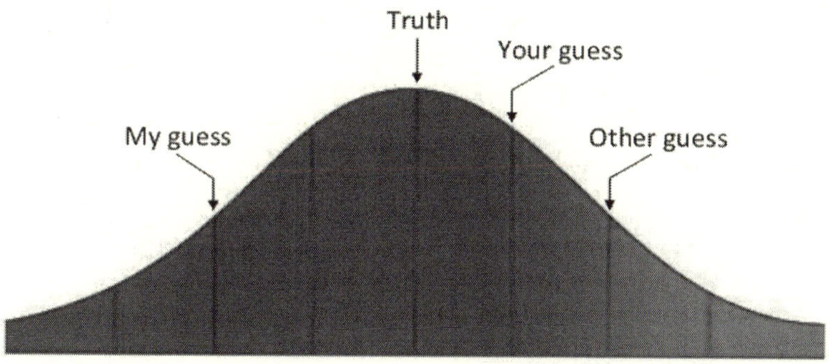

Figure 3.7 While each of our (educated) guesses may be significantly off from the true value, the *average* will get closer and closer to the true value as more and more people join in. This will hold true as long as we are all *independent* and *unbiased* in our guesses.

- The estimates were *unbiased*. Since everyone got a good look at the ox, there was no tendency across everyone to either guess too low or too high.
- The estimates were *independent*. None of the villagers saw anyone else's guess, and everyone made their submission in a "sealed envelope" way. Hence, nobody was influenced by anyone else.
- There were enough people participating.

We illustrate these factors in Figure 3.7, which form the basis behind Galton's remarkable result.

The wisdom of crowds

Returning now to Amazon, our hope is that when we average customer ratings for a product, the result will get close to the "right" rating. But can we even say that a "true" rating exists? Doesn't this depend at least somewhat on the specific customer?

In general, three factors are important in aggregating individual opinions.

- *Definition of the task*: Guessing a number is easy. Reviewing products on Amazon is a little more difficult.
- *Independence of the reviews*: Success in opinion aggregation stems not from having many smart individuals who are likely to guess correctly, but from the independence of each individual's view from the rest. So, are Amazon reviews independent of each other? Kind of. Even though you can see the existing reviews before entering

Figure 3.8 The concept of the wisdom of crowds is that the aggregate knowledge of "the many" will outperform the ability of "the individual" or "the few," provided we have independence and no bias.

your own, usually your rating will not be dramatically affected by the existing ones. But sometimes, reviews are indeed entered as a reaction to recent reviews posted on the website, either to counter-argue or to reinforce points made there. We will discuss *sequential* decision making more in Chapter 4.

- *Review population*: Galton's experiment would not have worked as well if there had been less people participating. And in general, the harder the task is and the more dependent the reviews are, the more people we need to obtain a trustworthy average. However, if we try to detect potentially inconsistent reviews, we can lower the necessary number of participants.

This leads us to a general result. Suppose we have a number of people, say 1000, playing a "guessing" game with a clear, well-defined task at hand. At the end, we collect each of the guesses and average them. As it turns out, we can expect that the error in this *average* will be reduced by a factor of 1000 from the error in *each separate guess*, *i.e.*,

$$\text{Error in average} \approx \frac{\text{Error in each guess}}{\text{Number of people}}, \tag{3.1}$$

provided that all the guesses are independent. We use an approximation symbol here because naturally, not each guess will have the same

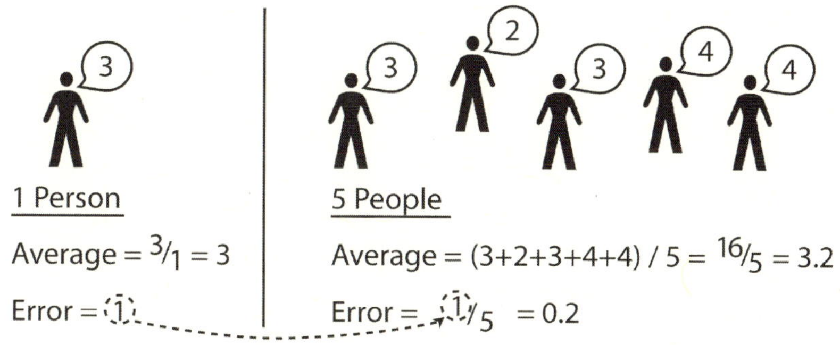

Figure 3.9 Simple illustration of Equation 3.1. We start with a single person, who's expected error is 1. When we move to 5 people, the error in the average of 3.2 is expected to reduce to 1/5, provided everyone's "guess" is independent and unbiased.

error. In fact, if they all had the same error, they would be completely *dependent* upon each other, and this result would not hold at all. But on average, we can expect this to be the case.

Equation (3.1) states mathematically what we have been discussing in this section, the **wisdom of crowds**. This is a remarkable result, that says even if no specific person in the crowd is particularly "wise," as long as they all guess independently of each other and have no bias, we can expect that their "collective guess" as a group will improve by the number of people in the crowd (see Figure 3.8). So if there are 5 people, it will be 1/5th (see Figure 3.9), if there are 10 people, it will be 1/10th, and so on.

So now, suppose we have a large number of reviewers for a product on Amazon. Does this discussion imply the average rating will be close to the "truth" we want? Not necessarily. As we pointed out, there are many complications. Let's explore a few of these next.

Challenges of rating aggregation

Here are three examples illustrating the key challenges in deciding when to trust ratings on Amazon.

Example 1: Number of reviews

Given that different products have different numbers of reviews, it's hard to determine which has the best quality. As was shown in Figure 3.3, two different HDTVs of comparable price have different average ratings: the Toshiba has 4 stars and the Panasonic has 4.5. But at the

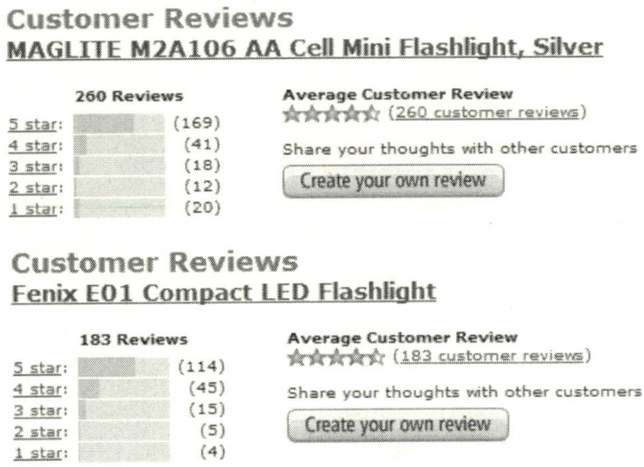

Figure 3.10 Two different flashlights on Amazon, both with an average review of about 4.5 stars. The MAGLITE (top) has 65.0% of reviews at 5 stars, and the Fenix (bottom) has 62.3%. On the other hand, the MAGLITE has 7.6% of reviews at 1 star, while the Fenix has 2.2%. So the MAGLITE has more review *variation*, making our choice harder.

same time, 8 people had rated the HDTV on the right, while 95 have rated the one on the left. The customer is faced with a tradeoff between choosing a product with a lower average rating and a larger number of reviews versus one with a higher average rating and a smaller number of reviews.

Example 2: Review variation
Consider two flashlights on Amazon, a MAGLITE and a Fenix, as shown in Figure 3.10. Both of them have roughly the same average customer review at 4.5 stars, and the number of reviews is on the same order. On the one hand, 65.0% of reviewers gave the MAGLITE 5 stars, and 7.6% gave it 1 star. On the other hand, 62.3% of reviewers gave the Fenix 5 stars, and 2.2% gave it 1 star. So the MAGLITE not only has a higher percentage of people giving it 5 stars, but also a higher percentage of people giving it 1 star. There is a larger *variation* in the ratings of the MAGLITE. Does that make the average rating more trustworthy, or less? This is a subjective question with no clear answer.

Example 3: Recent changes
In Figure 3.11, we compare the the 60 most recent ratings to the top

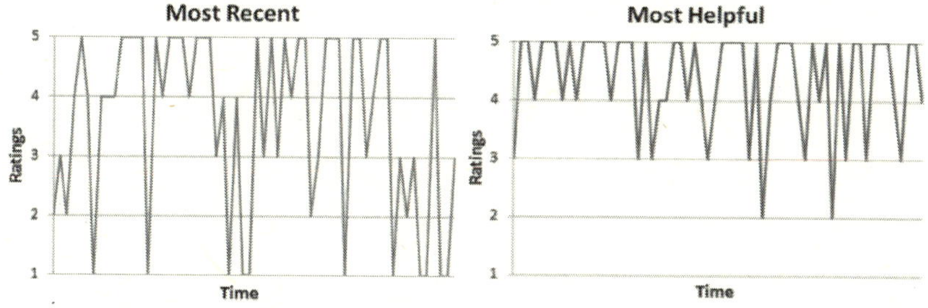

Figure 3.11 Illustration of the most recent (left) and the most helpful (right) ratings for an iPod touch on Amazon. Which of these two should the average rating be based off of? The two plots are quite different; the average on the left is 3.6, and on the right is 4.4.

60 "most helpful" ratings over time for an iPod4 Touch (16 GB). The mean of the most recent ratings is 3.6, while that of the most helpful is 4.4. Do the most recent ratings reflect a "real" change or just normal fluctuations? Which is a better fit for the average customer review? Again, the answer to this question is subjective.

3.1.4 Ranking methods

Naive averaging

Consider the table of ratings from Amazon shown in Table 3.1, for a set of 5 DVD players on some day in 2012. The manufacturer for each specific player is indicated, followed by the number of stars in each category and the total number of reviews.

First, let's calculate the average ratings of each of these products. We simply add together the total number of stars, and divide by the total number of reviews. For the Panasonic, we have five 5-star ratings, which is $5 \times 5 = 25$ stars in all; three 4-star ratings, which is $4 \times 3 = 12$; three 3-star ratings, or $3 \times 3 = 9$; and no two or one-stars. Hence, the total number of stars is $25 + 12 + 9 = 46$, and the total number of reviews is $5 + 3 + 3 = 11$. So, the average is $46/11 = 4.182$.

We can write the average as the following equation, where in the numerator we weight each rating by its number of stars:

$$\frac{(\text{Num. 5 star}) \times 5 + (\text{Num. 4 star}) \times 4 + (\text{Num. 3 star}) \times 3 + \cdots}{\text{Num. 5 star} + \text{Num. 4 star} + \text{Num. 3 star} + \cdots}. \tag{3.2}$$

The total number of reviews for each product is indicated in the last

DVD Player	5 stars	4 stars	3 stars	2 stars	1 star	Total
Panasonic	5	3	3	0	0	11
Sony	18	9	5	2	3	37
Phillips	23	15	11	5	13	67
Curtis	18	14	2	6	11	51
Toshiba	19	10	10	4	11	54
Total	83	51	31	17	38	220

Table 3.1 Ratings for a set of 5 DVD players on Amazon on some day in 2012. The manufacturer, followed by the number of stars in each category and the total number of ratings, is given for each. They are in order of their ranking based on simple naive averaging (Panasonic first). But if we decide to take into account the number of ratings for each, as in Bayesian ranking, the order will be quite different.

column of Table 3.1, where we have simply added across the rows. For instance, the Panasonic player has $5+3+3+0+0 = 11$ in all. We can use the numbers here as the denominator of Equation (3.2).

As we said, for the Panasonic player, we have

$$\frac{(5\times 5)+(3\times 4)+(3\times 3)+(0\times 2)+(0\times 1)}{5+3+3+0+0} = \frac{46}{11} = 4.182.$$

Similarly, we can do the same for the other four items:

$$\begin{aligned}
\mathbf{S}: &\ \frac{(18\times 5)+(9\times 4)+(5\times 3)+(2\times 2)+(3\times 1)}{18+9+5+2+3} = \frac{148}{37} = 4,\\
\mathbf{P}: &\ \frac{(23\times 5)+(15\times 4)+(11\times 3)+(5\times 2)+(13\times 1)}{23+15+11+5+13} = \frac{231}{67} = 3.448,\\
\mathbf{C}: &\ \frac{(18\times 5)+(14\times 4)+(2\times 3)+(6\times 2)+(11\times 1)}{18+14+2+6+11} = \frac{175}{51} = 3.431,\\
\mathbf{T}: &\ \frac{(19\times 5)+(10\times 4)+(10\times 3)+(4\times 2)+(11\times 1)}{19+10+10+4+11} = \frac{231}{67} = 3.407.
\end{aligned}$$

If we rank these from highest to lowest average rating, the order is as it appears in Table 3.1. But is this the "correct" ordering?

Back to our driving question: should a product with only two reviews, even though both are 5 stars, be played higher than a competing product with one hundred reviews that averages 4.5 stars? Intuitively this would be wrong. We are facing this problem with the DVD players in

Table 3.1: the Panasonic, which clearly has the highest average rating of 4.182, also has by far the smallest number of reviews.

Bayesian ranking

How can we deal with this issue? We should somehow weight the raw rating scores with the population sizes. Knowing how many reviews there are gives us *prior knowledge*, and we should take advantage of that.

So rather than considering each product *separately*, let's leverage the information we have about *all* of the products in question. Just as we computed the average rating for each, we can find the overall average across all items. If we add down the columns in Table 3.1, we get the total for each star-rating, as indicated in the last row. We can then apply Equation (3.2) with these totals to get the overall average:

$$\frac{(83 \times 5) + (51 \times 4) + (31 \times 3) + (17 \times 2) + (38 \times 1)}{83 + 51 + 31 + 17 + 38} = \frac{784}{220} = 3.564.$$

This average is based on the 220 total of ratings. Intuitively, we can use it as a "back up" for each of the individual products, as follows.

- The more ratings there are for a given product, the more trustworthy its individual average is relative to the overall. As such, we should place more emphasis on the individual.
- Similarly, the less ratings there are for a given product, the less trustworthy its average is. In this case, we should lean towards the overall.

We can think of this as a "sliding ruler" between the individual and

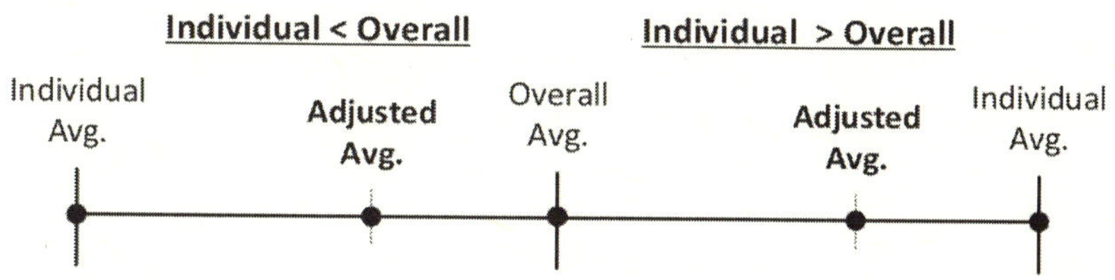

Figure 3.12 Illustration of the "sliding ruler" concept. The adjusted average rating sits somewhere between the individual and overall averages. Its exact location depends on the number of ratings the individual product has relative to the total. For clarity, we show on the left the case when a product's individual average is less than the overall, and right when it is greater.

the total, as shown in Figure 3.12. The adjusted rating lies somewhere between the two.

This can be mathematically crystalized in terms of the **Bayesian ranking** formula, is one of many applications of the work of the English mathematician Thomas Bayes in the early eighteenth century. For each product, the Bayesian-adjusted rating is

$$\frac{\text{Overall Num.} \times \text{Overall Avg.} + \text{Individual Num.} \times \text{Individual Avg.}}{\text{Overall Num.} + \text{Individual Num.}}, \tag{3.3}$$

where Num. stands for number of ratings.

Let's take a minute to examine this equation. We are weighting the two average ratings by their respective numbers, adding them, and dividing by the total number of ratings. Suppose we have 100 total, with an average of 2 stars. If one of the products has a 4-star average based on 5 reviews, then the Bayesian rating would be

$$\frac{100 \times 2 + 5 \times 4}{100 + 5} = \frac{220}{105} = 2.10,$$

which is much closer to 2 than it is to 4. The reasoning here is that the individual number of ratings is much *smaller* than the total, so we rely on the latter much more. On the other hand, if the 4-star average was based on 40 reviews, the Bayesian rating would increase to

$$\frac{100 \times 2 + 40 \times 4}{100 + 40} = \frac{400}{140} = 2.57,$$

which is moving away from 2. We show the "sliding ruler" for this in Figure 3.13.

Note that as an individual product's ratings make up a larger fraction of the total, the overall mean will already be closer to the individual. In these cases, Bayesian adjustment won't have much of an effect, because the product mean is already (presumably) trustworthy.

Returning now to our example in Table 3.1, we can apply Equation (3.3) to each of the items and see if the ordering changes. There are 220 total ratings, and we computed the overall average rating to be 3.564. For the Panasonic player, there are 11 ratings, and we computed an individual rating of 4.182. Hence, the Bayesian formula gives

$$\frac{220 \times 3.564 + 11 \times 4.182}{220 + 11} = \frac{784.08 + 46.002}{220 + 11} = 3.593.$$

Notice how much closer the Panasonic rating is to the mean, 3.564,

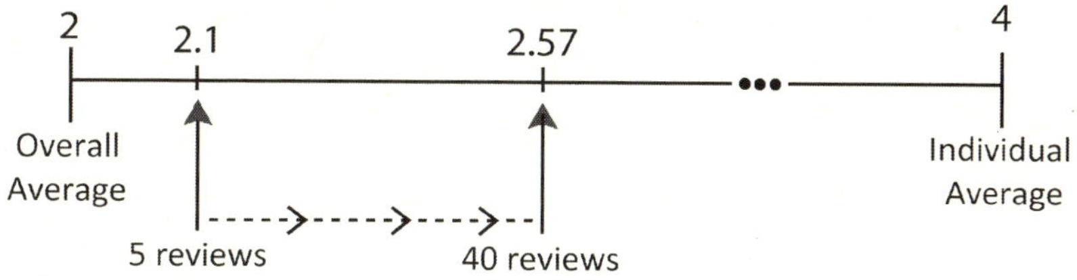

Figure 3.13 "Sliding ruler" for our quick example. Note that the overall average is smaller than the individual. Moving from 5 to 40 reviews causes the adjusted average moves from 2.1 to 2.57, closer to the individual.

after Bayesian adjustment. This is due to the fact that there is such a small number of ratings (11) for this specific player relative to the total (220), so Equation (3.3) weights the overall mean much more heavily. The computations for the other DVD players are similar:

$$\mathbf{S}: \frac{(220 \times 3.564) + (37 \times 4.000)}{220 + 37} = \frac{784.08 + 148}{220 + 37} = 3.626,$$
$$\mathbf{P}: \frac{(220 \times 3.564) + (67 \times 3.448)}{220 + 67} = \frac{784.08 + 231.016}{220 + 37} = 3.537,$$
$$\mathbf{C}: \frac{(220 \times 3.564) + (51 \times 3.431)}{220 + 51} = \frac{784.08 + 174.981}{220 + 51} = 3.539,$$
$$\mathbf{T}: \frac{(220 \times 3.564) + (54 \times 3.407)}{220 + 54} = \frac{784.08 + 183.978}{220 + 54} = 3.533.$$

So what is the result of this? Now, the ranked-order of the products, from highest Bayesian rating to lowest, is Sony, Panasonic, Curtis, Phillips, and Toshiba, respectively. The pre- and post- Bayesian ranking is shown in Table 3.2. As we can see, the Panasonic and Sony players, as well as the Phillips and Curtis players, swap places after Bayesian adjustment is applied. Beyond this, we note that all of the ratings are pulled closer to the mean, as we expect due to the "sliding ruler" concept.

In this example, the ordering of the list changes significantly. Note that sometimes, Bayesian ranking does not alter the ranked order of a set of comparable products.

Bayesian ranking in practice

In reality, quite a few websites adopt Bayesian ranking. For instance, the Internet Movie Database (IMDb)'s top 250 movies follows Equation (3.3) exactly. But sometimes, it's better to put some maximum on the

DVD Player	No. Ratings	Raw avg. Ranking	Raw avg. Rating	Bayesian adj. Ranking	Bayesian adj. Rating
Panasonic	11	1	4.182	2	3.593
Sony	37	2	4	1	3.626
Phillips	67	3	3.448	4	3.537
Curtis	51	4	3.431	3	3.539
Toshiba	54	5	3.407	5	3.533

Table 3.2 Comparison between the ranking based on a raw average and that of Bayesian adjustmement for the DVD players in Table 3.1. The Bayesian adjustment brings all the ratings closer to the mean. While the Panasonic is the highest based on raw averaging, the fact that it has a small number of ratings (11) relative to the total (220) demotes it considerably under Bayesian adjustment, bringing it down one ranking. Products 1 and 2, as well as 3 and 4, swap places.

value of "Overall Num." that we use when adjusting the ratings. The reason for this is that over time, the number of ratings for each product will keep increasing, and the range of the Bayesian formula can only decrease with this. Eventually, we could be at the point where we have so many ratings for so many products that we are essentially relying on the overall mean for each application of Equation (3.3). Beer Advocate's beer ranking, at `http://beeradvocate.com/lists/popular`, does just that: they choose "Overall Num." to be the minimum number of reviews needed for a beer to be listed there, rather than the total number of reviews.

In our analysis, we've assumed that there is a single, "true" value of a product's ranking. However, we discussed before how the "truth" can vary from person to person. Indeed, certain products simply create bipolar reactions: some love it and others hate it. To illusrate this, we show in Figure 3.14 a Sony DVD player on Amazon: most people give 4 or 5 stars, or 1 star. On the other hand, 2 and 3 stars are less frequent. In fact, sometimes, the only people who care enough to write reviews are those who have strong feelings in one of these two directions. We call such a collection of ratings a **multimodal** distribution, *i.e.*, one with many centers. Though we have focused **unimodal** distributions in this text, Bayesian analysis can be extended to the multimodal case as well.

We should caution that Bayesian adjustment applies only to adjust-

Figure 3.14 An example of a product on Amazon that created bipolar responses: some people liked it a lot, giving it 4 or 5 stars, while others hated it, giving it 1 star. 2 and 3 stars were less frequent. While we have focused on Bayesian analysis in the unimodal case, extentions to multimodal distributions are also possible.

ing ratings within a *comparable* family of products, *i.e.* a set of DVD players, a set of laptops, a set of shoes, etc. It cannot be readily used to adjust ratings of products that are likely to be different, e.g., a set of random electronic items. The family as a whole needs to have an overall average that is meaningful in order for Equation (3.3) to make sense.

3.1.5 What does Amazon do?

At this point, you're probably wondering what Amazon specifically does to rank its product lists. Amazon actually follows some secret formula that combines the average reviews with at least three additional elements:

- the Bayesian adjustment by review population,
- the recency of the reviews, and
- the reputation score of the reviewer.

The exact formula is not known outside of Amazon. But maybe we can reverse-engineer how these rankings are generated through an example. Let us consider the list of the top 20 thermoses in January 2013, shown in Table 3.3. It was otained from Amazon by applying the following sequence of search filters:

- Search query: Thermos
- Then select Home & Kitchen $\rightarrow$ Kitchen & Dining to enable sorting.

There are two rank-ordered lists.

Thermos	No. reviews	5 star	4 star	3 star	2 star	1 star	"Ave. review"	Ave. rating
1	210	179	20	7	3	1	4.8	4.776
2	87	74	8	3	1	1	4.8	4.759
3	1064	850	126	42	20	26	4.6	4.648
4	442	349	64	10	6	13	4.7	4.652
5	273	206	37	13	8	9	4.5	4.549
6	411	312	60	16	8	15	4.6	4.572
7	158	122	21	7	6	2	4.6	4.614
8	183	135	27	10	4	7	4.5	4.525
9	110	81	18	4	4	3	4.5	4.545
10	218	166	28	8	6	10	4.5	4.532
11	122	85	24	6	3	4	4.5	4.500
12	181	132	28	9	4	8	4.5	4.503
13	607	477	64	20	18	28	4.6	4.555
14	331	229	54	19	12	17	4.4	4.408
15	24	20	3	1	0	0	4.8	4.791
16	102	75	10	9	4	4	4.5	4.451
17	19	18	1	0	0	0	4.9	4.947
18	69	57	6	1	2	3	4.6	4.623
19	47	29	15	2	1	0	4.5	4.532
20	46	31	10	3	0	2	4.5	4.478

Table 3.3 A list of the top twenty thermoses on Amazon. Most of the items appear strangely out of order compared with the average reviews.

- The first is the ranking by Amazon, which orders the list in Table 3.3: 1, 2, 3, ...
- The second is the ranking by "average rating," or the average number of stars. Rather than showing this number explicitly, Amazon rounds it and gives an "average customer review." This leads to a different order: 17, 15, 1, 2, 4, 3, 18, 7, 6, 13, 5, 9, 10, 19, 8, 12, 11, 20, 16, 14.

Average rating has many differences from Amazon's ranking:

- Amazon places items 17, 15, 18, and 19 much lower than their average ratings would indicate.
- Amazon places items 5, 8, and 11, and 14 much higher than their average rating would indicate.

Let's try to reverse engineer what factors might have contributed to the actual ranking.

1. *Bayesian adjustment*: The population sizes of the ratings matter.

The raw rating scores must be weighted with the population size in some way.

2. *Recency of the reviews*: Perhaps some of the reviewers rated their thermoses as soon as they purchased them, and gave them high ratings because the products appeared durable, though sometimes items deteriorate quicker than anticipated.
3. *Quality of the reviewers or reviews*
   - (a) Reputation score of the reviewer: Reviewers with higher reputations should be given more "say" in the average customer review of a product.
   - (b) Quality of review: The quality of a review can be measured in terms of its length or associated keywords in the text.
   - (c) Review of review: Higher review scores indicate that customers found the review "helpful" and accurate.
   - (d) Timing of reviews: Review spamming from competing products can be partially detected from unusual patterns in review timing.

Bayesian adjustment

First, Bayesian adjustment uses Equation (3.3), so we need to know "Overall Num." and "Overall Avg." Here, the latter is the average rating of all the products (here assumed to be the top 20), while the former could be the total number of reviews, the *minimum* number of reviews necessary for a product to be listed (as in the Beer Advocate's website), or possibly some number in between. From Table 3.3, we can apply Equation (3.2) to obtain an average of 4.59.

Now, what should we choose for "Overall Num."? We compare the Bayesian rankings using 19, which is the lowest number of reviews for any product (number 17), 1064, which is the highest number of reviews for any product (number 3), 235, which is the average number of reviews per product, and 4704, the total number of reviews entered for the top 20 products. They lead to different Bayesian rankings, ordered from highest product to lowest:

- **19**: 17, 1, 2, 15, 4, 3, 18, 7, 6, 13, 5, 9, 19, 10, 8, 11, 12, 20, 16, 14.
- **1064**: 1, 3, 4, 2, 17, 15, 7, 18, 19, 9, 6, 20, 5, 11, 8, 10, 13, 16, 12, 14.
- **235**: 1, 3, 2, 4, 17, 15, 7, 18, 19, 6, 9, 20, 5, 13, 10, 8, 11, 12, 16, 14.
- **4704**: 3, 1, 4, 2, 17, 15, 7, 18, 19, 9, 20, 6, 5, 11, 8, 10, 16, 12, 13, 14.

Clearly, having the number too large or too small results in rankings that are far out of order. Both "1064" and "235" give results closer to Amazon's ranking, at least in terms of grouping clusters of products

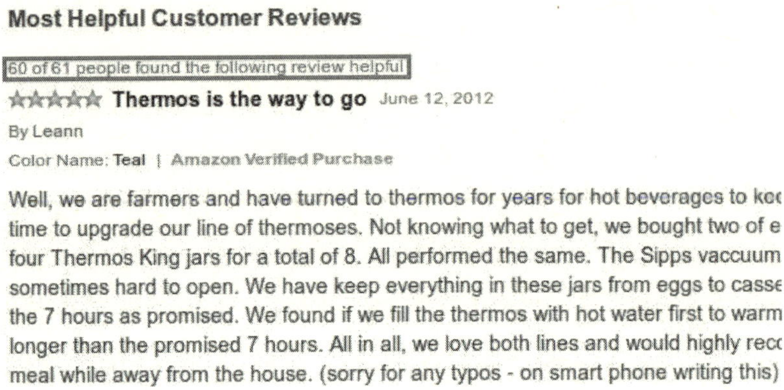

Figure 3.15 For the 15th thermos, 60 out of 61 people found the "most helpful" review helpful, which might help increase its rank.

together in the rank-ordered list. Still, there are some exceptions in each case. We will focus our attention on products 8, 15, 17, and 20.

Review recency and quality

We notice that products 15, 17, and 20 have the lowest number of ratings, specifically 24, 19, and 46, respectively. This helps explain why they are further down than expected. But, even so, why would they be placed *apart* from one another?

To answer this, we take into account the *review of reviews*: in the case of product 15, for instance, the "most helpful" review had 60 people find it helpful, as shown in Figure 3.15. On the other hand, products 17 and 20 had only seven and six, respectively, which could explain why 15 is ranked higher.

Then, we can look at the *recency of the reviews*: the "most helpful" review for product 17 was made in May of 2011. Product 17's "saving grace" is that its corresponding review was more recent, made in December of 2012, which would push it closer to product 15 in the ranking. On the other hand, Amazon may have deemed that product 20's reviews to not have enough overall helpfulness despite their recency.

Finally, product 15 had an extremely high quality of review in its descriptive listing of the pros and cons in each case. This may have also pushed its ranking higher.

Notice that despite being ranked so low, product 17 has the *highest* average rating. In addition to what we just discussed, there are two more factors that can help explain this. First, only half of the reviews

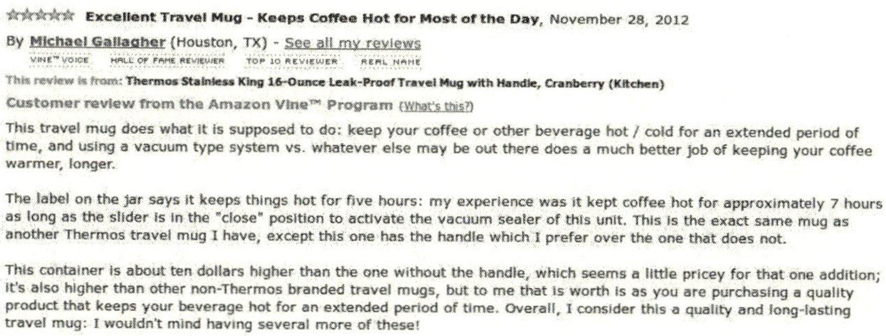

☆☆☆☆☆ **Excellent Travel Mug - Keeps Coffee Hot for Most of the Day,** November 28, 2012

By **Michael Gallagher** (Houston, TX) - See all my reviews

VINE™ VOICE HALL OF FAME REVIEWER TOP 10 REVIEWER REAL NAME

This review is from: **Thermos Stainless King 16-Ounce Leak-Proof Travel Mug with Handle, Cranberry (Kitchen)**

Customer review from the Amazon Vine™ Program (What's this?)

This travel mug does what it is supposed to do: keep your coffee or other beverage hot / cold for an extended period of time, and using a vacuum type system vs. whatever else may be out there does a much better job of keeping your coffee warmer, longer.

The label on the jar says it keeps things hot for five hours: my experience was it kept coffee hot for approximately 7 hours as long as the slider is in the "close" position to activate the vacuum sealer of this unit. This is the exact same mug as another Thermos travel mug I have, except this one has the handle which I prefer over the one that does not.

This container is about ten dollars higher than the one without the handle, which seems a little pricey for that one addition; it's also higher than other non-Thermos branded travel mugs, but to me that is worth is as you are purchasing a quality product that keeps your beverage hot for an extended period of time. Overall, I consider this a quality and long-lasting travel mug: I wouldn't mind having several more of these!

Figure 3.16 One of the reviews for the 8th thermos was from an Amazon "hall of fame" reviewer (7th overall), and he gave it five stars. This may have helped promote the ranking.

are from 2012. Second, average reviewer reputation is fairly low, with all but 2 having less than 10 helpful votes overall.

Now, why would Amazon decide to rank an item such as 8 so high, given that the Bayesian ranking places it around the 16th position? Well, it turns out that item 8's "most helpful" review had 79 out of 79 people find it helpful, and over 10 of Amazon's top 5000 reviewers reviewed the product (reviewer rankings extend beyond 3 million on Amazon). Further, the reviewers all have high ratings, and one of them is an Amazon "hall of fame" reviewer (ranked 7th overall), as shown in Figure 3.16. Quality of reviews overrides the quantity.

To summarize, the following set of guidelines is inferred from this (small) sample on how Amazon comes up with its ranking.

1. An initial Bayesian ranking is made, with "Overall Num." chosen to be somewhere around the maximum number of reviews entered for any product, or the average number of reviews across the products.
2. Products that have small numbers of reviews or low recency of their most helpful reviews are ranked separately amongst themselves, and are moved to lower locations (e.g., 15, 17, and 20).
3. Products that have very-high-quality, positive reviews from top reviewers are bumped up in the ranking (e.g., 8).
4. Products for which there are few reputable reviewers (e.g., 17), are severely demoted in the ranking.

In this section, we've seen how under certain assumptions rating aggregation can be useful in attempting to find the "truth" about a product. Unfortunately, there's still a gap between the theory behind Bayesian rating and what is practiced by Amazon's secret formula. In the next section, we will turn to Netflix recommendation, where rather than extracting one rating for the mass we will predict a mass of ratings for each person.

3.2 Movie recommendation on Netflix

We continue our theme of opinion aggregation, but now turn to *recommendation*, where we use the knowledge we have about product ratings to advise customers on what to purchase next. Specifically, we will look at Netflix (see Figure 3.17), and some of the main ideas behind how it recommends movies to its subscribers. We will see how this requires us to predict unknown ratings from the sea of raw data they have aggregated. The wisdom of crowds is implicit here, because the more data we have access to, the better our predictions get.

3.2.1 Internet video streaming

Back in 1997, a man named Reed Hastings had rented the movie Apollo 13 from a video store. Being that he was already six weeks late on returning it, he had racked up a large sum of late fees. Bothered by this, he became convinced that a different payment model would be better: Rather than charging people per rental, they should be charged a flat rate per month, relying on other mechanisms to incentivize customers to return DVDs.

In the same year, Hastings became one of the founders of Netflix, and the company started its DVD rental business. For the first two years,

Figure 3.17 Illustration of Netflix's trademark red envelope it packages its DVDs with. This logo alone has made them stand out and increased their customer base. At the same time, it makes their DVDs the target of theft.

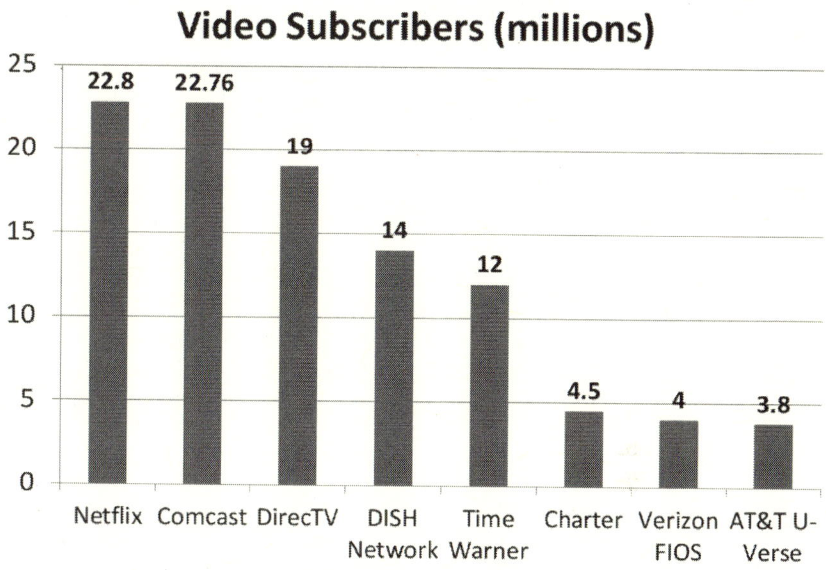

Figure 3.18 Number of video subscribers in the US by different companies in March 2011. Netflix had just surpassed Comcast, the world's leading cable provider, aggregating roughly 23 million customers.

Netflix operated as the more traditional pay-per-rental type store, until it began offering two new features.

- In 1998, they opened an online store, allowing people to wait for DVDs to arrive by mail rather than walking or driving.
- In 1999, they introduced the monthly subscription model, whereby customers pay Netflix a monthly fee. Then, a subscriber could keep a rented movie out for as long as he wants, but there is a limit on the number that he can have out simultaneously.

Netflix operated with great *scalability* (meaning that the cost for additional customers is much lower as the number of customers increases), and *stickiness* (meaning users are reluctant to change services). By 2008, there were roughly 9 million Netflix users in the US and Canada, which was a ten-fold increase over the number of subscribers it had at the turn of the century! Today, around 35,000 different film titles are contained in the 1 million DVDs Netflix sends out every day.

But Netflix moved on to another form of delivering entertainment: streaming movies and TV programs through the Internet to Internet-connected devices. These devices include TVs, set-top boxes, game con-

Rank	Upstream		Downstream		Aggregate	
	Application	Share (%)	Application	Share (%)	Application	Share (%)
1	BitTorrent	52.01	Netflix	29.70	Netflix	24.71
2	HTTP	8.31	HTTP	18.36	BitTorrent	17.23
3	Skype	3.81	YouTube	11.04	HTTP	17.18
4	Netflix	3.59	BitTorrent	10.37	YouTube	9.85
5	PPStream	2.92	Flash Video	4.88	Flash Video	3.62
Total	Top 5	70.64	Top 5	74.35	Top 5	72.59

Figure 3.19 Amount of Internet traffic consumed by the top 5 applications in March 2011 in the US (report by Sandvine). The **downstream** (from application to customer) and **upstream** (from customer to application) percentages are measured at peak times, while the aggregate is over the course of the month. Netflix held almost 30% of peak downstream and 25% of aggregate Internet traffic in this month.

soles, smartphones, and more. This form of rental service has become so dominant that traditional video stores are largely out of business: Blockbuster filed for bankruptcy in 2010 and has moved on to mail delivery.

Video streaming caused Netflix's subscriber base to skyrocket; nearly tripling to 23 million in 2011. They are considered to be the world's leading provider in video distribution today: Figure 3.18 shows that in 2011 they surpassed Comcast in terms of the number of video subscribers for the first time. In fact, Netflix video streaming generated so much traffic in March 2011, that one in every four bits on the Internet was their own. And at times of *peak* Internet usage, they accounted for almost 30% of all downstream traffic. Some other statistics put the percentage at an even higher number. In Figure 3.19, we show fractions of Internet traffic by application for that month.

In future chapters, we will look at the technical network aspect of video-distribution services. But for the time being, we will focus on the social-network dimension of Netflix.

3.2.2 The recommendation problem

Netflix's website allows you to browse for movies by genre, look at Critic's Picks, or use the Netflix Top 100 to decide which movies you want to watch. As you browse, you have three options for each movie: you can rate it if you've already seen it, add it to one of your queues, or tell Netflix you're not interested in it. Believe it or not, these three actions all have a tremendous impact on the movies that Netflix will recommend for you, as well as for others. This brings us to the main question we will explore in this section:

- How does Netflix recommend movies for you to watch?

It's as if they are reading your mind to predict your movie rating. An effective **recommendation system** is important to Netflix because it enhances user experience, increases loyalty, and helps with inventory control.

Even beyond video distribution, recommendation systems are important for many applications. Here are some representative cases:

- *YouTube*
  YouTube makes recommendations of videos to you at the end of each one you watch. This is dependent upon the type of content you are viewing as well as what other people who viewed that video have also watched.
- *Amazon*
  In the last section, we saw how Amazon comes up with its average ratings and rankings for specific products. A different variation is how Amazon *recommends* products to you on the basis of your purchase and viewing history, adjusting it each time you browse. This is similar to Netflix: they both involve recommending specific items (movies or products) to specific people.
- *Pandora*
  Pandora's online music selection has experts that develop recommendations. You have the ability to give a thubs-up or thumbs-down to the songs they recommend, serving as a feedback signal.

Netflix instead has championed a recommendation system that uses the rich history of *all* the user behaviors to profile *each* user's taste in movies. This system has the following inputs and outputs, as summarized in Figure 3.20.

- **Input**
  Each time a user rates a movie, their rating is stored in the Netflix database. Each of these data points consist of four numbers of interest:
  (1) The user's ID;
  (2) The movie's title;
  (3) The number of stars (an integer from 1 to 5); and
  (4) The date of the rating.
  This is a very large data set: Netflix has over 20 million users, and

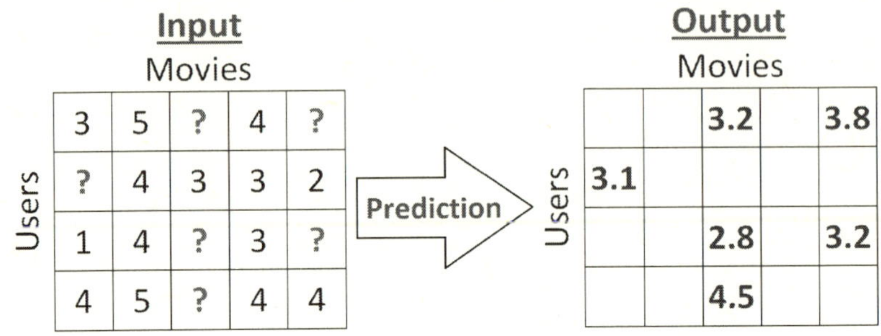

Figure 3.20 Simple diagram of the Netflix prediction process. The known ratings of users for movies are inputed into the predictor, which outputs a prediction of all the unknown values. Realistically, the input data could have 23 million users and 75 thousand movies.

roughly 75 thousand movie titles. But only a small fraction of users will have watched a given movie, and only a fraction of this fraction will have bothered to rate it. Still, this data set consists of billions of points. We illustrate this input conceptually as a table, with users along the rows and movies along the columns (see the left side of Figure 3.20).

- **Output**

  The output is, first of all, a set of predictions for each movie that a user hasn't watched yet (*i.e.* the unfilled entries in the table). These can be real numbers, not just an integer between 1 and 5. We can interpret a predicted rating of, say, 4.2 as saying there is a 20% chance that the user would give the movie 5 stars, and an 80% chance he would give it 4 stars (since he's 80% towards 4, and 20% towards 5).

  But these predicted ratings are not shown to the viewers: the final output is a series of short, rank-ordered lists of movies recommended to individual users. We could obtain this list from the predictions using different criteria: for instance, we could recommend the top five predictions for movies a user hasn't seen yet, or all the movies with a prediction of over 4 stars.

In the end, a prediction system must be evaluated for accuracy. The true test is how many users actually *like* the movies that Netflix recommends to them. But this information is hard to collect. So, Netflix needs a proxy as the metric to evaluate its performance.

To this end, predictions can be made not only on the unknown ratings, but also on some of the *known* ratings. Then, they can evaluate the error on how well these predictions match the known ratings. Netflix quantifies the prediction error in terms of **Root Mean Square Error** (RMSE). Its definition is a mouthfull: you "take the square root of the mean of the squared errors of all points in the data set." Before diving into this in more detail, let's see how RMSE played an important role in Netflix history.

3.2.3 The Netflix Prize

Netflix's original prediction algorithm, CineMatch, worked well: it was within half a star 75% of the time. But recognizing how important it was to have the best prediction algorithm possible, they launched a challenge in October 2006. Their open, online, international competition, called the **Netflix Prize** (see Figure 3.21), awarded $1 million to the team that could develop an algorithm which would improve RMSE by 10% over CineMatch.

Logistics

At the beginning of the competition, Netflix released a set of over 100 million ratings to the public, as part of its records from 1999 to 2005. That amount of data could fit in the memory of standard desktops in 2006, making it easy for anyone in the world to participate in the competition. The rating data came from more than 480,000 users and 17,770 movies. On average, each movie was rated by more than 5000 users and each user rated more than 200 movies.

At a first glance, it seems like this would be ample data to make an accurate prediction for each user. But the real difficulty is that many

Figure 3.21 The Netflix Prize began in October of 2009 and ran until July 2009. The objective was for teams to concoct an algorithm to outperform Netflix's own recommendation system by at least 10%.

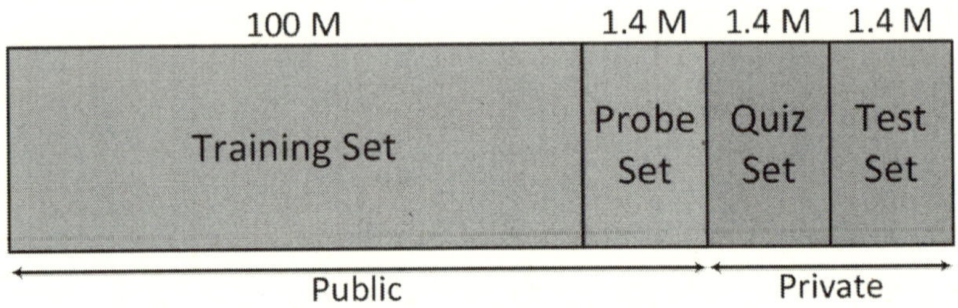

Figure 3.22 The Netfix Prize's four data sets. The training set (100 million data points) and probe set (1.4 million) were publicly released, whereas the quiz set and test set were hidden from the public and known only to Netfix. The probe set could be used to evaluate RMSE as often as anyone wanted, whereas Netflix only allowed each team to evaluate on the quiz set once a day. The final decision was based on RMSE evaluated on the test set.

users rated only a few movies, and a very few users rated a large number of movies (for instance, one user rated over 17,000 movies), so for the majority of users not a lot of information on their tastes were available.

The exact distribution of the Netflix Prize data is shown in Figure 3.22. Two portions of it were released to the public, and two were private.

1. Roughly 100 million ratings were made public, which constituted the *training set*. This is what competitors would use as input data.
2. About 1.4 million additional ratings were made public, and had similar properties to the test and quiz sets described next. This set of ratings was called the *probe set*, and competitors in the Netflix prize could test their performance on this whenever they wanted.
3. Another 1.4 million ratings were hidden from the public, referred to as the *quiz set*. Again, this set had similar properties to the test set. Competitors were allowed to run their algorithms on this at most once a day. The RMSE scores continuously updated on the leaderboard of the Netflix Prize's website were based on this set's data.
4. Another 1.4 million ratings were also hiden from the competitors and constituted the *test set*. This was the real test. The RMSE scores on this set would determine the winner.

Sectioning the sets like this was the best way for Netflix to be sure that the winning algorithm could beat CineMatch by 10%, on more than just this data. Had they released the test set to the public, many teams would have tried to **reverse-engineer** that *specific* set of data,

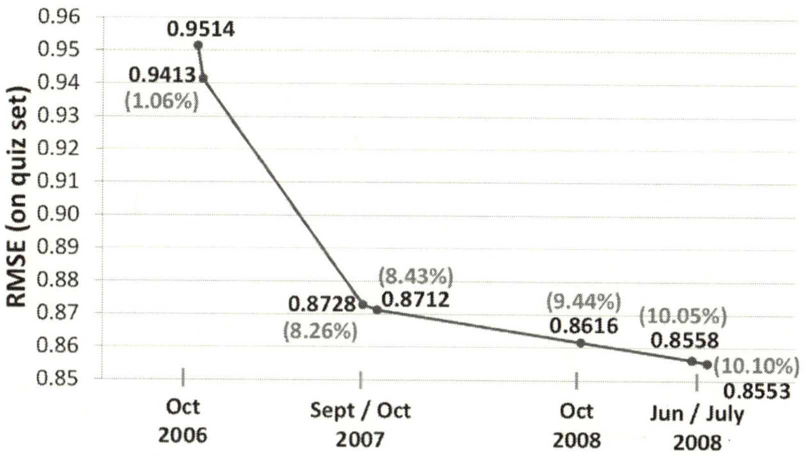

Figure 3.23 Key events of the Netflix Prize competition, which lasted for almost three years and came down to a difference of 20 minutes between two leading submissions. Here, we show progress in reducing the RMSE on the *quiz set*. Also included is the percent improvement over that of CineMatch.

adding as many parameters to their algorithm as necessary to get zero prediction error. But then the algorithm would not work on other data sets, and this was *not* what Netflix wanted. If they were going to replace CineMatch with the winning predictor, it needed to show promise of outperforming CineMatch in the most general case possible. In a similar manner, they only allowed the teams to test on the quiz set once a day. The probe set, on the other hand, was released to the public to help them evaluate performance.

Netfix's CineMatch gave an RMSE of 0.9525 on the test set. This put the end goal to achieve an RMSE of $(1-0.1)\times 0.9525 = 0.8573$ on the test set. Improving it by even 0.01 can sometimes make a significant difference in the movies recommended to a user.

Let the competition begin

This competition sparked some of the most intense activity that has been seen in certain research communities in recent years. Over 5000 teams worldwide entered more than 44,000 submissions. Figure 3.23 highlights some of the important RMSE improvements on the quiz set over the almost three-year period, as we will elaborate on now.

- **Start**

  The competition began on October 2, 2006. Within a week of the start of the competition, CineMatch was beaten: one team had obtained over a 1% improvement in RMSE.

- **2007 Progress Prize**
  Each year, the competition had a Progress Prize of \$50,000, which would be awarded to the leading team, as long as it had achieved an improvement of at least 1% from the previous year (of course, in 2007 there was no "previous year.")

  On September 2, 2007, the competition entered the "last call" phase for the 2007 Progress Prize, meaning everyone had 30 days to make their final submissions for that year. By this time, a team of researchers calling themselves BellKor had the lead with an RMSE of 0.8728, which was an 8.26% improvement over CineMatch. First place changed hands a few times over this period, but BellKor retained the lead in October, reaching an improvement of 8.43% and winning the prize.
- **2008 Progress Prize**
  In 2008, teams started merging. BellKor and another leading team, BigChaos, combined efforts as BellKor in BigChaos and received the 2008 progress prize, having pushed the RMSE improvement to 9.44%. This meant that 2009 would be the last year, since the minimum increase of 1% would push the RMSE to over 10%, the goal of the competition.
- **The Grand Finale**
  BellKor in BigChaos further merged with Pragmatic Theory. In June 2009, the new team – BellKor's Pragmatic Chaos – became the first team to achieve more than 10% improvement, beating CineMatch by 10.06% on the quiz set. At this point, the competition entered the last call phase for the grand prize.

  At the end of this period, two teams had beaten CineMatch by more than 10% on the quiz set: BellKor's Pragmatic Chaos had 10.09%, and The Ensemble had 10.10%, slightly better. But the winner was determined by comparing the RMSE improvement on the *test set*. While both teams beat CineMatch by more than 10% on the test set, and actually got the same improvement of 10.06%, BellKor's Pragmatic Chaos had submitted their algorithm 20 minutes earlier. Hence, they were declared the winner of the \$1 million grand prize.

3.2.4 Key ideas

You must be wondering what BellKor's Pragmatic Chaos's winning algorithm consisted of. As it turns out, obtaining the last few per-

	I	II	III	IV	V
A	5	?	4	?	4
B	4	3	5	**3**	4
C	**4**	2	?	?	3
D	2	**2**	3	1	2
E	4	?	**5**	4	5
F	4	2	5	4	**4**

Figure 3.24 Example of a data set with 6 users (A - F) and 5 movies (I - V). The training data are the 20 points in the table that are not bolded or labeled with a question mark. The test set consists of the 5 bolded entries, since we know them and can "test" our predictor on them. The question marks denote ratings that were never made by the respective users.

cent improvement in RMSE requires hundreds of ingredient algorithms blended together and thousands of model parameters tuned just right. It is not our intention to explain the details involved in such procedures. Instead, we will focus on two key ideas that are the first steps in obtaining such algorithms: the baseline predictor and the neighborhood model.

Our example

Consider the example in Figure 3.24. The data set consists of 6 users (A - F), one for each row, and 5 movies (I - V), one for each column. The entries that are not bolded or question marks have been designated as the *training data* for us to base our predictor off of. The 5 bold entries have been made the *test data*, so that we can test our algorithm's performance on this set. And the 5 question marked ratings are data that we seek to predict in the end.

Clearly, this example is nowhere near the scale of Netflix's dataset: they have tens of millions of users and tens of thousands of movies. Also, Netflix's data is much more **sparse** than ours, meaning that a

	I	II	III	IV	V
A	5	?	4	?	4
B	4	3	5	**3**	4
C	**4**	2	?	?	3
D	2	**2**	3	1	2
E	4	?	**5**	4	5
F	4	2	5	4	**4**

Mean ⇒

	I	II	III	IV	V
A	3.5	**3.5**	3.5	**3.5**	3.5
B	3.5	3.5	3.5	**3.5**	3.5
C	**3.5**	3.5	**3.5**	**3.5**	3.5
D	3.5	**3.5**	3.5	3.5	3.5
E	3.5	**3.5**	**3.5**	3.5	3.5
F	3.5	3.5	3.5	3.5	**3.5**

Figure 3.25 Original ratings (left) and predictor based on a simple arithmetic average (right). This is a very unsophisticated and lazy predictor that gives an RMSE of 1.1619 on the training set and 1.0247 on the test set.

great portion of the entires are not available: in our example, we have 83% of the entires in the table filled. For illustration purposes, our example will suffice.

Baseline predictor

Now, how to develop a predictor? For starters, we could try to take the **mean**, or arithmetic average, of all the entries in the training set, and attribute this to all the unknown entries. The mean is given as:

$$\text{Mean} = \frac{5+4+2+4+4+3+2+2+\cdots}{20} = 3.5,$$

since we are averaging over the 20 numbers in the training data. The result is shown on the right side of Figure 3.25, and the original table has been repeated on the left for convenience.

Let's evaluate the accuracy of this predictor using the RMSE proxy. As we stated before, we must "take the square root of the mean of the squared differences." We will focus here on the test set, but the training set is treated exactly the same way.

To evaluate RMSE, we move backwards in the acronym step-by-step:

1. *Square the difference*: For each element in the test set, we find the difference between the raw rating and the predicted rating, and square it. Take User B and Movie IV. The prediction is 3.5, and the raw is 3, so the squared difference is $(3 - 3.5)^2 = 0.25$. For User D, Movie II, it's $(2 - 3.5)^2 = 2.25$.

	I	II	III	IV	V
A	5	?	4	?	4
B	4	3	5	**3**	4
C	**4**	2	?	?	3
D	2	**2**	3	1	2
E	4	?	**5**	4	5
F	4	2	5	4	**4**

Figure 3.26 Illustration of the need to take user-movie interactions into account: there are reviewers as well as movies with *bias*. Here, user 4 appears to be a harsh critic, while Movie III seems to be better received by the public. Their predictions should reflect this.

2. *Take the mean*: Once we have obtained the squared-differences for each point in the test set, we simply take the mean of all these values.
3. *Take the square root*: Finally, since we squared the differences in the first place, we take the square-root of the result for consistency.

Applying this to the test set using Figure 3.25, we have

$$\sqrt{\frac{(4-3.5)^2+(2-3.5)^2+(5-3.5)^2+(3-3.5)^2+(4-3.5)^2}{5}}$$

$$\begin{aligned} &= \sqrt{\frac{0.25+2.25+2.25+0.25+0.25}{5}}, \\ &= 1.0247. \end{aligned}$$

We can repeat this exact logic to find the RMSE on the training data to be 1.1619. But if you haven't already guessed, this is an extremely lazy and inaccurate prediction system. We need to dive deeper and try to take *user-movie interactions* into account:

- *Easy / harsh reviewers*
  Some users naturally rate more critically than others. For instance, take a look at User D, highlighted in Figure 3.26. The highest rating

he gave was a 3, and this was to a movie (III) that received all 4's and 5's from other users. He also gave two 2's, and was actually the only person to give a 1. He therefore seems like a harsh critic, and the value we predict for his rating of Movie A should reflect this. By the same token, some users may be lenient critics.

- *Good / bad movies*
  Let's face it: some movies are received by the mass as better than others. For instance, take a look at Movie III. With the exception of the one 3 it received (which was by the harsh critic D), it received all 4's and 5's. As a result, we expect this movie to be more well-liked by those who haven't rated it yet, and this should be reflected in its prediction for User C's rating. At the same time, we expect some movies to be less well-received.

Each rating is going to be affected by the *bias* of the corresponding user and movie. To account for this, we add two bias terms to the mean: one for the user, and one for the movie. For each rating, we obtain the following:

$$\text{Rating} = \text{Mean} + \text{User's Bias} + \text{Movie's Bias}. \tag{3.4}$$

This is the general form of the **baseline predictor**. In general, to find the best possible bias terms, we would need to solve an optimization problem with $6 + 5 = 11$ variables (one for each user and movie). Rather than doing that, let's take a simpler approach: for a user, find his average rating across all movies he critiqued in the training data, and compare this to the overall mean. If it's higher, this is an indication of how lenient he is relative to everyone else. And if it's lower, it gives us an idea of how much more critical he is. Similarly, for each movie we find its average over all the users that critiqued it, and compare that to the overall mean.

We can compute these biases from Figure 3.24 by taking the average of each row/column (using only the training data) and subtracting the mean. For the harsh User D, there are 4 ratings in the training data: 2, 3, 1, and 2. Therefore,

$$\text{Bias D} = \frac{2+3+1+2}{4} - 3.5 = -1.5.$$

This is significantly below zero, as we would expect: his average rating is much less than the overall average. As for the good Movie C, there are also 4 ratings: 4, 5, 3, 5. Therefore,

	I	II	III	IV	V	
A	5	-	4	-	4	**0.83**
B	4	3	5	-	4	**0.50**
C	-	2	-	-	3	**-1.00**
D	2	-	3	1	2	**-1.50**
E	4	-	-	4	5	**-0.83**
F	4	2	5	4	-	**0.25**
	0.30	**-1.17**	**0.75**	**-0.50**	**0.10**	

Figure 3.27 Here, we show the user and movie bias terms, at the end of their respective rows/columns. These are obtained by taking the average across each row/column, and subtracting the overall mean. For instance, User D's bias is −1.50, and Movie III's bias is 0.75.

$$\text{Bias IV} = \frac{4+5+3+5}{4} - 3.5 = 0.75.$$

This is significantly above zero, again as expected. The remaining bias terms are computed in exactly the same manner, and are shown at the ends of the rows (for the users) and columns (for the movies) in Figure 3.27. The results are rounded to two decimal places.

With these terms in hand, we can use Equation (3.4) to determine the baseline predictor. For instance, we would say that User D would rate Movie III at

$$\text{Mean} + \text{Bias D} + \text{Bias III} = 3.5 - 1.5 + 0.75 = 2.75,$$

which is only 0.2 away from his actual rating. Taking a rating we don't already have, we would predict User A to rate Movie II with

$$\text{Mean} + \text{Bias A} + \text{Bias II} = 3.5 + 0.83 - 1.17 = 3.16.$$

Applying this logic to all 30 user/movie combinations, we arrive at the baseline predictor shown on the right side of Figure 3.28. We can evaluate its accuracy by computing its RMSE. For the test set, we have:

	I	II	III	IV	V
A	5	?	4	?	4
B	4	3	5	**3**	4
C	**4**	2	?	?	3
D	2	**2**	3	1	2
E	4	?	**5**	4	5
F	4	2	5	4	**4**

Baseline →

	I	II	III	IV	V
A	4.63	**3.17**	5.00	**3.83**	4.43
B	4.30	2.83	4.75	**3.50**	4.10
C	**2.80**	1.33	**3.25**	**2.00**	2.60
D	2.30	**1.00**	2.75	1.50	2.10
E	4.63	**3.17**	**5.00**	3.83	4.43
F	4.05	2.58	4.50	3.25	**3.85**

Figure 3.28 Baseline predictor (right) computed using Equation (3.4). Any value computed greater than 5 is assumed to be 5, and any less than 1 is assumed 1. The bias terms for each of the users and movies are shown in Figure 3.27.

$$\sqrt{\frac{(4-2.80)^2+(2-1.00)^2+(5-5.00)^2+(3-3.50)^2+(4-3.85)^2}{5}}$$

$$= \sqrt{\frac{1.44+1+0+0.25+0.0225}{5}}$$

$$= 0.7365.$$

Compared to the lazy, average predictor's test set RMSE of 1.0247, we have made an improvement of $(1-0.7365/1.0247)\times 100\% = 28.13\%$. We can equivalently evaluate the baseline predictor's RMSE on the training data, which comes to be 0.4709. This represents an improvement of 59.47% over 1.1619. Not too shabby.

Neighborhood model

Let's add a little more detail to our baseline predictor to squeeze in some more RMSE improvement. So far, we have been taking averages along single rows or columns of the rating data. But what about the similarities among *different movies* and *different users*? This is the essence of the **neighborhood model**, where we call two users "neighbors" if they share similar tastes in movies, and two movies "neighbors" if they were rated similarly by different users. This is the simpler of the two methods in **collaborative filtering**, where we try to discover *patterns* across the table, rather than looking at users or movies in isolation.

- *User-user similarity*

Figure 3.29 We say that two users are *similar* when they give the same feedback on a number of movies. Here, Anna and Ben have similar taste. As a result, we would predict that Ben will like "Jurassic Park", since Anna does. Similarity can be present among movies as well.

Consider two users, Anna and Ben, as shown in Figure 3.29. If Anna and Ben both like "Good Will Hunting" and "A Beautiful Mind" but do not like "Lion King" and "Aladdin," then knowing Anna likes (does not like) "Jurassic Park" means that Ben probably likes (does not like) it either. And this works in reverse, too: if Ben does not like the first two, and likes the second two, then if Anna *likes* (does not like) "Jurassic Park," we would expect that Ben *does not like* (likes) it. This illustrate the difference between **positive correlation** and **negative correlation** among neighbors.

- *Movie-movie similarity*
  Consider the two movies "Good Will Hunting" and "A Beautiful Mind." Suppose Anna and Ben rate both of them high, and Charlie rates them both low. Then, if Dana rated "Good Will Hunting" high (low), we would expect her to also rate "A Beautiful Mind" high (low), since the others' ratings have indicated that the movies are probably similar in type, quality, etc. This also works in reverse, as shown in Figure 3.30: if Anna, Ben, and Charlie had rated "A Beautiful Mind" opposite to what they rated "Good Will Hunting," then we would expect Dana to rate the movies opposite as well.

We compute the similarity among different users or movies using a metric called **cosine similarity**, which can be be visualized geometrically. Consider Figure 3.31, where we plot four different cases of the angles between two line segments. If the angle is *very small*, as in (a), then the two line segments are pointing in similar directions, and they exemplify a strong **positive correlation**. On the other hand, if the angle is *very large*, as in (d), then the two line segments are pointing in opposite directions, and they have a strong **negative correlation**.

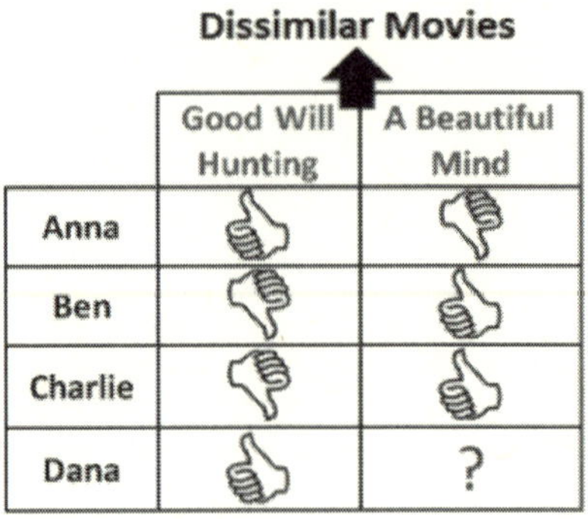

Figure 3.30 We say that two movies are *dissimilar* when they receive opposite feedback from a number of users. Here, "Good Will Hunting" and "A Beautiful Mind" have dissimilar ratings from each user. As a result, we would predict that Dana will not like "A Beautiful Mind," since she likes "Good Will Hunting." Dissimilarity can be present among users as well.

Intermediate values of the angle, as in (b) and (c), indiate that the segments are not correlated.

Now, let each line segment represent one of two Movies, I and II. AI is an indication of the rating that a User A gave Movie I, BI of a rating that a User B gave Movie I, AII of A to II, and BII of B to II. We can directly apply the correllation logic of the previous paragraph here. In (a), User A rated both I and II positively, as did User B: this shows a strong similarity, or positive correllation, between the movies. And in (d), User A rated Movie I positively and II negatively, and User B also rated them opposite to one another: this shows a strong dissimilarity, or negative correllation, among the movies.

We can quantify the similarity / dissimilarity between movies by taking the *cosine* of the angle between the segments. The cosine is given as

$$\frac{\mathrm{AI} \times \mathrm{AII} + \mathrm{BI} \times \mathrm{BII} + \cdots}{\sqrt{\mathrm{AI}^2 + \mathrm{BI}^2 + \cdots} \times \sqrt{\mathrm{AII}^2 + \mathrm{BII}^2 + \cdots}} \tag{3.5}$$

In the numerator, each term is the product of how a user rated each movie. If he rated them both "good" or both "bad," the term will be positive, otherwise it will be negative. Then, these products are summed. For strong positive correllation, we need at least mostly positive terms, and for strong negative correllation, we need at least mosly negative terms. The denominator simply normalizes the sum by the length of the movie "segments," thereby restricting Cosine to lie be-

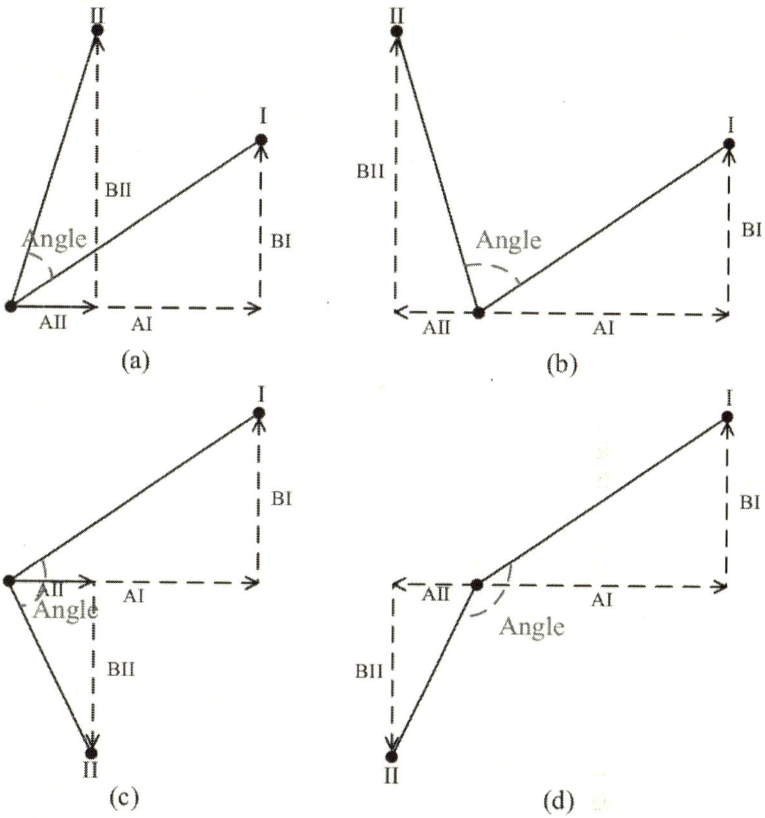

Figure 3.31 Illustration of four possibilities of the angle between two line-segments. For our purposes, if we abstract the segments to two Movies I and II, then AI becomes an indication of the rating User A gave to Movie I, BI of B to I, AII of A to II, and BII of B to II. The smaller the angle between the two segments, the more *similar* the movies are, and the larger, the more *dissimilar*. In this way, we are interested in cases where the angle between the movies is close to 0° (as in (a)) or 180° (as in (d)), but not intermediate values (as in (b) and (c)).

tween -1 and 1. This equation is easily extended to more than two users: just add more terms in the numerator and denominator, as indicated.

You may be wondering how some of the values of AI, BI, AII, BII, ... can be negative if Netflix requires a rating scale of 1 to 5. Well, we won't be computing similarity based on the *absolute* ratings, but rather based on the *error* between the baseline predictor in Figure 3.28 and the actual values. For each user / movie pair, we subtract the baseline from the raw ratings, and the result is shown in Figure 3.32.

	I	II	III	IV	V
A	0.37	?	-1.00	?	-0.43
B	-0.30	0.17	0.25	-	-0.10
C	-	0.67	?	?	0.40
D	-0.30	-	0.25	-0.50	-0.10
E	-0.63	?	-	0.17	0.57
F	-0.05	-0.58	0.50	0.75	-

Figure 3.32 Table of baseline-prediction errors, computed by subtracting the predictions from the actual ratings in Figure 3.28.

Let's compute the similarity values. Take, for instance, Movies I and II. For I, we have entries in the training data for Users A, B, D, E, and F, and for II, we have it for B, C, and F. We only make use of those that rated *both* Movies: B and F. Applying Equation (3.5), we have

$$\begin{aligned}\frac{\mathrm{BI} \times \mathrm{BII} + \mathrm{FI} \times \mathrm{FII}}{\sqrt{\mathrm{BI}^2 + \mathrm{FI}^2} \times \sqrt{\mathrm{BII}^2 + \mathrm{FII}^2}} &= \frac{-0.30 \times 0.17 + -0.05 \times -0.58}{\sqrt{0.30^2 + 0.05^2} \times \sqrt{0.17^2 + 0.58^2}} \\ &= \frac{-0.0220}{0.3041 \times 0.6044} \\ &= -0.11.\end{aligned}$$

As -0.11 is neither close to +1 or -1, we conclude that Movies I and II are not very correllated. Let's take another example: Movies III and V. This time, we have three Users – A, B, and D – that rated both movies. Hence, Equation (3.5) gives the following:

$$\begin{aligned}&\frac{\mathrm{AIII} \times \mathrm{AV} + \mathrm{BIII} \times \mathrm{BV} + \mathrm{DIII} \times \mathrm{DV}}{\sqrt{\mathrm{AIII}^2 + \mathrm{BIII}^2 + \mathrm{DIII}^2} \times \sqrt{\mathrm{AV}^2 + \mathrm{BV}^2 + \mathrm{DV}^2}} \\ &= \frac{-1.00 \times -0.43 + 0.25 \times -0.10 + 0.25 \times -0.10}{\sqrt{1.00^2 + 0.25^2 + 0.25^2} \times \sqrt{0.43^2 + 0.10^2 + 0.10^2}} \\ &= 0.79.\end{aligned}$$

	I	II	III	IV	V
I	---	-0.11	-0.82	0.01	-0.74
II	-0.11	---	-0.74	-1.00	0.88
III	-0.82	-0.74	---	0.50	0.79
IV	0.01	-1.00	0.50	---	0.48
V	-0.74	0.88	0.79	0.48	---

Figure 3.33 Table of movie-movie similarity values computed using Equation (3.5). For each movie in its respective column, we have highlighted its closest neighbor, which is the highest value in magnitude. So Movie I's closest neighbor is III, Movie II's is IV, and so on. These will be used to augment the baseline predictor shortly.

As 0.79 is close to +1, these two movies are positively correlated.

Similar computations are carried out for other movie pairs, and the results are shown in Figure 3.33. A few things to mention: first, the table is **symmetric**, meaning the similarity of A to B is the same as that from B to A. Second, we don't consider the similarity of a movie to itself, as this would not really tell us anything.

The values given in Figure 3.33 are exactly what we use in the development of the neighborhood model. In general, we can use one of various metrics to determine the neighbors of a particular movie. For instance, we could go down each column and select the three movies with the highest similarity values. Or, we could say that any movie is a neighbor if it has a similarity value higher than a specific number (when we say "highest" here, we mean highest in absolute terms: it could be either very strongly positive or negatively correlated). In general, there's no telling what the best thing to do for the particular situation is: that can only be determined *a postieri* after we train with the test set data.

For simplicity, we choose the best, single neighbor for each movie, as highlighted in Figure 3.33. The highlighted box indicates that the

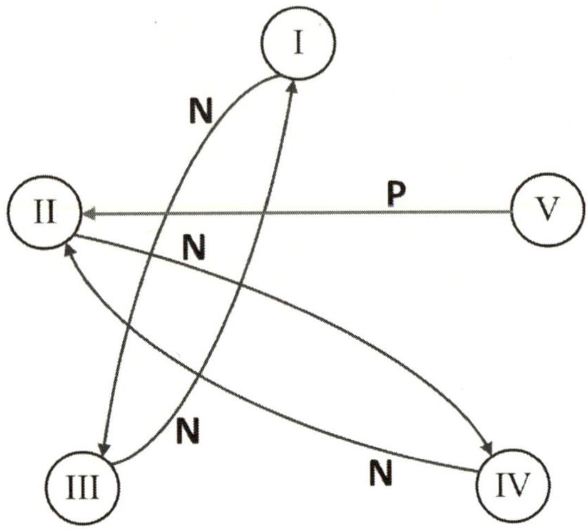

Figure 3.34 Graphical representation of nearest neighbor movie relations. The link from a node indicates who the origin's closest neighbor is. Additionally, "N" means the correlation is negative, and "P" means it's positive. For instance, V points to II with label "P," so II is V's choice, and the correlation is positive.

movie in the column will select the movie in the row as its neighbor, *but not the other way around*: just because V chooses II as its neighbor, for instance, does not imply that II will choose V, as its correlation with IV is higher. We show these relations graphically in Figure 3.34.

Now, what do we do with these neighbors? We want to leverage such information to increase the accuracy of our predictor. We do this by making use of the error we obtained using the baseline predictor, in Figure 3.32.

Let's take Movie V as an example. V's closest neighbor is II, which has a high positive correlation (+0.88). User C rated Movie II, and the predictor made an error of +0.67 in that rating. This tells us that we were *too low* by 0.67 in making that prediction. And since V is similar to II, chances are we were too low in predicting User C's rating of Movie V, too. So, we add 0.67 to the baseline prediction of 2.60 to obtain

$$2.60 + 0.67 = 3.27.$$

Compared to the actual rating of 3, the error here is smaller: we are too high by 0.27 rather than too low by 0.40.

Let's take another example, of User B on Movie IV. IV's closest neighbor is II, which has a perfectly negative correlation (-1.00). The error in predicting User B's rating of Movie II was +0.17, telling us we were too low by 0.17. But since IV is dissimilar to II, chances are we were too *high* in predicting User B's rating of Movie IV. So, we subtract 0.17 from the baseline prediction of 3.50 to obtain

$$3.50 - 0.17 = 3.33.$$

Again, the error here is smaller (0.33 rather than 0.5).

In general, if the correlation is positive, then we *add* the neighbor's baseline error, and if it's negative, then we *subtract* the error, and obtain the following:

$$\text{Rating} = \text{Baseline} \pm \text{Neighbor's Error}. \tag{3.6}$$

Equation (3.6) is a simple form of a **neighborhood predictor**, where we only use the closest neighbor. We could also use two, three, or more neighbors, by averaging the error terms, but we won't get into that here.

It's important to know that we don't expect the error to get lower for *every* rating. The neighborhood method is just a proxy we use based on intuition, and it won't work in all cases. As an example, take User F's rating of Movie V. As we said, V's closest neighbor is II, and they are similar. For Movie II, the baseline predictor's error was −0.58, indicating it was too high. By Equation (3.6), we should add −0.58 to the baseline of 3.85, to obtain

$$\text{Rating} = \text{Baseline} + \text{Neighbor's Error} = 3.85 + (-0.58) = 3.27.$$

But 3.27 is actually further from the true rating of 4 than 3.85 was!

Even though the neighborhood predictor will be off in some cases, we expect that it will help more than it will hurt the *overall* RMSE. Let's find out what that is now.

We can carry out the computations for the rest of the neighborhood predictions just as we have. The results are shown on the right side of Figure 3.35. Note that some of the values are the same as in the baseline predictor: this happens when either the user did not rate the neighboring movie, or the neighboring rating is not part of the training data. For instance, take User C's rating of Movie I. Movie I's neighbor is III, and C did not rate that one. So, we don't change the baseline

	I	II	III	IV	V
A	5	?	4	?	4
B	4	3	5	**3**	4
C	**4**	2	?	?	3
D	2	**2**	3	1	2
E	4	?	**5**	4	5
F	4	2	5	4	**4**

Neighbor →

	I	II	III	IV	V
A	5.00	**3.17**	4.63	**3.83**	4.43
B	4.05	2.83	5.00	**3.33**	4.27
C	**2.80**	1.33	**3.25**	**1.33**	3.27
D	2.05	**1.50**	3.05	1.50	2.10
E	4.63	**3.00**	**5.00**	3.83	4.43
F	3.55	1.83	4.55	3.83	**3.27**

Figure 3.35 Table of neighborhood predictions (right) computed using Equation (3.6). The original ratings have been repeated on the left for convenience.

prediction (2.80). As another example, take User B's rating of Movie II. II's closest neighbor is IV, and B's rating of IV is part of the test set. As a result, we don't change the baseline prediction (2.83) here, either.

Using Figure 3.35, we can evaluate the RMSE of the neighborhood model. Focusing on the test set, we have

$$\sqrt{\frac{(4-2.80)^2+(2-1.50)^2+(5-5.00)^2+(3-3.33)^2+(4-3.27)^2}{5}}$$

$$= \sqrt{\frac{1.44+0.25+0+0.11+0.53}{5}},$$
$$= 0.6826.$$

Similarly, we can evaluate the RMSE on the training data to be 0.4660. These represent improvements of 33.26% and 68.48% over our original, simple average predictor's performance on the test and training sets, respectively. Additionally, this is an extra 5 - 10% over the improvement percentages we saw for the baseline predictor.

3.2.5 Wrap up

In Figure 3.36, we compare the RMSE values on the training and test data in our example for various predictors. The first three are those we discussed in this section. The fourth shows what RMSE we would have

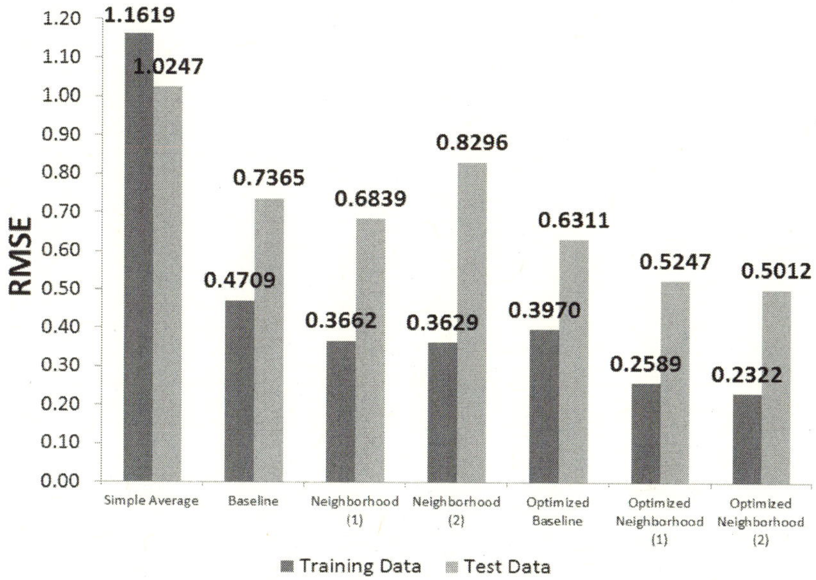

Figure 3.36 Bar graph of the RMSE errors obtained on the training and test data for different predictors. The first three columns are those we illustrated in this section. The fourth is the neighborhood method had we used the nearest two neighbors instead. The last three are the results from optimizing the 11 baseline bias variables and applying the neighborhood predictor on this method instead, which reduces the RMSE a little more.

obtained if we had used the two closest neighbors for each movie, rather than just the highest. As it turns out, the RMSE on the test set stays roughly the same, while on the training data it increases significantly. One neighbor was actually the best choice here.

We stated previously that choosing the movie/user bias values in the baseline predictor to simply be the column/row averages, less the total average, was suboptimal. The fifth set shows the RMSE that we would have obtained if we had optimized the 11 bias terms: it lowers the RMSE by 0.08 to 0.1 over what we have in the second column. We also plot the RMSE that would have been obtained if we used the neighborhood method with the optimized basline. As it turns out, using *two* neighbors here is actually a better choice: in the end, that leads to an improvement of 51% and 80% over the RMSE of the simple average on the test and training sets, respectively.

Another option that we didn't explore in this section was a neighborhood model where *users* are the "neighbors," rather than movies.

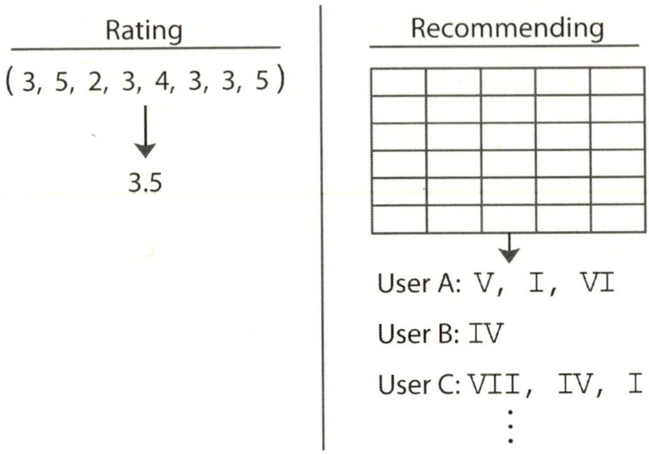

Figure 3.37 Summary of the three different case studies of consensus we've looked at in this chapter – rating products on Amazon, and recommending movies on Netflix. In the former, we are dealing with a set of numbers (e.g., star ratings) and generating a single summary value (e.g., by raw averaging or Bayesian adjustment). In the latter, we start with a table of numbers (e.g., user–movie ratings) and generate a rank-ordered list of items for each user (e.g., by the neighborhood and/or baseline predictors).

But this is a simple extension, with all mathematical development remaining the same.

As we said, it wasn't our objective to explain all the details of the winning algorithm of BellKor's Pragmatic Chaos in the Netflix Prize. This would require highly advanced mathematical techniques and blending of hundreds of ingredient algorithms. But we have illustrated the basic ideas behind how Netflix extracts structure from the slew of information it has in its database, and individualizes recommendation from data about the crowds.

Summary

In this chapter, we have looked at different flavors of the opinion aggregation problem, focusing on online products. We have seen that under the right circumstances, the wisdom of crowds is an effective tool in generating an accurate estimate of the underlying "truth" from a set of user ratings. And when the individual sample size isn't large enough, techniques such as Bayesian adjustment can be applied to weigh it against the larger population. These tools were introduced as we discussed some of the principles behind how Amazon ranks products, though their actual formula that goes from "rating" to "ranking" remains a secret.

We then turned to the recommendation problem, specifically for Netflix. We saw how they play a "mind–reading" game to predict unknown user–movie ratings. Part of their technique involves forming a baseline from the past history of that user and that movie, which is subsequently adjusted based on similarities across different users and/or different movies. We summarize these two case-studies in Figure 3.37.

Our discussion here has been similar in virtue to the problem we explored in Chapter 2 of finding consensus within a crowd. We can unite them under the general theme of rating, ranking, and recommendation. And as with consensus formation on Wikipedia, we see a significant gap between the theory and practice in opinion aggregation on Amazon.

4 Crowds Are Not So Wise

In the last chapter, we explored the "wisdom of crowds", and how we can aggregate the opinions of the mass in an attempt to discover the overall "truth". This idea applied directly to rating products on Amazon, and even seeped its way into recommending movies with Netflix.

Inherent to the wisdom of crowds is an assumption that everyone's opinions are *independent*. So what happens if they aren't? Clearly, what other people think will have some influence on your actions. Like that YouTube video you watched because everyone was talking about it, or that iPad you bought because everyone else had one. And the more people that have followed through with such an action, the more the temptation increases for you to do the same.

In light of this, our focus in this chapter will be on *dependence* of opinions. First, we will look at viralization on YouTube, and then we will talk about social influence on Facebook and Twitter. By the end of the chapter, you should be convinced that in certain situations, it can be relatively *easy* to influence crowds.

4.1 Viral videos on YouTube

In the space of user-generated video content, YouTube (see Figure 4.1) has become the dominant sharing site. We can find videos ranging from lectures on academic subjects to the music of self-proclaimed "YouTube artists". More recently, the site has also started a movie rental and purchase business, to compete against iTunes and the studios.

In this section, we will explore the concept of viralization. In doing

Figure 4.1 YouTube's trademarked logo.

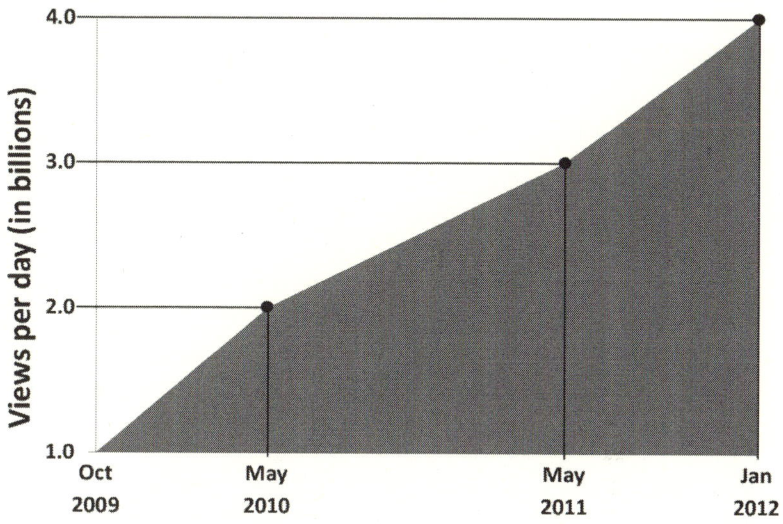

Figure 4.2 Depiction of the average number of YouTube video views per day, from 2009 to 2012. In October 2009, the number hit 1 billion, and by the following May, that had doubled. By January 2012, YouTube hit four billion views per day.

so, we will recognize how YouTube is a perfect example of dependencies created by information spread.

4.1.1 YouTube and viralization

YouTube was founded in February 2005, at the hands of three former PayPal employees. From its inception, the site grew at a rapid rate: by July 2006, YouTube was getting 65,000 new videos and 100 million views daily. And just a few months later, in November of 2006, the company was acquired by Google for $1.65 billion.

Within several years, people were watching videos on YouTube so much that it became a search engine second in size only to Google itself (even though we normally would not think of YouTube as a "search engine"). We show the progression of YouTube's average daily view count in Figure 4.2: in October 2009, YouTube reached one billion daily views on videos, and by January 2012, this number had increased by a factor of four. By Winter 2013, YouTube had over one billion unique monthly viewers, with over four billion views per day. Further, there

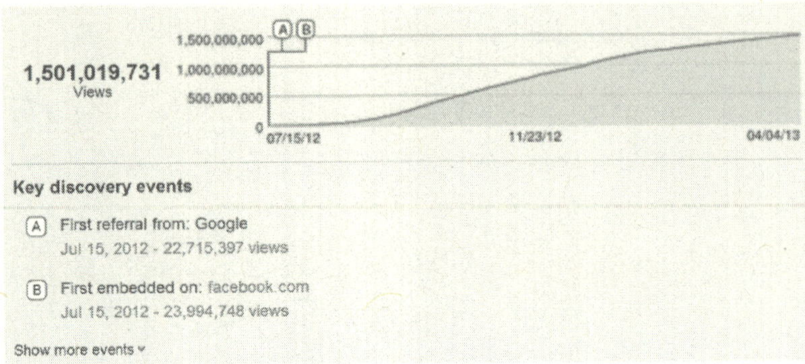

Figure 4.3 Aggregate views over time for the music video Gangham Style by PSY, on YouTube. This is the first video ever to reach 1 billion views, and by April 2013, it had received over 1.5 billion. In addition to statistics, YouTube provides various analytic measures for its content.

were roughly 72 hours of new content uploaded each minute, which is almost 12 years worth of videos each day!

Alexa has ranked YouTube as the third most visited website on the Internet, behind Google (most visited) and Facebook (second most visited). After all, the short-clips and recommendation sidebar make the site almost addictive for viewers. At the end of the day, YouTube has really become somewhat of a "viral" phenomenon itself.

Viral Style

YouTube offers interesting tools, such as "YouTube Insight," that highlight the overall viewing behavior on individual videos. Using these analytics, we can look at the trends behind those videos that have gone "viral." The most extreme example is "Gangham Style," a four-minute music video by PSY which is the first and only video to have reached over one billion views. Released on July 15, 2012, it broke that barrier in just five months, and as of April 2013 it had amassed over 1.5 billion views. The graph of its views over time, provided by YouTube's analytics, is shown in Figure 4.3.

This brings us to the driving question behind this section:

- How do I viralize a YouTube video?

In fact, there has been a lot of research on how to make a video go viral. The most logical starting point is to look at the four main paths by which a viewer is led to a YouTube clip in the first place:

- *Web search*, through sites like Google.

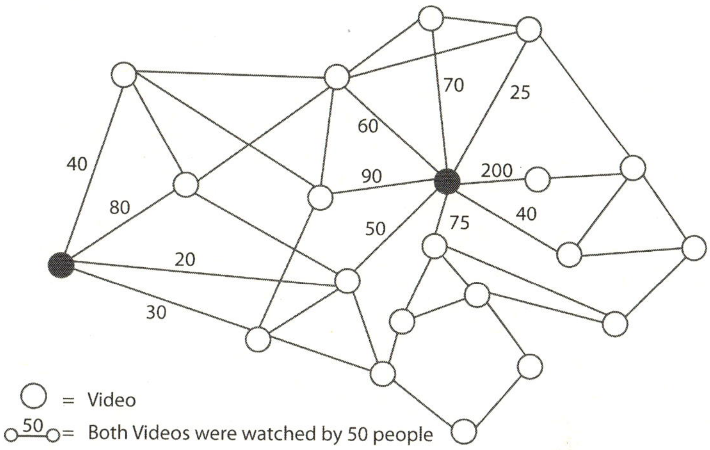

Figure 4.4 YouTube recommendation leverages the co-visitation account between pairs of videos over some time. We can view this as a graph, where each node is a video, and a link's weight is the number of people who watched both videos over that duration. For instance, the left black node has been viewed by 40, 80, 20, and 30 people who have also viewed the respective neighboring video. Similarly, the right black node has a co-visitation count of 200, 40, 75, and so on with its respective neighbors. Naturally, the higher the co-visitation between two videos, the more likely a user watching one will be recommended to watch the other.

- *Referral*, through a variety of means such as email, Twitter, and Facebook.
- *Subscription* to a YouTube channel.
- *Browsing* through the YouTube recommendation page.

Often, subscription and recommendation play a bigger role in a video's popularity than the number of likes and dislikes it has. It is also interesting to see that YouTube does not use a PageRank-style algorithm for ranking the video clips, since linking the videos by tags is pretty unreliable. Nor does it use a sophisticated recommendation engine like Netflix, since viewing data for short clips is too variable, and YouTube videos often have short lifecycles. Instead, YouTube recommendation simply leverages video association through **co-visitation count**: how often each pair of videos is watched together by a viewer over, say, 24 hours. So if 100 people have watched both Video A and Video B in the last day, we could establish a "link" between these two videos with a value of 100. Applying this procedure to each pair of videos gives a link relationship between them, and then we can make

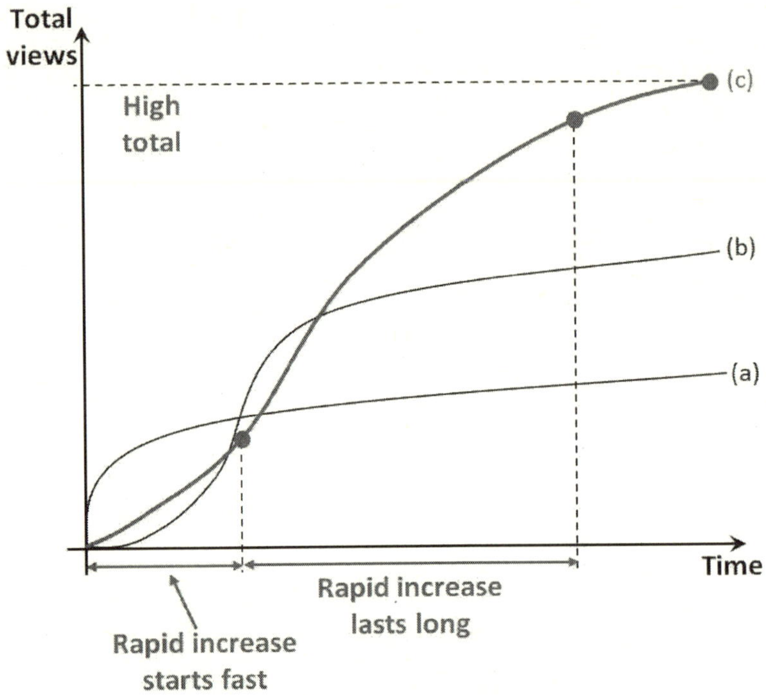

Figure 4.5 A few typical shapes of the total number of views over time. Curve (a) stays at a low level; Curve (b) rises very quickly but then flattens out rapidly too; and curve (c) has a reasonably rapid increase of long duration, and stays that way for a while. Of course, we can also have combinations of these curves, *e.g.* a curve that has a sharp rise to a high level and stays there.

a graph with videos as nodes, as shown in Figure 4.4 (as opposed to webpages as nodes in Chapter 2). YouTube combines this information with matching of the keywords in the video title, tags, and summary to generate your recommendation page upon searching.

It has also been observed that often only those videos with a watch-count number similar to, or slightly higher than, that of the current video are shown in the recommendation page. This makes it easier for widely watched videos to become even more widely watched, possibly becoming viral.

Now, how do you even define "viral"? There is no commonly accepted definition, but the notion of "viral" typically implies that the total number of views over time exhibits three features, like curve (c) shown in Figure 4.5 (or the Gangham Style curve in Figure 4.3):

- A high total view count.
- A "rapid" increase of sufficient duration.
- A short time before the rapid increase begins.

Unfortunately, there is no "formula" you can follow to guarantee your video will become viral. It's as much of an art as a science; otherwise, we would have many more "Gangham Styles" on YouTube. However, researchers have developed some sophisticated models to analyze the spread of an "item" throughout a population, in which the case-study "item" has been anything from a physical product to a disease. In this section, we will focus on one simple yet elegant model that can give us some understanding of phenomena like viralization.

Popularity

So, what makes an item popular to begin with? We can break that down into two distinct factors.

1. *Intrinsic value*

   You may enjoy certain music or buy a particular product simply because you like or need it, regardless of whether the rest of the world agrees or not.
2. *Network effect*

   But in many instances, your decision depends on what others do, for one of two reasons.

   - First, the fact that others like a product naturally gives you information about its quality, appeal, and other factors. This could lead to what we call an **information cascade**, an example of the **fallacy of crowds**. In these situations, the assumption of independence we made in Chapter 3 completely breaks down.
   - Second, the value of the service or product may depend on the number of people who use it. For instance, the fax machine becomes more useful as more people accept faxes, and Wikipedia becomes richer with knowledge as more entries are written. This is an example of the positive network effect.

Our focus here will be on the fallacy of crowds. Specifically, we will study an example of information cascade, to see why it may occur. But keep in mind that we do not have a definite answer as to which model, if any, fits reality sufficiently well. Once again we will see a large gap between theory and practice.

4.1.2 Information cascade

An information cascade is precisely what we assumed would not happen when we discussed the wisdom of crowds in Chapter 3. This occurs when people observe the actions of others and follow the crowd, irrespective of their own instincts. Fortunately or unfortunately, this occurs quite often, in various situations. Let's take a look at some practical examples.

- *Homework assignment*

  Suppose you and your friends are working on a homework assignment together. One of the questions concerns voting theory, and asks you whether Condorcet voting can lead to a cyclic output or not. You have a gut feeling that the answer is "yes," but your friends all insist that it's "no" and will not budge. Not wanting to be the odd man out, you also put "no." Unfortunately, it turns out that your original instinct was right.
- *Looking up*

  On a street corner, if one person looks up to the sky, you will probably think she has a nose bleed or just suddenly turned philosophical, and you would go on with your business. But if ten people look up at the sky together, you may think there is something wrong up there, and stop and look up to the sky too. Now that makes the crowd even bigger, so the next person sees 11 people, which is even more convincing (see Figure 4.6). Finally, someone shouts "Hey, these guys have nose bleeds, that is why they stopped and tilted their heads!",

Figure 4.6 The idea behind an information cascade is that if you see a crowd of people performing some action, you may ignore your own instinct, thinking there's a good reason since there are a lot of people doing it. Here, everyone is looking up at the sky, which would make you think there might be something wrong up there, and hence you would look too.

and you think everyone else, just like you, was misled. Suddenly, the whole crowd disperses.

- *YouTube*

  On YouTube, if many others before you watched a video, you are more likely to watch it too. You might decide to stop viewing it if you don't like it, but this will still count towards the viewing number shown next to the video, and partially determines its place on the recommendation page. Hence, other people viewing the video may cause you to watch it, which in turn influences more people, and this accumulation keeps on building.

We could probably write a whole book on all the places where you would see such information cascades: from stock market bubbles and fashion fads, to the emergence of pop stars and the collapse of totalitarian regimes throughout history. And at least to some extent, all such examples involve people *following the crowd*, rather than everyone listening to their own gut feeling.

Let's generalize some of these ideas. In a situation of **sequential decision making**, where people are making decisions one at a time, we can point out several key observations.

- Each person gets a *private signal* (my nose starts bleeding) and releases a *public action* (let me stop walking and tilt my head to the sky). Subsequent users can observe the public action, but *not* the private signal.
- If we come to a point where there are *enough* public actions of the same type (ten people looking at the sky), then all later users will ignore their own private signals and simply follow what others are doing. A cascade starts, and the *independence assumption* behind the wisdom of crowds that we discussed in Chapter 3 breaks down. What constitutes *enough* pulic actions will depend on the situation at hand. For instance, it's presumably much harder to get everyone to watch your YouTube video than it would be to get people to look up at the sky.
- A cascade can start *easily* if people are ready to rely on others' public actions in their reasoning. It can accumulate to a *large size* through what's known as **positive feedback**: when more people are displaying the same public action, it gives the next person more incentive to follow along, and so on, as shown in Figure 4.7. Notice that positive feedback contrasts our discussion of negative feedback in Chapter 1. Rather than moving in the *opposite* direction of an

Figure 4.7 Illustration of positive feedback in an information cascade. As the number of people performing some public action increases, this increases the chance you will follow along. And if you follow along, this increases the number of people performing the action, increasing the chance the next person will follow along, and so on.

error to correct for it, we are moving in the *same* direction as the influence, thereby creating more of it.

And it's important to realize that cascades can be *wrong*: after all, there's nothing wrong with the sky!

- But a cascade is also *fragile*. Even if a few private signals are released to the public, the cascade can quickly disappear or even reverse direction, precisely because people have little faith in what they are doing even when there are many of them doing the same thing. They are *malleable*. For instance, if the people who had nose bleeds shouted this out, the crowd would immediately dissipate.

Number-guessing thought-experiment

There are many dimensions to understanding the above observations. Let's turn our focus to a particular model from 1992 for sequential decision making: the number-guessing experiment.

Consider a group of people lined up to play a game in which they will guess a number. The moderator has picked some number, either 0

Figure 4.8 Number guessing process: is the true number "0" or "1?" A number of people are lined up, and one at a time they are called up to write their guess on the chalkboard. Here, the third player sees that the first two people wrote down "0" and "1." Between this information and her private signal, she tries to guess what the true number is.

or 1, to be the true one. Each person knows that the number is either 0 or 1. One at a time, they come up to a blackboard, where they are to write down what they think the number is (see Figure 4.8).

The moderator has two cards, one of which has a 0 written on it, and one of which has a 1. When each person comes up, the moderator chooses one of these cards, and shows it to them: this serves as the person's private signal. There's no guarantee that the number she shows will be right, but everyone is told that there's more of a chance that the card they are shown is *right* than *wrong* (*i.e.*, more people will be shown the right number than the wrong number).

When a person goes to make a guess, she gets to see the public actions of everyone who went up before her. But she does not get to see the private signals they were shown. As an example, Figure 4.9 shows the situation when the third person goes up to the board. She is shown a private signal on a notecard, say `PRV III` (`PRV` for "private," `III` for person 3) and sees the public actions of the first two users, say `PUB I` (`PUB` for "public," `I` for person 1) and `PUB II` on the blackboard. Using this information, her task is to make her guess, `PUB III`, and write it on the board.

If you are the first person, what should you do? Well, there's nothing currently on the blackboard, so all you have to go by is what you are

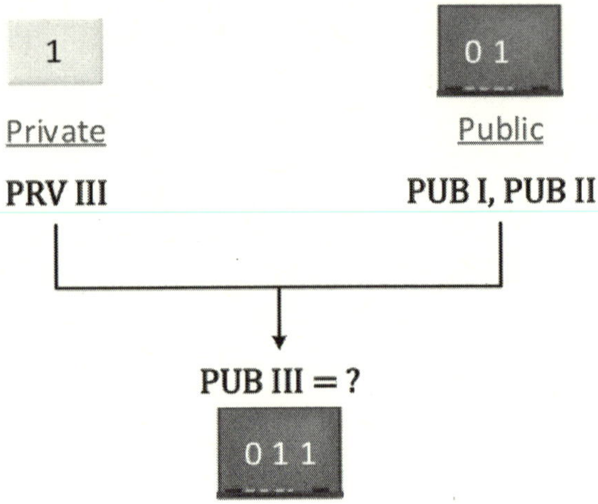

Figure 4.9 When the third person comes up to the board to guess, she sees the public actions of the first two people (`PUB I` and `PUB II`), and receives her own private signal (`PRV III`). From this information, she writes down her guess on the chalkboard (`PUB III`), which becomes her public action for succeeding guessers to see.

shown on the notecard. And you know that what you see if more likely to be right than wrong. So if you see a "0," you will write down a "0," and if you see a "1," you will write down a "1."

Now suppose I'm the second person to go up. I see the public action of the first person, `PUB I`, and my own private signal, `PRV II`. But in reality, I know how the first person *reasoned*: even though I cannot see her private signal, it *must* be the same as `PUB I`, because she had no other information and thus would have just written down her private signal. So I really know *two* private signals, `PRV I` and `PRV II`.

- If they are both "0," then obviously I will write down "0": this is just a stronger case than what the first person was facing, and I can be more confident that "0" is correct.
- If they are both "1," then similarly, I will write down "1."
- But if they are *different*, then I really have no information: I can't be sure which number is more likely to be correct. In this case, I will flip a coin and randomly write down "0" or "1."

Now here comes the first chance of an information cascade starting. If you're the third person, what should you do? As shown in Figure 4.9, you have one private signal, `PRV III`, and two public actions, `PUB I` and `PUB II`.

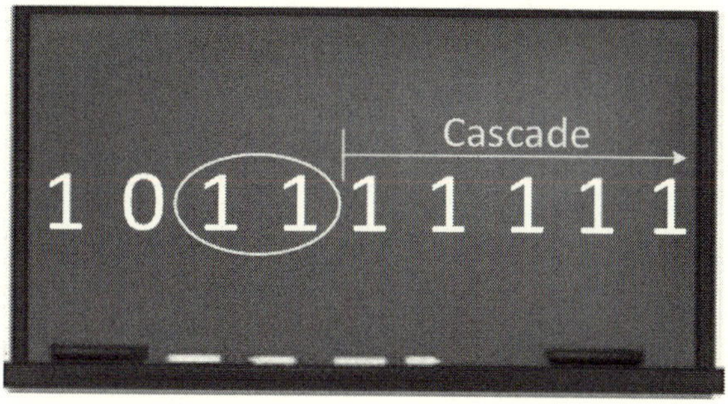

Figure 4.10 Example of a cascade of 1's. The first and second people write down two different public actions, and as such, no cascade is triggered yet. But the third and fourth both show 1's, and as a result, it is in everyone else's best interest to guess a 1.

If `PUB I` and `PUB II` are different, then you know what dilemma the second person faced. Her private signal was different from the first person's, and therefore she must have guessed randomly (otherwise, she would have written down the same thing the first person did). So, you are in exactly the same shoes as the first person was, and you should just rely on your own private signal `PRV III`. And the fourth person will be in the same shoes as the second user.

But consider the case that `PUB I` and `PUB II` are the same. If your `PRV III` is the same as these guesses, then obviously you should pick that number for `PUB III`. But it turns out that even if your private signal is *different*, you should *still* write down what they did. So if the first two users both wrote down the same guess, then an information cascade starts: it's not logical for you to follow your private signal. It's in your best interest to just keep with the crowd.

Supposing no cascade starts after the first two people, then it could just as well start after the next two people, and then the next two, and so on: all it takes is some even-numbered person to show the same public action as the odd-numbered person right before her. From this point forward, every person will copy the same public action. In Figure 4.10, a cascade of 1's is triggered after the third and fourth people show public actions of 1.

The first pair

Now, let's run through a few calculations to see how these factors interplay for the first pair of people. Assume that the moderator has decided on "1" as the correct number, and that the chance that she shows each

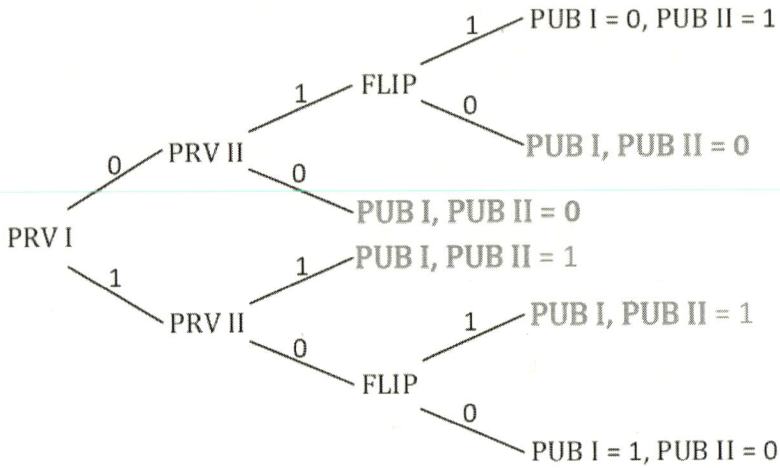

Figure 4.11 Illustration of all the possible sequences of events arising from the first two people, as a tree. For instance, if the first person is shown PRV I $= 1$, we are on the lower-half of the diagram. Then, if the second person is shown PRV II $= 0$, we arrive at FLIP, which denotes a coin flip. Depending on this outcome, the second person might choose "1," so that PUB I, PUB II $= 1$, or "0," so that PUB I, PUB II $= 0$. We have colored / shaded (depending on color / black and white version) the paths that would lead to correct and incorrect cascades.

person a "1" is 80%. The second assumption implies that "0" is given 20% of the time.

First, what is the probability that we have *no* cascade after the first pair? In Figure 4.11, we illustrate all the possible sequences of events arising from the first two people, as a tree diagram. For no cascade, we need the first two public actions to be *different*. This means PUB I $= 0$ and PUB II $= 1$ or PUB I $= 1$ and PUB II $= 0$, which will happen, respectively, as follows:

- PRV I $= 0$, PRV II $= 1$, FLIP $= 1$

  Here, the first person gets a private signal of 0, which has a probability of 20%. Then, the second person gets a 1, which happens with a 80% chance. Since PRV I and PRV II are different, the third person flips a coin, which turns up heads and tells her to guess "1." This coin flip has 50% chance of turning up heads. Since each of these three events must happen together, we *multiply* them to get the probability: $0.20 \times 0.80 \times 0.5 = 0.08$, meaning this has an 8%

chance of happening.

- PRV I $= 1$, PRV II $= 0$, FLIP $= 0$
  This is exactly the reverse of the logic we just saw. The first person sees a "1," with probability 80%, the second sees a "0" with probability 20%, and the coin flip for the third person comes out tails. Multiplying this gives the same as above: $0.20 \times 0.80 \times 0.5 = 0.08$, or 8%.

Since either of these will lead to no cascade, we *add* them to find the total probability of *not* having a cascade after the first two users: $0.08 + 0.08 = 0.16$, or 16%.

Now, if a cascade were to occur, it could either be correct (1's) or incorrect (0's). In order to get a correct cascade, we need to have the first two public actions be 1. Referring back to Figure 4.11, this can happen in one of two ways:

- PRV I, PRV II $= 1$
  Here, we simply have the first two private signals being 1's. The probability of each case is 80%. Since we need both to happen, we multiply: $0.8 \times 0.8 = 0.64$, or 64%.

- PRV I $= 1$, PRV II $= 0$, FLIP $= 1$
  In this case, the first user receives a private signal of 1, with probability 75%. Then, the second user sees a 0, with a probability 25%. When the third person flips a coin, it turns up heads, telling her to go against her private signal and guess a 1. Again, the coin flip is 50/50. Therefore, the overall probability is $0.8 \times 0.2 \times 0.5 = 0.08$, or 8%.

Adding these two probabilities together, we get $0.64 + 0.08 = 0.72$, meaning there is a 72% chance of having a correct cascade after the first pair. That's pretty high, due to the fact that the probability of the private signal being correct is high (80%).

Now, what is the chance of having an *incorrect* cascade? Well, there are only three outcomes after the first pair: we have no cascade, we have a correct cascade, or we have an incorrect cascade. These probabilities must add up to 1, because exactly one of them must happen. Subtracting 0.16 and 0.72 from 1, we get $1 - 0.72 - 0.16 = 0.12$, or 12%.

Let's verify that through calculation as we did for the other two cases. For an incorrect cascade, we must have the first two public actions being 0. Referring back to Figure 4.11, we see this can happen in one of two ways:

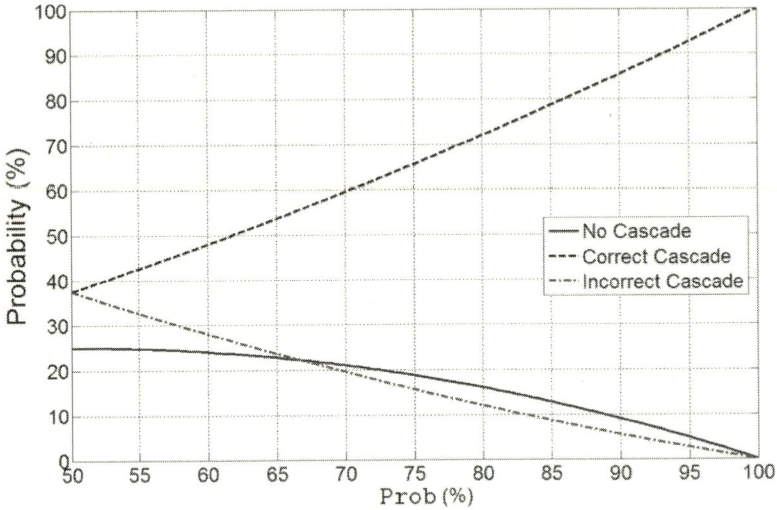

Figure 4.12 Probabilities of none, correct, and incorrect cascades after the first pair of people, as a function of the probability that the private signal is correct, `Prob`. When `Prob` $= 0.5$, we have the same chance of correct as incorrect cascade, and less of a chance of no cascade. As `Prob` increases to 1, the probability of correct cascade increases to 1, and the probabilities of no and incorrect decay to 0.

- `PRV I`, `PRV II` $= 0$
  Here, the first two private signals are 0's. The probability of each case is 20%. So we have $0.2 \times 0.2 = 0.04$, or a 4% chance.

- `PRV I` $= 0$, `PRV II` $= 1$, `FLIP` $= 0$
  In this case, the first user receives a private signal of 0, with a probability of 20%. Then, the second user sees a 1, with a probability of 80%. The coin flip turns up tails, telling the third person to go against her private signal and guess a 0. Multiplying, we have $0.2 \times 0.8 \times 0.5 = 0.08$, or an 8% chance.

Adding these together, we have $0.04 + 0.08 = 0.12$, or 12%, just as we expected.

What we've illustrated here are the probabilities of no cascade, correct cascade, and incorrect cascade after the first pair, for a certain chance that each private signal is correct (80%). In Figure 4.12, we go one step further and plot the probability of each outcome as this chance (denoted `Prob`) is varied from 50% to 100%. We see that increasing it reduces both the probability of not having a cascade and of having an incorrect cascade, which eventually drop to 0%. On the other hand, the probability of having a correct cascade increases, eventually reaching 100%. This should make sense, because as the chance the private signal is right increases, we are more likely to have the first two people see 1's, which makes a correct cascade more probable.

Future pairs

Figure 4.12 show us the different probabilities after the *first pair* of people have guessed the true number. But what happens after the second, third, and future pairs?

Let's keep the chance of the moderator showing a "1" at 80%. After a given number of pairs have guessed, what is the probability of having *no cascade*? Well, for that to happen, each of the pairs needs to not get stuck in one. We need the first not to cause one, the second not to, and so on. And since we need these events to all happen, we multiply the probabilities.

Before, we calculated the probability that a pair does not cause a cascade to be 16%. So, for two pairs, it would be $0.16 \times 0.16 = 0.0256$, or 2.6%. That's a very small chance. For five pairs, it becomes $0.16 \times 0.16 \times 0.16 \times 0.16 \times 0.16 = 0.000105$, or less than one-hundreth of a percent. For 50 pairs, the decimal probability has 37 trailing zeros before the first significant digit.

Clearly, this probability is going to zero, and very quickly at that. This means that after a few pairs, we are more or less *guaranteed* that we will have a cascade. Is this a good thing? Not necessarily. We have only shown that the probability of having *some* cascade is 100%. Remember, this could be either a correct *or* an incorrect cascade. Unfortunately, as more and more pairs of users pass by, while there's greater than a 50% chance of it being correct, there is still a significant chance it will not be. We can see this in Figure 4.13, where we have plotted the probability of having a *correct* cascade for different numbers of pairs, which we call `NumPairs`, as the chance of showing the correct number varies. While there's a significant difference between the curves for 1 and 2 pairs, increasing the number above 5 really does not increase the probability at all, which is especially alarming when the chance of showing the correct number is closer to 50%. Indeed, if the chance is 50%, then we have an equal chance of having a correct as an incorrect cascade, even with an infinite number of pairs!

This is somewhat couter-intuitive, because when the number of pairs is large, we expect a large amount of information to be available (i.e., more public actions), and that a correct cascade should happen with

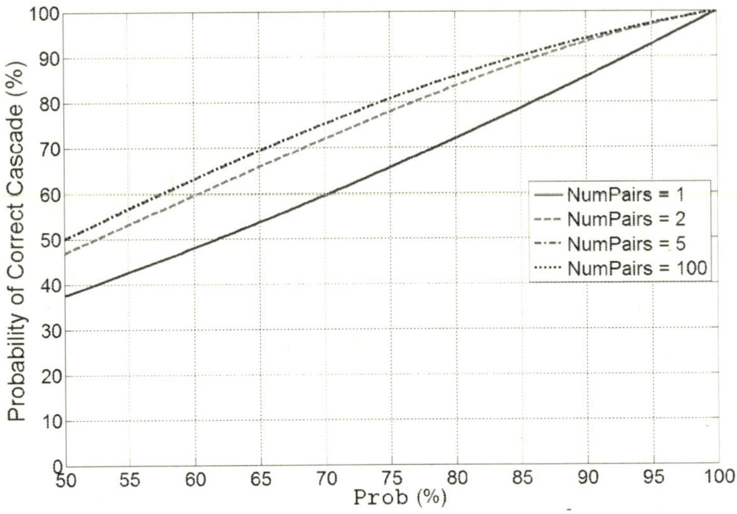

Figure 4.13 Probabilities of winding up in a correct cascade as functions of `Prob`, for various numbers of pairs that have guessed so far, or (`NumPairs`). We see that increasing `NumPairs` from 1 to 2 does increase the probability of a correct cascade, but beyond this, more pairs do not increase the probability much. We see this because when `Prob` is below 1, there's still a pretty good chance of having an *incorrect* cascade.

high probability. But what happens here is that cascades block the aggregation of independent information, which was so important to us in the wisdom of crowds in Chapter 3. Once we get a pair that displays two 0's, we are stuck with 0's forever, because people will trust the crowd's guess over their own.

As it turns out, more people really aren't doing much for us here: they are just guaranteeing that we will have *some* cascade. We just have to hope the moderator has decided that she will show the right value a lot more than the wrong one, so that when a cascade is triggered, it will be correct.

As one final point, note that we did not compute the correct and incorrect cascade probabilities for multiple pairs. These calculations are not as simple as just multiplying repeatedly, as we can do to find the chance of no cascade. Though that math is beyond us, explaining the key points doesn't require it.

Emperor's New Clothes

We have just seen how easy it is to start a cascade (depending on the situation). So, how long will a cascade last? Well, forever, unless

Figure 4.14 The Emperor's New Clothes effect is the counterpart of information cascade. In this story, the emperor is told that he has the finest clothing which is invisible to all those who are "unfit." The crowd, one after another, begins complimenting his attire. This triggers a cascade effect, because each person trusts the crowd's opinion over their own eyes, fearing reprisal for not following. All it takes is one child crying out that the emperor is naked for the cascade to stop.

there is some kind of *disturbance*, e.g., a release of private signals. And how many such disturbances do we need? Even a few will often suffice, because, despite the number of people in the cascade, they all know they are just playing "follow the leader" to maximize their chance of doing the right thing.

This is the counterpart of information cascade: the **Emperor's New Clothes effect**. This name arises from the 19th century short story (see Figure 4.14) in which a vain emperor is told this his new "clothing" is of the finest, best suit of fabric that is invisible only to those who are unfit for their positions. In reality, there are no clothes at all. While everyone plays along (their public actions), including the emperor himself, not wanting to seem "unfit" (their private signals), it only takes one kid's shouting "hey, he is wearing nothing at all!" until everyone becomes more confident that he is, in fact, not wearing any clothes.

So how do we break a cascade? Suppose a cascade of 1's has started. Now when one person gets a private signal of 0, she shouts *that* out, instead of her public action, which is still to guess that the correct number is 1. If the next person gets a private signal of 0, she would think that, on the one hand, now there are two private signals of 0s. On the other hand, there may have been only two private signals of 1's to begin with, since that would have been enough to kick off a cascade of 1's (or even just one private signal of 1, since the following user receiving a private signal of 0 may break the tie by randomly choosing

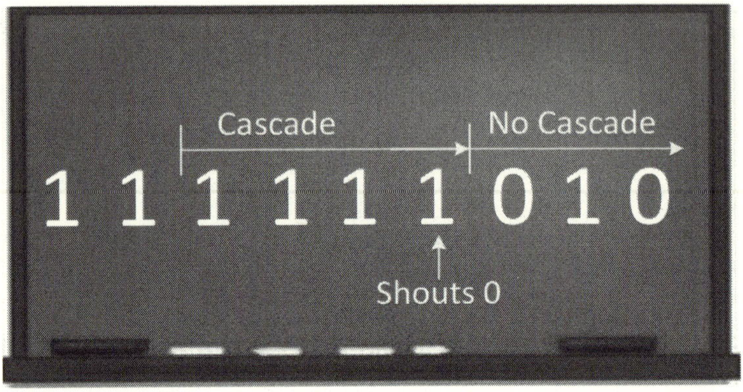

Figure 4.15 After the first two people, a cascade of 1's is triggered. But the sixth person decides to shout out her private signal of "0," when she writes her guess of "1." Then, assuming the seventh person also has a private signal of "0," she will guess "0," since all she can be sure of is that the first person had a private signal of "1." This breaks the cascade (until another one starts).

1 as the public action). So she will decide to guess 0. That breaks the cascade. We show an example of this in Figure 4.15. Remember, a cascade represents only what happened with a few people right around the time it started. If everyone knows that, another block of a few people may be able to break it.

Sometimes it takes the apparent conformity of more users, e.g., ten people, instead of just two, standing on a street corner looking up at the sky. For something with a low probability (e.g., something is wrong in the sky, wrong enough that I want to stop walking and take a look), you need more public actions to override your private signal. In a crowd on a street corner, a passer-by often needs more convincing than in a simple game where she is guessing the right number with higher than 50% probability. That is why she needs to see a bigger crowd all tilting their heads before she is willing to stop and do the same. A more curious and less busy person may require a lower crowd size before she would stop and do the same.

In this analysis, we have assumed that everyone acts rationally and can decide what the best guess to make is depending on their current information. What each person *should* do can be quite different from what she *actually* does in reality. Researchers have observed that the number-guessing experiment does not play out in reality as the theory predicts, perhaps because we are not all able to go through the probabalistic thought processes on the spot.

Implications

There are many implications of this information-cascade model, since (1) various social phenomena exhibit the features of a crowd making a decision that ignores each individual's own private information, and yet (2) the chain reaction is fragile. Information cascade is part of the reason why US presidential primary elections use Super Tuesday to avoid sequential voting, why teenagers tend to obtain information from the experiences of peers, and why, once people suspect the underlying true signal has changed (whether it has actually changed or not), the ongoing cascade can quickly reverse.

Now, how can we translate this theory to figuring out how to viralize a YouTube video? Not easy. But the main idea should be clear: you want your video to undergo an information cascade, so that when a person sees or hears about it (the "public action"), they will most likely watch it, irrespective of whether or not she is interested in it (her "private signal"). But how many public actions does it take before a person will watch your video automatically? Does such a number even exist? Clearly, if it does, it will be different for everyone, depending on how malleable a person is. All of these are open research questions with no definitive answers.

So far in this chapter we have focused on a single influence model – information cascade – to show how people's actions may be highly dependent upon others'. Specifically, we have assumed that everyone has access to all the public actions before them, as shown in Figure 4.16. Information cascade is *population-based*, meaning that it assumes the population interacts as a whole, rather than taking into account social

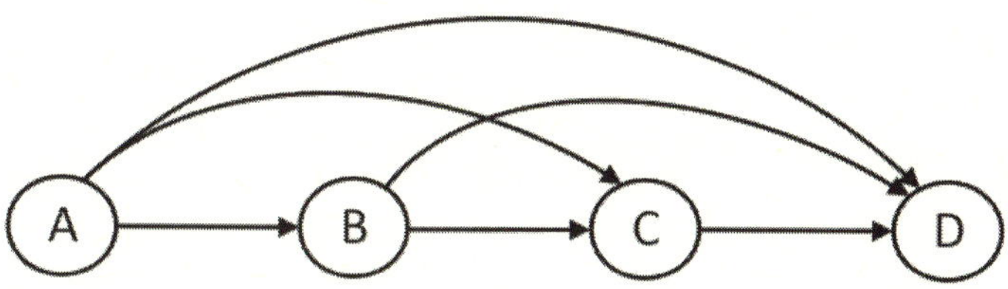

Figure 4.16 Line-type relationship between nodes (*i.e.*, people) we assumed in the number-guessing game. A link from node A to node B means that B obtains information (*i.e.*, public action) from A. So the second node is influenced by the first, the third by the first and second, and so on.

relationships among people. Naturally, such relationships are important in the study of influence: you will be more affected by the decisions your best friends make than of some random person you see walking down the street. In the next section, we will study influence models that are *topology-dependent* (*i.e.*, take social graphs into account) in the study of behavior models for Facebook and Twitter.

4.2 Influencing people in social networks

Let's turn now to considering how the people you are *connected* to can influence you. In doing so, we will focus on topology, rather than just the overall populations as we did in the last section.

4.2.1 Social influence

By far the largest social networking site today is Facebook (see Figure 4.17). Though it wasn't formally founded until mid-2004, its roots can be traced back to January of that same year. At that time, Mark Zuckerberg, then a student at Harvard University, launched a social networking site called "thefacebook," which had been inspired by a previous one he had created called Facemash.

Initially restricted to only students at Harvard, this site quickly attracted more than half of the undergraduate population there. In March of 2004, less than three months after it launched, it expanded to other Ivy League schools, and gradually gained the attention of most universities in the US and Canada. In 2005, the site officially changed its name to Facebook when the corresponding domain `facebook.com` was purchased. Around this time, it opened up to high school students, and then employees of certain companies. Finally, in September 2006, Facebook began allowing anyone 13 years or older with a valid email address to join, and it remains that way to the present day.

Figure 4.17 Famous logos, one of Facebook (left) and Twitter (right), the two social networking sites we will be focusing on in this section.

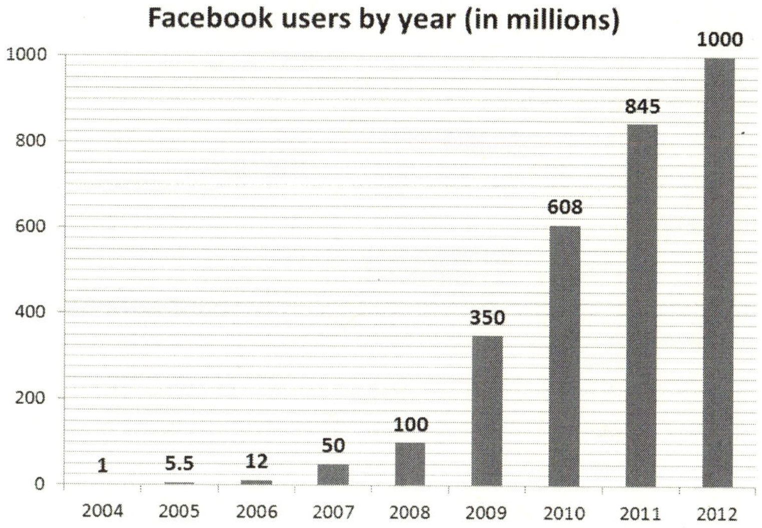

Figure 4.18 Number of active Facebook users by year since its inception in 2004. In 2012, it hit 1 billion.

Present through a wide range of demographics, Facebook has become almost ubiquitous in today's society. Figure 4.18 shows the rapid growth in the number of active Facebook users per year since its inception. In October 2012, it hit 1 billion users worldwide. In fact, in that same year, it filed for an IPO (initial public offering) which priced Facebook at $104 billion, the highest valuation to-date of any newly public company. Though the US has more than double the number of Facebook users than any other country – over 160 million registered as of February 2013 – the site has a broad international foothold; for instance, in that same month, there were over 60 million users in each of Brazil and India.

Many links and connections have been formed among the people using Facebook. If we took all the users on the site, represented them as nodes in a graph, and drew links between them if they were friends, just imagine how large and complex that graph would be. But one of the questions that will come up in this section is, what really constitutes a "link" between two people on Facebook? Is it simply being "friends?" Or is it some stronger notion, like a certain amount of *mutual* activity on each other's walls?

Not far behind Facebook in popularity is Twitter (see Figure 4.17), the social networking site that enables users to send and receive text-

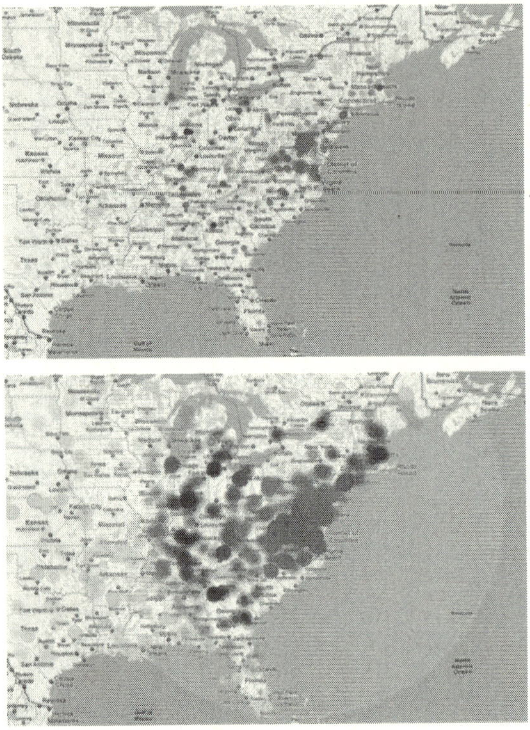

Figure 4.19 Twitter activity 10 seconds (top) and 80 seconds (bottom) after the earthquake hit the east coast of the US in summer 2011. Red is most intense, while green is least intense.

based "tweets." Twitter was created in March 2006 by Jack Dorsey of Odeo and launched in July of that same year. In April 2007, it officially became an independent company.

In just six years, Twitter attracted more than 500 million users. On its seventh anniversary in March 2013, it was handling more than 400 million tweets each day, up 60 million from the year before. Twitter has become highly popular among celebrities, likely due to the one-way nature of "following" relationships (i.e., you can follow them without them having to follow you); as of May 2013, the pop-singer Justin Bieber had the most followers at just under 40 million. In addition to following, Twitter combines two more features into a single platform: microblogging (with no more than 140 characters), and group texting.

Facebook and Twitter are two of the most influential modes of communication, especially among younger generations. Both of them, though especially Twitter, tend to spike in usage during major events. Here are two interesting examples:

- During the east-coast earthquake in Summer 2011, tweets traveled faster than the earthquake itself from Virginia to New York. In Figure 4.19, we show the level of Twiter activity on the east coast just 10 and 80 seconds after the earthquake hit (notice the volume of increase).
- In February 2013, Twitter announced a record-breaking 24.1 million tweets during the Super Bowl of that year.

Who is "important?"

With the proliferation of online social networks, there have been all kinds of attempts at figuring out the answers to the following questions:

1. How do we analyze people's "opinions" on Facebook or Twitter?
2. How can we measure the *influential power* of individuals on Facebook or Twitter?
3. How would we leverage the knowledge of who's influential and who's not to actually influence people online?

For example, on question (1), we recently studied all the tweets about Oscar-nominated movies during the month of February 2012. Overall, the opinions on Twitter tended to be a lot more positive than peoples' opinions on other forums. We found it necessary to combine tweet analysis with data from other venues like IMDb or Rotten Tomato to get a more accurate reading.

Question (2) is an analysis (i.e., breaking down) problem and question (3) a synthesis (i.e., applying what we learned) problem. Neither is easy to answer; and there is a significant gap between theory and practice, perhaps the biggest such gap you will find in this book.

But difficulties have not prevented people from trying:

- Regarding (2), there are many companies charting the influential power of individuals on Twitter. There are several ways to approximate that: by the number of followers, number of retweets, or number of URL repostings a person has. There are also many companies attempting to put together the entire social graph of Facebook: with access to that, they would be able to see which nodes were most "important" (more on this next).
- As to (3), Facebook uses simple methods to recommend friends, often based on email contact lists or common backgrounds. Marketing firms also use these sites to stage marketing campaigns. Some "buy

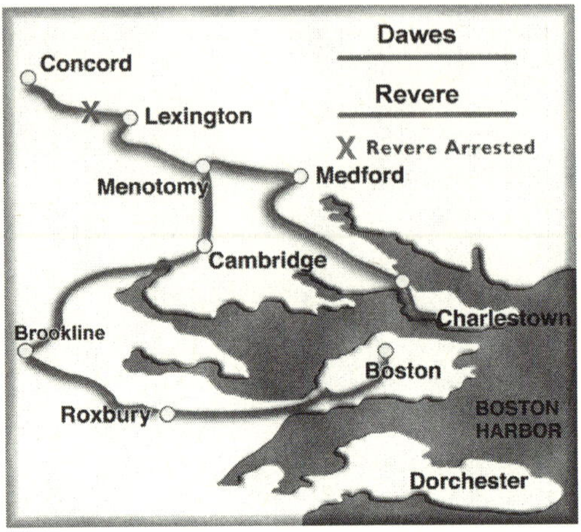

Figure 4.20 Night rides by Dawes and Revere the night before the American Revolutionary War started in 1775. Revere was more effective in spreading the word than Dawes because he took a path with many more influential militia leaders.

off" a few influential individuals on these networks, while others buy off a large number of randomly chosen, reasonably influential individuals, and seed them with the product they want everyone to buy. At the end of this section, we will explore how finding the "best" people to seed, even when we have the graph, can be difficult.

It is important to figure out who is influential, and in turn to leverage that information, in many situations. We can cite this in a wide range of historical examples. For instance, consider the "midnight rides" of Paul Revere and William Dawes on the brink of the American Revolution in 1775. That night, starting near Boston and Cambridge, both of them were to alert Americans that the British were planning an assault on Lexington and Concord. They each took different paths towards Lexington, before meeting there and riding towards Concord, as traced out in Figure 4.20. Revere was able to alert powerful militia leaders on his route to Lexington, and as a result he was more effective than Dawes in spreading the word (though Revere was also arrested by the British military). This helped the Americans win the first battle that started the war the following day.

Now, how do we *quantify* which nodes are more important? Not easy. Throughout the years, many different definitions of what it means to be "important" in a social network have emerged. Throughout this section, we will examine some of these interpretations, and in doing so will see the benefits and weaknesses of each.

4.2.2 Measuring node importance

Consider the social graph depicted in Figure 4.21. This graph defines the connections between six different people, represented here as nodes. For instance, Cara is connected to Anna, Dana, and Evan. Our main task will be to look at how "important" each of the people are, based on the connectivity of the graph.

Before looking at the graph in any detail, you will probably wonder how someone would come up with such a model for social relationships. Indeed, how do we define a "link" between two people? As we said, the notion of a "link" in this context can sometimes be ambiguous, but let's assume it means the two connected nodes "know" each other. So Dana knows Cara, Evan, and Frank. On the other hand, Ben only knows Anna.

The first thing you likely notice about the graph in Figure 4.21 is that the links are different from the ones we looked at in Chapter 2 when we discussed PageRank: there are no arrows at the end of the links. The reason for this is that we assume the graph is **undirected**, which

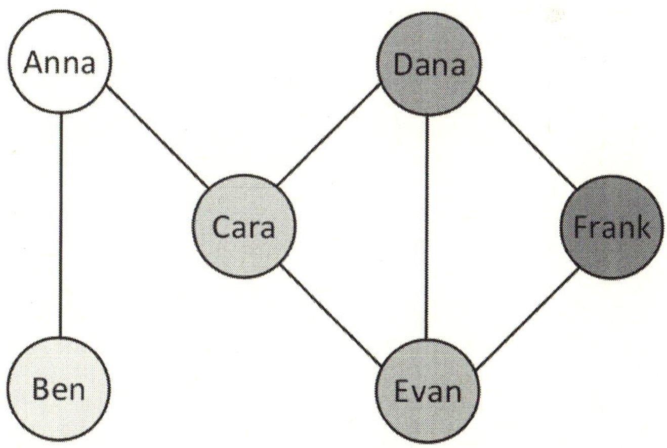

Figure 4.21 Social graph we will use to illustrate different centrality measures. There are six people, represented as nodes. The links between them indicate that they are connected. Notice that the links are directed, meaning that they go both ways (e.g., Facebook "friends" as opposed to Twitter "followers").

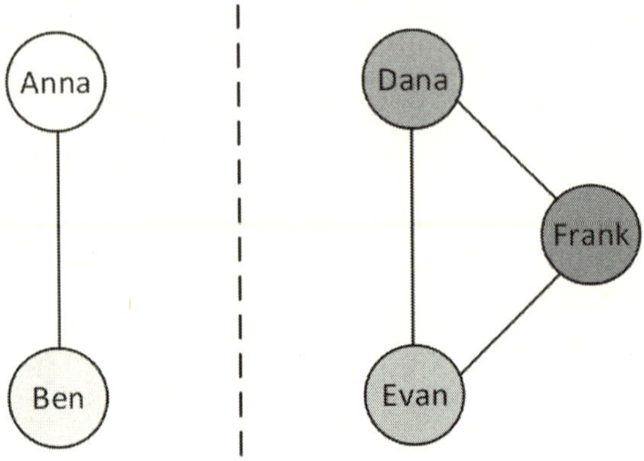

Figure 4.22 If we remove Cara, then Anna and Ben will not be connected to the rest of the nodes anymore. This should say something about the centrality of Cara. This same logic applies to Anna, whose removal would cause Ben to be split from the rest.

means the links are **bidirectional**: if Person "A" has a connection to Person "B," then Person "B" has a connection to Person "A." This is how Facebook's graph of friends work: if you're my "friend," then I must be your "friend," too. On the other hand, Twitter's graph of following relationships would be **directed**: even if you "follow" your favorite musician, chances are he or she does not "follow" you.

Which node is most important? Well, that depends on the definition. Indeed, there are at least three different approaches to measure the importance, or **centrality**, of a node.

Degree centrality

The first obvious choice is **degree centrality**: the number of nodes connected to the node. As we saw in Chapter 2, for a directed graph, we can count both the in-degree and out-degree. But for undirected graphs, there's no distinction here, so our task is even simpler.

What does this imply in Figure 4.21? Anna's degree is 2: she is friends with Ben and Cara. Ben's is 1: he is only friends with Anna. Similarly, those of Cara, Dana, Evan, and Frank are 3, 3, 3, and 2, respectively.

By degree centrality, then, we have Cara, Dana, and Evan as the most important, and further, they are *equally* as important. Similarly, Anna and Frank are the second most important, and Ben is the least.

But is this really reasonable? Well, Ben should probably be the least important: his only connection is to Anna. We can also probably agree that Dana and Evan should be equally as important, as they connect to the same nodes.

But consider Cara. There's something special about her: she provides the *only* connection that Anna and Ben have with the rest of the graph. Without her, the network would **partition** into two groups, as shown in Figure 4.22. Additionally, Ben would not be connected to the graph without Anna. If we are discussing "centrality," then, shouldn't we reward Ben and Cara more since their connections are more vital to the graph's connectivity?

As we saw when looking at PageRank, there are many issues with using the degree of a node as its centrality measure. We just pointed out a specific example. Another is that if you are connected to many important nodes, you should be rewarded more than if you were connected to many unimportant nodes. When ranking webpages, we solved this problem by having each node "spread" its importance to its neighbors. And indeed, we could also use the PageRank algorithm to compute the centrality of these nodes. But rather than using the PageRank algorithm, let's take other, more conventional approaches.

Closeness centrality

One factor we did not take into account with degree centrality was the *distance* between nodes in the graph. Here, the distance is simply the number of links we traverse over the **path** to get from one point to another. A path consists of a set of links, and we will refer to a path by the nodes we visit in traversing it, from start to end. In doing so, we will abbreviate the node names by their first letters, for convenience. So from Figure 4.21, the link from Ben to Anna is denoted as `(B, A)`. On the other hand, the path from Ben to Frank that goes through Anna, Cara, Evan, and then Dana is `(B, A, C, E, D, F)`. The distance of this path is 5, since it consists of five links.

But notice there are more ways to get from Ben to Frank than by path `(B, A, C, E, D, F)`. For one, we could have visited D before E: `(B, A, C, D, E, F)`. But we could have also gotten there using *fewer links*, since there's no need to go through both Dana and Evan. Specifically, we could have taken `(B, A, C, D, F)`, or `(B, A, C, E, F)`, which only uses four links. This is the **shortest path** between Ben and Frank, as illustrated in Figure 4.23. If we find the length of the shortest paths between each pair of nodes, the largest of these is known as the **diameter** of the network. In fact, 4 is the diameter of the network in Figure 4.21.

This brings us to our second interpretation of importance, which is

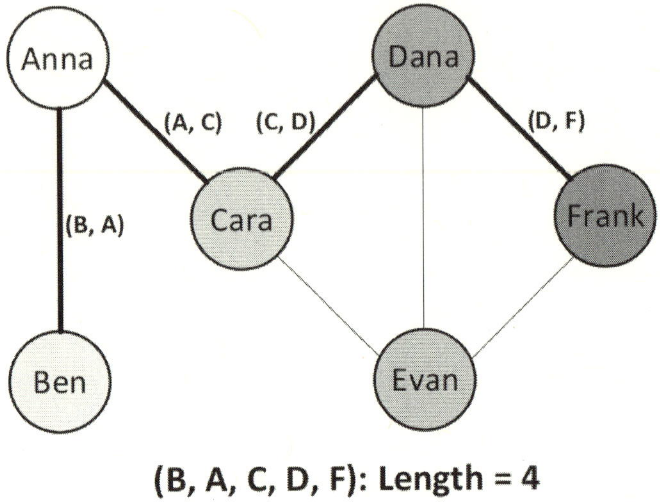

Figure 4.23 Here, we have highlighted one of the shortest paths between Ben and Frank, `(B, A, C, D, F)`, which is composed of the links `(B, A)`, `(A, C)`, `(C, D)`, and `(D, F)`. The total length is 4. Note that we could have gone through Evan rather than Dana and still kept the same length. As 4 is the *longest* of the shortest paths between any pair of nodes, it is also the diameter of the network.

to look at how "far" a node is from its neighbors. Specifically, we find the shortest path lengths to each of the other nodes, and then take the *average* of all these (rather than the largest, as we would for the diameter). Nodes for which this is small have shorter path lengths to their neighbors on average, and hence, are more "important." This is the idea behind **closeness centrality**.

Let's illustrate this for Cara. First, we need to find the shortest paths between her and other nodes in Figure 4.21.

- Cara to Ben: The shortest path is the only path, `(C, A, B)`, which consists of links `(C, A)` and `(A, B)`. This has length 2.
- Cara to Anna: The shortest path consists of one link, `(C, A)`. This has length 1.
- Cara to Dana: The shortest path is link `(C, D)`, which has length 1.
- Cara to Evan: The shortest path is link `(C, E)`, which has length 1.
- Cara to Frank: The shortest path is either `(C, D, F)` or `(C, E, F)`, as shown in Figure 4.24. Either has length 2.

Now, we find the *average* of these five shortest distances for Cara:

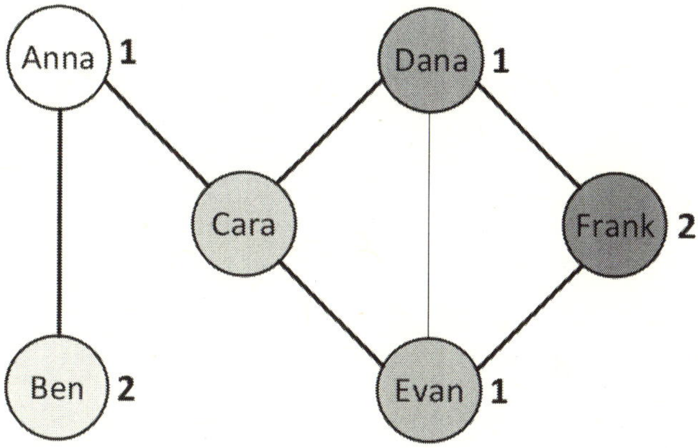

$$\text{Average C} = \frac{1+2+1+1+2}{5} = \frac{7}{5}$$

$$\text{Closeness C} = \frac{1}{\text{Average C}} = \frac{5}{7}$$

Figure 4.24 Here, we illustrate the shortest paths and closeness centrality for Cara. To Anna, Ben, Dana, Evan, and Frank, the shortest distances are 1, 2, 1, 1, and 2, respectively. Averaging these values gives 7/5, and taking the reciprocal gives us a closeness centrality of 5/7.

$$\text{Average C} = \frac{2+1+1+1+2}{5} = \frac{7}{5}.$$

We are just about there, except for one minor detail. We want a *smaller* average to give a *higher* centrality, since that indicates the person is "closer" to the rest of the nodes. To account for this, we simply take the reciprocal, so that a higher average gives a smaller value. Hence, the closeness centrality for Cara is:

$$\text{Closeness C} = \frac{1}{\text{Average C}} = \frac{5}{7} = 0.714.$$

We illustrate these steps in Figure 4.24.

Now, let's repeat the process for Dana. We start by finding the shortest path lengths to the other nodes.

- Dana to Anna: The shortest path is `(D, C, A)`, which consists of two links, `(D, C)` and `(C, A)`. Hence, the length is 2.

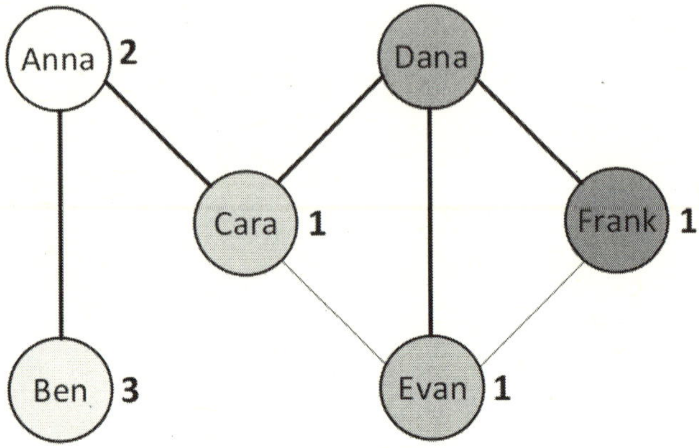

$$\text{Average D} = \frac{2+3+1+1+1}{5} = \frac{8}{5}$$

$$\text{Closeness D} = \frac{1}{\text{Average D}} = \frac{5}{8}$$

Figure 4.25 Here, we illustrate the shortest paths and closeness centrality for Dana. To Anna, Ben, Cara, Evan, and Frank, the shortest distances are 2, 3, 1, 1, and 1, respectively. Averaging these values gives 8/5, and taking the reciprocal gives us a closeness centrality of 5/8.

- Dana to Ben: We just add one more link to make `(D, C, A, B)`, which consists of three links, making the length 3.
- Dana to Cara: The shortest path is simply link `(D, C)`, which has a length of 1.
- Dana to Evan: The shortest path is again one link, `(D, E)`, with length 1.
- Dana to Frank: Again, the length of the single link `(D, F)` is 1.

Taking the average, we have

$$\text{Average D} = \frac{2+3+1+1+1}{5} = \frac{8}{5},$$

and then the reciprocal gives the closeness centrality:

$$\text{Closeness D} = \frac{5}{8} = 0.625.$$

These paths and lengths are summarized in Figure 4.25.

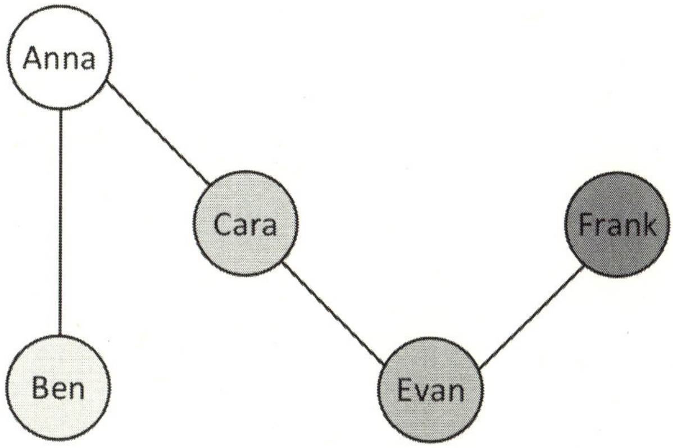

Figure 4.26 We can remove Dana or Evan without causing the rest of the nodes to disconnect. Hence, our intuition tells us that Cara, and probably even Anna, should be more important than these two.

We can similarly compute the shortest paths and associated closeness centrality measures for the other nodes. Note that we can simply flip the denominator and the numerator in the average equation to get closeness, as follows:

$$\begin{aligned}
\text{Closeness A} &= \frac{5}{1+1+2+2+3} = \frac{5}{9} = 0.556, \\
\text{Closeness B} &= \frac{5}{1+2+3+3+4} = \frac{5}{13} = 0.385, \\
\text{Closeness E} &= \frac{5}{2+3+1+1+1} = \frac{5}{8} = 0.625, \\
\text{Closeness F} &= \frac{5}{3+4+2+1+1} = \frac{5}{11} = 0.455.
\end{aligned}$$

So, with closeness centrality, Cara is the highest, followed by a tie between Dana and Evan, and then Anna, Frank, and finally Ben. Compared to degree centrality, a lot of the ties have been broken: Cara is first and Anna now beats Frank.

Closeness centrality is quite intuitive: the more people that are close to you, measured by path length, the more central you are to the network. But is this the best we can do? Clearly, it's more reasonable than degree centrality: in our example, Cara and Anna have both been promoted, which is justified since they are both vital to the graph's connectivity. But there are still some issues with this.

- First, Dana and Evan really don't hold the graph together. We show

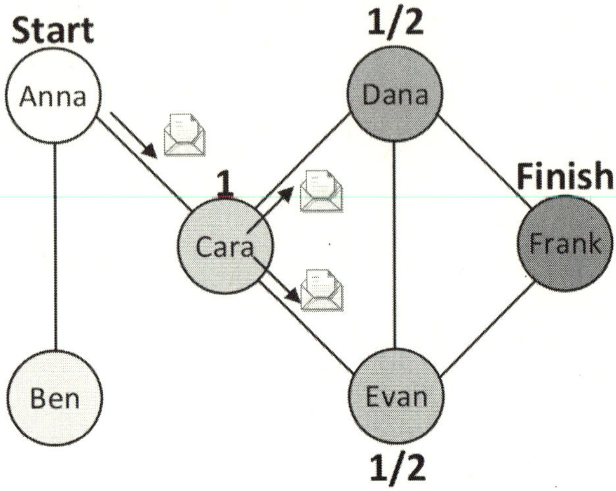

Figure 4.27 If Anna wants to send a message to Frank, she has one of two choices: she can tell Cara to pass it to Dana or Evan. Intuitively, since Dana and Evan each lie on one of the paths, they should be awarded half of what Cara is for this situation. Applying this logic to all node pairs and adding up the points for a person gives that person's betweenness centrality.

in Figure 4.26 that removing Dana will not cause the other nodes to disconnect. So, why should Anna be less important than Dana and Evan?

- Second, Cara isn't *that much* more "central" than Dana and Evan, since 0.714 is only 13% larger than 0.625. Is it possible to increase this by taking more into account?

There's another notion of centrality that's just as useful, and perhaps can fit our intuition more "correctly."

Betweenness centrality

How can we explicitly bring "connectivity" into importance? Let's start with a message-passing example.

Suppose for a moment that Anna wants to send a message to Frank, using the shortest distance possible. Clearly, she must tell Cara, and then have Cara forward it on. But Cara has two possibilities: she can tell Dana or Evan, either of whom can get it to Frank directly. What does this tell us? Anna really has two ways to get her message to Frank using the shortest possible distance. Both of the paths involve Cara, so if we are assigning points to the intermediate messengers, she should get the most. And only one of the two paths involve Dana or Evan, so each of these people should get half of what we gave Cara. This is illustrated in Figure 4.27.

Now, take another case. Suppose Ben wants to send a message to Dana. Here, there's only one shortest path: send to Anna, who forwards to Cara, who passes it to Dana. If we are rewarding points to messengers, Anna and Cara should get the same amount. But how much should they get relative to what we gave Cara, Dana, and Evan in the previous case? Well, whereas previously there were two possible paths, now there is only one. Without Anna or Cara, Ben couldn't message Dana, and at the same time, without Cara, Anna couldn't message Frank. So, Anna and Cara should get the *same* number of points for this path as Cara did for the other one, which is twice as many as Dana and Evan.

This is the idea behind **betweenness centrality**. We consider each pair of nodes, find the shortest paths between them, and award the other nodes points. The number of points is fraction of the shortest paths between these two nodes that the one in question lies on. So for the pair Anna and Frank, we would award Cara 2/2 (she lies on both paths), and Dana and Evan each 1/2 (they lie on one of the two paths). As for the pair Ben and Dana, Anna and Cara would each get 1/1 (they each lie on the only path). Once we do this for all pairs, we add up the number of points each node has, and that's its betweenness centrality.

Let's find the betweenness centrality for Cara. To do this, we will have to consider each pair separately, and ask the following two questions:

1. How many shortest paths are there between the pair of people?
2. How many of these shortest paths contain Cara?

This will allow us to compute the desired fractions. In computation, we don't have to consider any pairs that contain Cara, and we don't have to double-count pairs (e.g., once we consider Anna and Evan, we don't have to consider Evan and Anna). As a side note, we expect Cara to have a high betweenness centrality, because moving from Anna or Ben to Dana, Evan or Frank requires passing her (see Figure 4.28).

- Anna and Ben: How many shortest paths are there? Just one, link `(A, B)`. Clearly, it doesn't include Cara, so she gets $0/1 = 0$ points.
- Anna and Dana: There's only one shortest path, `(A, C, D)`. Since this includes Cara, she gets $1/1 = 1$ point.
- Anna and Evan: Again, there's one shortest path, `(A, C, E)`. Since it contains Cara, she again gets $1/1 = 1$ point.
- Anna and Frank: Now we have two shortest paths, `(A, C, D, F)`

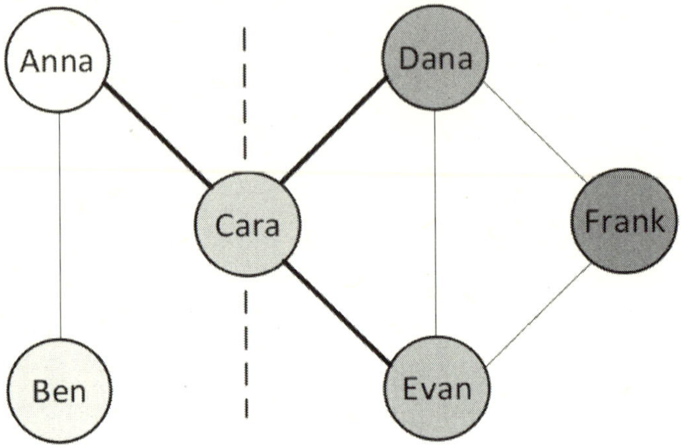

Figure 4.28 Cara lies on any of the shortest paths that contain `(A, C, E)` or `(A, C, D)`. Most do, because they are required in moving between one of the two subgraphs separated by the dotted line. Hence, we expect Cara to have a high betweenness centrality.

and `(A, C, E, F)`. Since both of them include Cara, she gets $2/2 = 1$ point.

- Ben and Dana: There's one shortest path, `(B, A, C, D)`. And it contains Cara, so she gets $1/1 = 1$ point.
- Ben and Evan: Again, there's only one shortest path, `(B, A, C, E)`, and it contains Cara, so she gets $1/1 = 1$ point.
- Ben and Frank: There's two shortest paths, `(B, A, C, D, F)` and `(B, A, C, E, F)`. Since both contain Cara, she gets $2/2 = 1$ point.
- Dana and Evan, Dana and Frank, Evan and Frank: Each of these pairs has one shortest path, `(D, E)`, `(D, F)`, and `(E, F)`. Since none of these contain Cara, she gets $0/1 = 0$ points for each.

Now, we add up all of the points Cara got to get her betweenness centrality:

$$\text{Betweenness C} = 0 + 1 + 1 + 1 + 1 + 1 + 1 + 0 + 0 + 0 = 6.$$

Take Dana as another example. Note we have already found most of the shortest paths, except those involving Cara. As shown in Figure 4.29, we don't expect Dana to have a high betweenness centrality value because most shortest paths do not include her.

- Anna and Ben, Anna and Cara, Anna and Evan: Dana is not on any

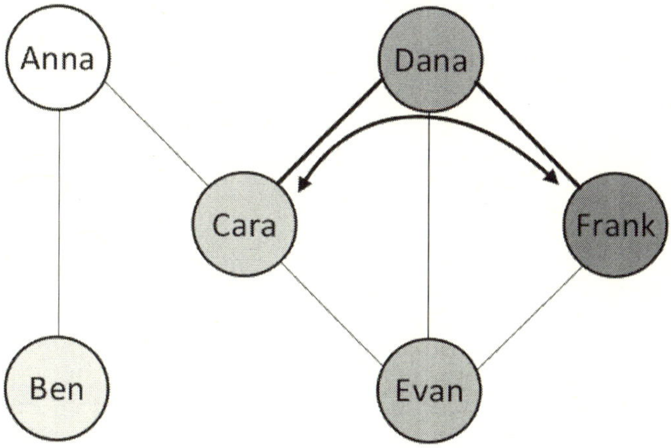

Figure 4.29 Dana only lies on the shortest paths between pairs that traverse path (C, D, F). The only paths that use these are those between Anna and Frank, Ben and Frank, and Cara and Frank. Hence, we expect Dana not to have a high betweenness centrality.

of the shortest paths between these pairs, so she gets $0/1 = 0$ for each.

- Anna and Frank: Out of the two shortest paths, Dana is on (A, C, D, F). Hence, she gets $1/2 = 0.5$ points.
- Ben and Cara, Ben and Evan: Dana is not on any of these shortest paths, so she gets $0/1 = 0$ for each.
- Ben and Frank: Dana is on one of the two shortest paths, (B, A, C, D, F), so she gets $1/2 = 0.5$ points.
- Cara and Evan: Cara is not on link (C, E), so she gets $0/1 = 0$ points.
- Cara and Frank: There are two shortest paths, (C, D, F) and (C, E, F). Since Dana is on one of them, she gets $1/2 = 0.5$ points.
- Evan and Frank: Cara is not on link (E, F), so she gets $0/1 = 0$ points.

Adding these values together, we get:

$$\text{Betweenness D} = 0 + 0 + 0 + 0.5 + 0 + 0 + 0.5 + 0 + 0.5 + 0 = 1.5.$$

We can continue this process for the other nodes:

- Anna will only lie on shortest paths that involve Ben. Further, she

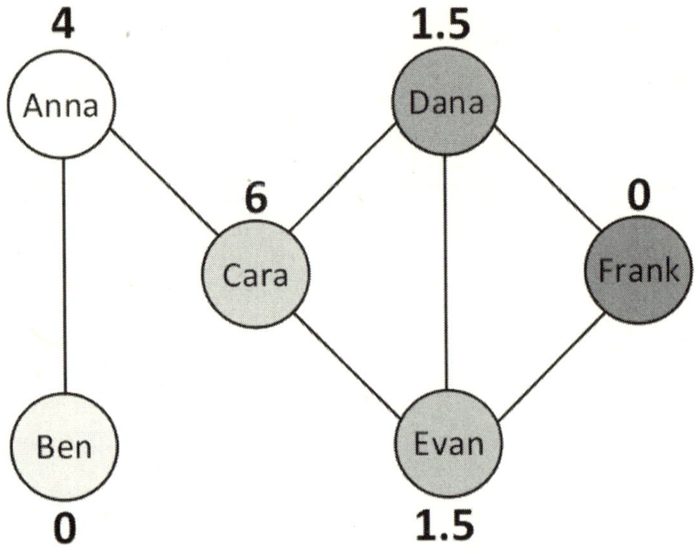

Figure 4.30 Betweenness centrality values for each person in our social graph.

will lie on all such shortest paths. As a result, she gets 1/1, 1/1, 1/1, and 2/2 for Cara, Dana, Evan, and Frank, respectively, and thus Betweenness A $= 1 + 1 + 1 + 1 = 4$.

- Ben doesn't lie on any shortest paths (other than ones where he is an endpoint), since he is only connected to Anna. Hence, Betweenness B = 0.
- Evan will have exactly the same centrality as Dana, since the structure of the graph is the same from both of their perspectives. Hence, Betweenness E = 1.5.
- Frank, like Ben, lies on no shortest paths. Hence, Betweenness F = 0.

We summarize the betweenness centrality values for each node in Figure 4.30. By this metric, we have Cara as the most important, followed by Anna, a tie between Dana and Evan, and then another tie between Ben and Frank. We make two observations:

- *Now, Cara is by far the most important.* Her value is 50% higher than Anna's, and 4 times larger than Dana and Evan's.
- *Now, Anna is more important than Dana and Evan.* We indicated before that both of these characteristics are likely desirable because

of Cara and Anna's contribution to the overall connectivity of the graph.

In Table 4.1, we summarize the different centrality measures we have seen: degree, closeness and betweenness.

Node	Degree		Closeness		Betweenness	
	Value	Rank	Value	Rank	Value	Rank
Anna	2	2nd	0.39	5th	4	2nd
Ben	1	3rd	0.56	3rd	0	4th
Cara	3	1st	0.71	1st	6	1st
Dana	3	1st	0.63	2nd	1.5	3rd
Evan	3	1st	0.63	2nd	1.5	3rd
Frank	2	2nd	0.45	4th	0	4th

Table 4.1 Summary of the different centrality measures used in this example: degree, closeness, and betweenness. Betweenness is perhaps the closest to what our intuition would tell us.

It's also possible to compute betweenness centrality values on the *links*, rather than the nodes. We will not have time to detail this mathematically, but the procedure is exactly the same as we did with nodes, except we look at the number of shortest paths containing each *link*.

4.2.3 Contagion

We just discussed the basic metrics of a graph, and how they can make certain nodes more "central" than others. With this in mind, we will now continue with our discussion of influence models, but this time taking social graphs into consideration. More specifically, we will try to understand how a person's social relationships can influence her to adopt a product or good, depending on how many of her neighbors have done so.

Consider the network in Figure 4.31, where we show eight nodes connected to one in the middle. Each of the nodes connected to the one in the middle are in one of two states: the four labeled "N" are in state-0, and the four labeled "Y" are in state-1. These states are a generalization of not having adopted a product ("N"), and having adopted it already ("Y"). The question we will ask is, do the four

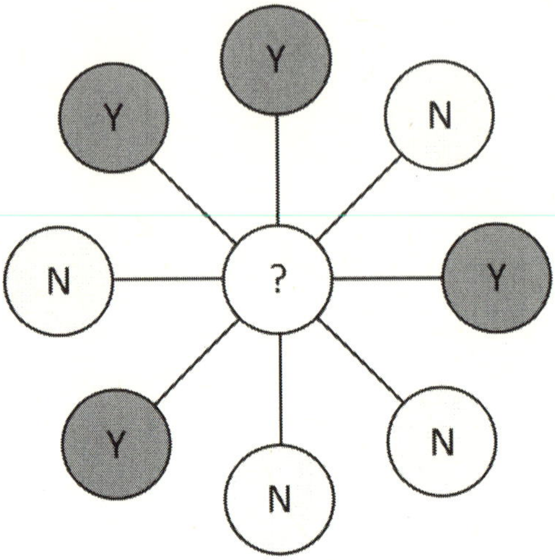

Figure 4.31 Here, the center node has eight nodes that it is connected to. Four of them have switched to state-1 ("Y"), and four of them have not ("N"). Will the middle node flip to state-1 or not? If the flipping threshold is 50% or less, than it will, but otherwise, the social influence is not strong enough.

nodes that are in state-1 constitute enough *social influence* to cause the center node to flip (adopt the product) as well?

As an example, suppose you are the person in the middle, and the edge nodes represent your friends. Some of your friends have recently purchased the newest brand of iPad (state-1), while others have not (state-0). You can imagine that the more people you see that have the product (assuming they are satisfied with it), the more influence that will have over you, and the more likely you will be to buy an iPad too. Is there a way for us to tell whether or not you will follow suit?

In order to answer this question, we take the approach of defining a *flipping threshold* for each node. This is the fraction of the node's neighbors that must have already flipped before that node will as well. In Figure 4.31, 50% (4/8) of the center node's neighbors have flipped, and back to our iPad example that means 50% of your friends have the new iPad. So if your threshold is less than 50%, you will buy one. But if it is something higher, then you will not, because there is not enough social influence. For instance, if your threshold was 80%, then you would need at least 7 of your friends to have adopted it (since $7/8 > 0.8$, but $6/8$ is not).

Realistically, the flipping threshold is hard to estimate:

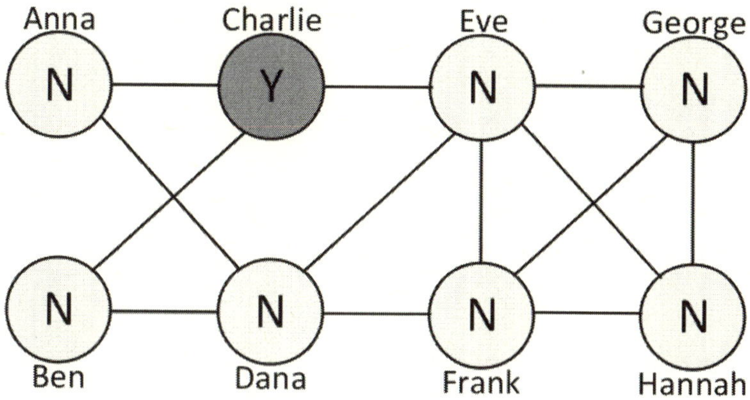

Figure 4.32 Eight-node social graph that we will use to illustrate contagion. You can think of this as a group of people purchasing a product. Initially, Charlie has the product and "flipped," while everyone else has not.

- It will depend on the *product* we are considering. For example, some products are cheaper than others, and some are naturally attractive than others. Either of these would intuitively lower the threshold.
- Additionally, it will depend on the *person* in question. I might be easy to sway, and if I see one of my friends adopting something, I will do so as well. You, on the other hand, may never budge.

It will depend on many other factors as well, like strength of social ties, and what we mean by a "link," among others. But for our purposes, we will assume that we know the flipping threshold, and that it's the same for each node.

Consider the social graph in Figure 4.32, consisting of eight people and their relationships to one another. We have initialized Charlie to have flipped, while everyone else has not. Assuming a threshold of 50%, let's trace out the flipping behavior over time. In each iteration, we go through person by person to see if their flipping thresholds have been met, and if so, they will become adopters. This process is known as **contagion**.

Time 1

- Anna: Initially, one of her neighbors is flipped (Charlie), and the other is not (Dana), meaning 50% of them have. Since the threshold is exactly 50%, Anna has been presented with just enough social influence, and as such, she will flip.

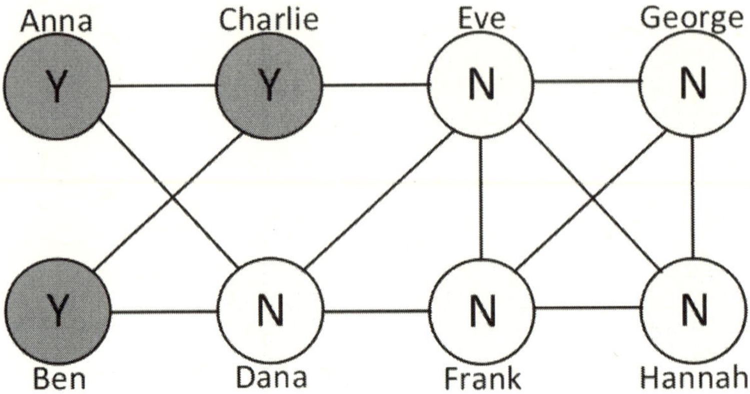

Figure 4.33 Contagion graph at Time 1: Anna, Ben, and Charlie have flipped.

- Ben: At this point (before Anna flips), one of his neighbors has (Charlie), and the other has not (Dana). Similar to the previous case, 50% is enough, so he will flip.
- Dana: None of her neighbors have flipped (Anna and Ben don't switch until this round). Clearly then, Dana will not either.
- Eve: One of her neighbors has flipped (Charlie), and the other four (Dana, Frank, George, and Hannah) have not. So, she has 1/5, or 20% of her neighbors flipped, Since this is less than 50%, it is not enough influence.
- Frank, George, and Hannah: None of their neighbors have flipped, and hence, they won't either.

The resulting graph at Time 1 is shown in Figure 4.33. Now, let's proceed to Time 2 using this updated graph, where we will consider all the unflipped nodes.

Time 2

- Dana: Now, two of her neighbors (Anna and Ben) have flipped, and two have not (Eve and Frank). 50% is just enough to cause her to flip.
- Eve: The situation hasn't changed from last time. Still, only one neighbor has flipped, making 20%. As such, she will not budge.
- Frank, George, and Hannah: Again, none of their neighbors have flipped, so they won't either.

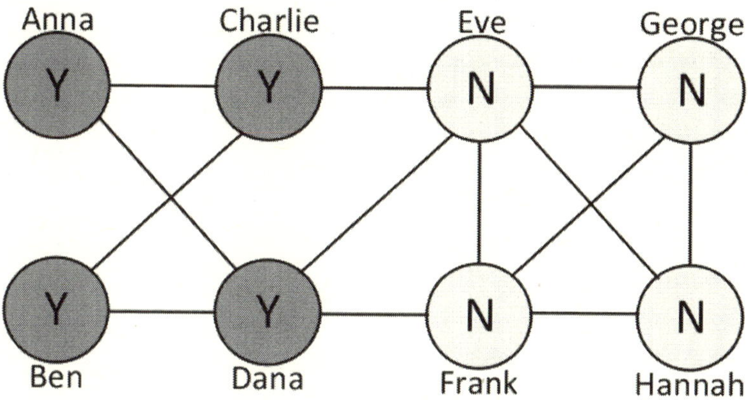

Figure 4.34 Contagion graph at Time 2. Anna, Ben, Charlie, and Dana have flipped.

Figure 4.34 shows the state of each node at the end of Time 2. We have the left half of the network having flipped, and the right half having not. Are we on track to have everyone flip eventually? Let's see.

Time 3

- Eve: Two of her neighbors have flipped, and the other three haven't, making 40% that have. Not quite enough.
- Frank: One of his neighbors has flipped, and the other three haven't. This makes 25%. Again, not enough.
- George and Hannah: None of their neighbors have flipped, so they won't either.

So, none of the nodes switch state at Time 3. Why is this? Each of person has *too large* a fraction of their neighbors to friends that have not flipped, and not enough to those that have. Out of Even, Frank, George, and Hannah, Eve has the highest fraction of links to those in state-1. But this is only 40%, not quite high enough.

In general, we can take any group of nodes that have connections among themselves and call them a **cluster**. Taking a group of unflipped nodes, if each of them have too many neighbors inside the cluster than outside, then it will be impossible to flip it.

For instance, consider the cluster in Figure 4.35. Eve has 60% of her neighbors inside, Frank 75%, and George and Hannah each have 100%. The *smallest* of these fractions, 60%, is known as the *density* of the cluster. Unfortunately, this number is too high: we need at least 50%

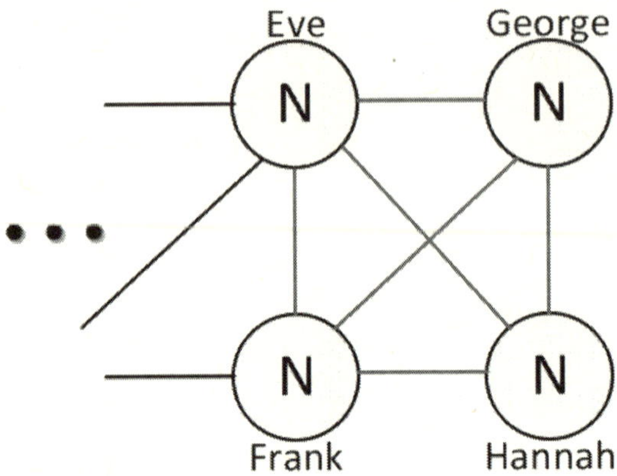

Figure 4.35 The cluster of Eve, Frank, George, and Hannah is too *dense* to be penetrated by the outside social force. Eve has 60% of her links inside the cluster, Frank 75%, and George and Hannah 100%. The smallest of these, 60%, is the cluster density, which is too high for any node inside to be flipped, given the threshold of 50%.

(the threshold) of the density to be *outside* of the cluster in order to penetrate it.

Marketing strategies

Now suppose we are on some marketing campaign, and want to make sure *everyone* buys the product. What can we do to make this happen? Given we know the social graph, there are a few possibilities:

1. *Lower the threshold.* If we lower the flipping threshold, even for just some nodes, naturally we will have a better chance of having everyone end up in state-1.
2. *Break the cluster.* If we could cut some social ties within the cluster, this would also work.
3. *Seed one of the nodes in the cluster.* Finally, we could try to influence one of the nodes in the cluster indirectly.

The third option is the most plausible one: we could *pay* one of the nodes in the cluster to adopt our product. At this point, seeding *any* of Eve, George, Frank, or Hannah will do the job.

More generally, suppose each node in a graph can be influenced by paying them a specific amount of money. Presumably, those nodes that perceive themselves as more influential will charge a higher price (e.g., a designer company might pay a famous celebrity a large sum of money to wear their clothing brand). Under a total budget constraint, which

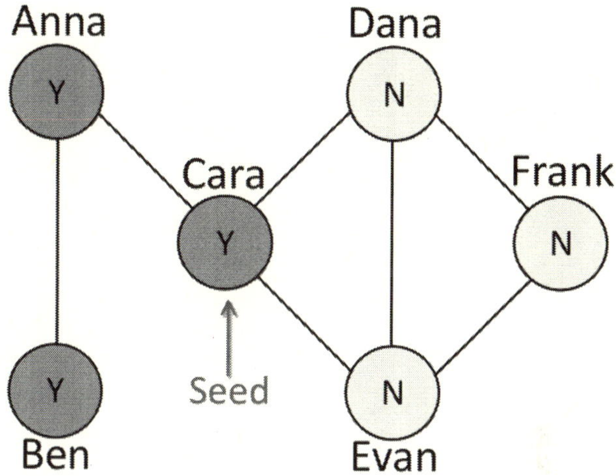

Figure 4.36 If we can only choose one person to seed initially, then the most central, Cara, is the best bet. At equilibrium (after two iterations), Anna, Ben, and Cara will each have flipped. Seeding any other person will cause at most one (in addition to the seed) to switch.

nodes would you seed (i.e., pay to change their states) in order to maximize the extent of flipping at equilibrium, and perhaps minimize the time it takes to reach this equilibrium?

While this question is hard to answer, some intuitions are clear. If you can seed just one node, it should be the most important one. And how do we define the most important node? By the centrality measures we discussed before. But once we can seed more than one node, it becomes a little more complicated. You don't just want to choose the two most "central" nodes: you have to make sure their *combined* influential power is large enough.

For example, consider again our original social graph of five nodes, as shown in Figure 4.36. As you may recall, Cara was the most important nodes both in terms of closeness and betweenness centrality, and hence we expect her to have to most social influence. If we choose to seed Cara initially, we will get the greatest magnitude of change by the end: Anna, Ben, and Cara will all have flipped to state-1, after two iterations. On the other hand:

- Seeing Anna would cause only Ben to flip, just as Ben would cause only Anna.
- Seeding Dana or Evan would only cause Frank to flip.

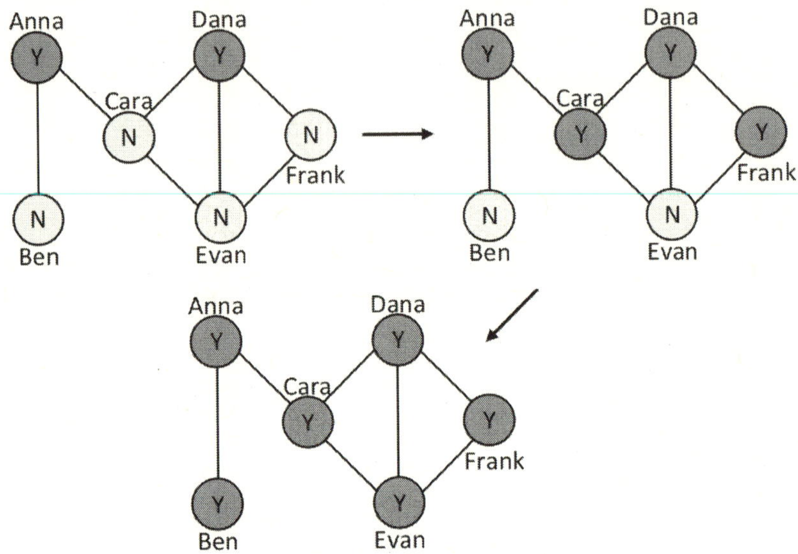

Figure 4.37 While when choosing only one person to seed, the most important is probably our best bet, this is not necessarily the case when choosing two people. Here, seeding Anna and Dana will cause the entire network to flip after two iterations, whereas seeding the most important two (Cara and Anna) will only cause the left half to flip in the end.

- Seeding Frank would cause no one to flip.

Clearly then, if we can only seed one node, we would want it to be Node 3. But what if we could seed *two* nodes? We might figure, let's just take the two most important by betweenness centrality: Cara and Anna. But it turns out that flipping these nodes together does *nothing* additional for us, besides decreasing the time by one iteration.

So, this is may not be the best strategy. For instance, consider picking Cara and Frank: after one iteration, this would cause Anna, Evan, and Frank to flip, and after two, Ben would flip, meaning everyone will. And Frank is one of the *least* important nodes here.

But do we even need to pick Cara at all? It turns out not: in Figure 4.37, we show that by seeding Anna and Dana:

- After one iteration, we cause everyone to flip except Evan.
- After two iterations, Evan also flips, meaning we have achieved our goal.

The important thing is to make sure the nodes we choose will *collectively* have the most influence.

Summary

Throughout this chapter, we have examined ways in which people are influenced by one another, motivated by examples from our everyday, networked lives. First, we looked at how population-based models such as information cascade can shed light on why some YouTube videos become viral while others do not. Then, motivated by platforms like Facebook and Twitter, we took a topology-based approach, where we defined what it means to be "influential" before looking at the contagion model.

Overall, we see here the complete opposite of what we discussed in Chapter 3. Through influence, people's actions can become entirely *dependent* on one another, which causes the underlying assumption behind the the wisdom of crowds to shatter. Crowds become less wise, and we can take advantage of them to spread information in various contexts.

5 Network Is Expensive

In this chapter, we will switch gears and look at network pricing methods. In particular, we will focus on cellular, and why providers such as AT&T and Verizon started charging their customers for *data* usage (see Figure 5.1). We will look at different methods of doing so, and discuss their implications on user consumption and the corresponding costs incurred by the provider.

5.1 Pricing data

How do our cellular providers determine what to charge us? In this section, we will see that the commonly-employed usage-based pricing scheme, where we are charged based on how much we actually consume, sends us better pricing signals than flat-rate, "buffet" schemes. In doing so, we will resort to some basic theories in network economics.

5.1.1 Our mobile data plans

Almost all of the utility bills we pay – electric, water, gas, etc. – are based on the *quantity* of the service that we use. For instance, our elec-

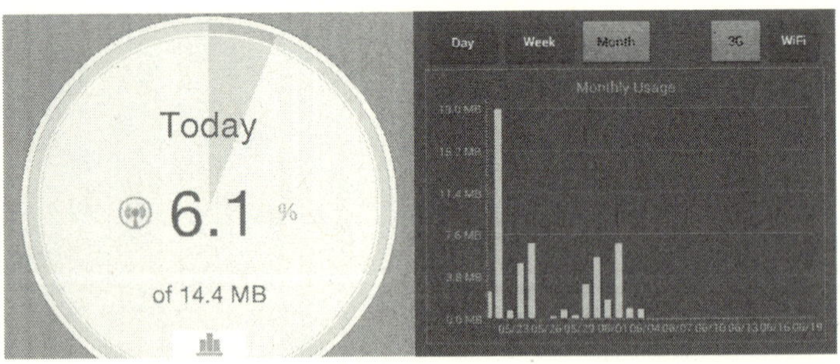

Figure 5.1 Several mobile apps can track your cellular (and WiFi) data usage, such as DataWiz pictured here. If you have a monthly consumption cap, they can recommend a maximum daily usage amount (here, 14.4MB) to make sure you don't exceed it. How does your provider determine what to charge you based on this usage?

Figure 5.2 A restaurant buffet illustrates the principle behind flat-rate pricing. You pay some initial cost, say \$20, to enter, and then you can consume as much as you want. There's no difference between one, two, five, or more plates, in terms of price. But if customers kept increasing their appetites everytime they came back, could restaurants handle it?

tric bills are based on a certain amount we pay for every unit of energy, measured in killowatt-hours (KWH), that we consume in a month. So if you pay \$0.10 per KWH, and you use 500 KWH in a month, your bill will be $\$0.1 \times 500 = \$50.00$.

With the introduction of data plans, first for texting and now for surfing the web, streaming videos, and so on, our cellular service providers have started charging us for data usage as well. Indeed, they incur some capacity cost in the network everytime you use one of these services. But even though wireless cellular capacity is expensive to provide and difficult to crank up, consumers in some countries like the USA used to have the luxury of only paying a pre-determined, fixed data cost each month for many years. Such a scheme is known as **flat-rate** pricing.

Are buffets OK?

What do we mean by a flat-rate price? It's one that doesn't depend on how much you actually consume, much like when a restaurant offers a buffet (see Figure 5.2). You pay some fixed amount, say \$20, to enter, and once you are in, you can eat as much as you want. And it's in your best interest to follow suit: you won't pay a penny more depending on whether you have one, two, or five plates. May as well satisfy your appetite while you have the chance.

Such a buffet is great for you, especially if you're hungry. But it may not be so great if you don't want to eat much: would you pay \$20 for a piece of bread? Absolutely not. Even beyond this, consider it from the perspective of the restaurant: what happens if people consume more and more food each time they come back? Can a restaurant keep

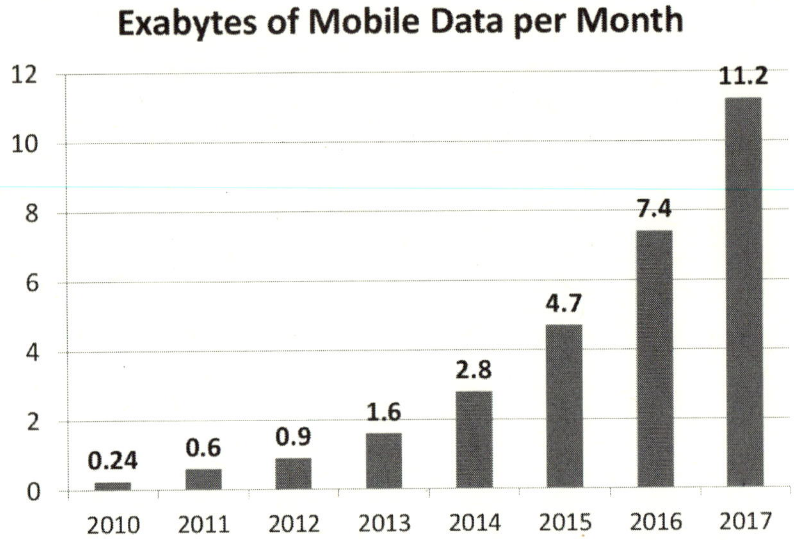

Figure 5.3 Plot of yearly mobile data volume, with projections by Cisco out to 2017. Here, data sizes are measured in *Exabytes*: one Exabyte is *one billion* GB. Demand increased 3.75 times between 2010 and 2012, and is expected to increase another 12 times by 2017.

offering buffets with the same price if customers keep doubling their appetites every year? Or will it have to stop at some point?

Flat-rate pricing was common practice in the cellular industry for years for data, but not voice call. That is because wireless operators in the past decades saw voice calls and messages as the main use of cell phones, rather than mobile data. Pay some rate, say \$30 each month, and the network became a buffet for you, allowing you to consume as much data as you wanted. And for a long time, such a scheme made sense for providers, because the amount of data consumed by cellular devices was low. The primary purpose of cell phones was voice, and data was viewed as a "secondary" feature that you could add on.

But as smartphones became popular, this changed very quickly. The demand for data began to rise rapidly as these devices became capable of surfing the web, streaming videos, and supporting a wide variety of other data-intensive applications. More importantly, increasingly many of these applications began to run in the background, not even requiring a human being on the other side. For instance, the introduction of the first generation iPhone in 2007 caused a 50-fold jump in cellular data demand. Such increases are also fueled by the sheer number of people

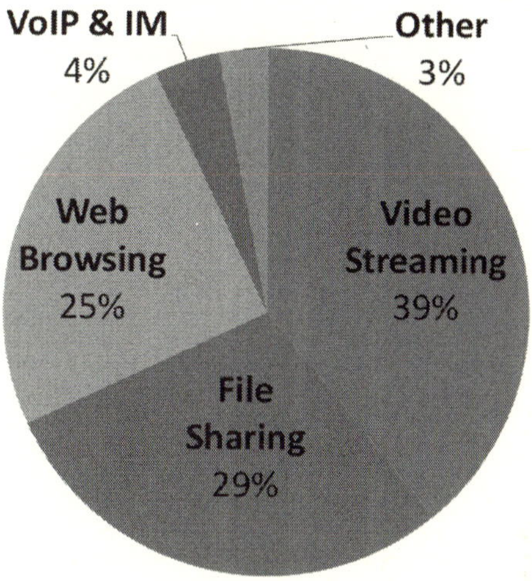

Figure 5.4 Fraction of data consumption by application on cell phones, in 2011.

using smartphones, and the upcoming machine-to-machine or device-to-device communications.

In Figure 5.3, we show just how high mobile data traffic has become, and how much it is expected to increase over the next few years. To understand the magnitude of this graph, we first have to explain how data size is quantified. The fundamental unit for measuring the size of data is the **byte** (B), which is 8 bits. But this quantity is rather small: an audio file typically consists of millions of bytes, and video can be somewhere in the range of billions. Hence, we will find it more convenient to speak of data sizes in terms of megabytes (MB), or 10^6 bytes, and gigabytes (GB), or 10^9 bytes.

So *how much* is 1 GB of content? If you watch 15 minutes of medium-resolution YouTube videos a day, and do nothing else with your Internet access, that is about 1 GB a month. Also, half of a typical standard definition movie consumes 1 GB already. Referring to Figure 5.3, we are measuring monthly data size in *Exabytes*: one Exabyte is *one billion* GB, or 10^{18} bytes. So in 2012, there was almost one Exabyte of mobile data running through the Internet per month. And by 2017, this number is expected to increase by more than 12 times.

These increases have also been fueled by new, data-intensive applications on mobile devices. To highlight this, we break down the fraction of consumption by application in Figure 5.4 as of 2011. Video streaming, at 40%, accounts for the most, followed by file sharing at 29%.

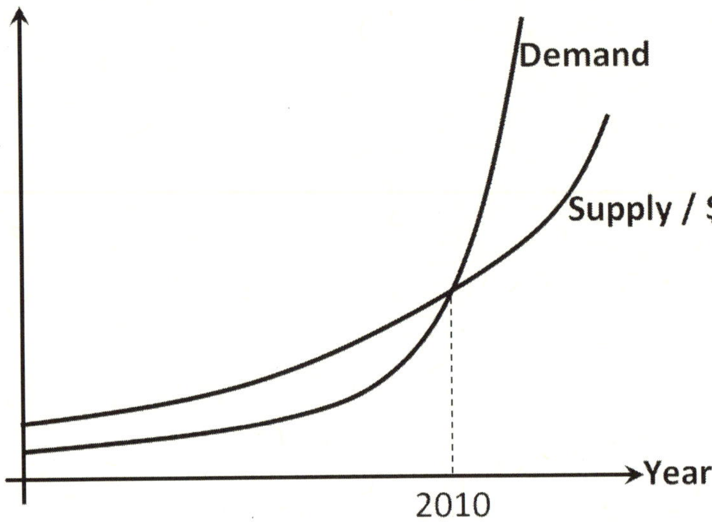

Figure 5.5 The trend of demand and supply/$ of wireless cellular capacity over time. Demand caught up with supply (per dollar of cost) around the turn of the decade, through turning points such as the introduction of iPhone by AT&T in the USA in June 2007, which caused a 50-fold jump in cellular data demand. More importantly, demand is projected to keep growing at a faster pace than supply/$.

With the exception of limited web-browsing, these features were not available before smartphones or tablets.

So, why do such extraordinary levels of demand pose a problem? Well, networks have a *limited capacity*, just like buffets only have a certain amount of food to serve to customers. Around 2007, a prediction was made that the growth in demand was beginning to outpace the growth in the amount of capacity that could be provisioned with each additional dollar (e.g., supply/$). Further, around the turn of the decade, demand would become *greater* than supply/$, with the gap widening in the years beyond that. The keyword here is *growth rate*: over time, the difference gets *larger and larger*, as shown in Figure 5.5.

We call this "Job's Inequality of Capacity", after the late Steve Jobs. The problem is that once the device suppliers and application communities, such as Apple and iOS app developers, figured out how to make it easy and attractive for users to consume mobile Internet capacity, innovation in those spaces proceeded faster than the supply side can keep up with. In Figure 5.3, we saw that Cisco predicts that the mobile Internet demand will keep doubling every year. That's more than 64

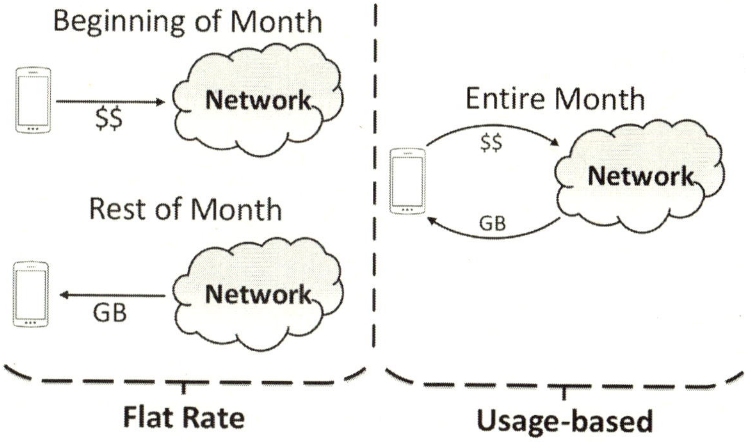

Figure 5.6 Flat rate and usage-based pricing send different pricing signals to consumers. With flat rate, after the initial payment (e.g., $30) is made, the customer can consume as much data as she wants for the rest of the month without incuring any additional cost. On the other hand, with usage-based, the monthly cost is accrued as the person uses data, and every piece of consumption is accounted for in the total. This sends a continuous feedback signal to the customer.

times after five years. No technology can double the supply/$ each year forever.

Migrating to usage-based pricing

So, what was the solution that Internet Service Providers (**ISPs**) came up with? Rather than a flat-rate, they began switching to **usage-based** pricing schemes, in which customers are charged based on how much data they use each month, much like a utility bill. Charging this way sends a different *pricing signal* to consumers: instead of a single initial charge each month, they incur a *continuous* cost for each additional piece of data they use, as illustrated in Figure 5.6. In this way, each person is deducted for the congestion she creates in the network by being there in the first place. We will investigate these differences in detail throughout this section.

In any market, there are two typical precursors to the introduction of usage-based pricing:

1. We need network usage to have surged throughout the market, and for its projected demand to climb even *higher and faster.*

   This happened with the introduction of iPhones, Android smartphones, and iPads. With more widespread users and more capable devices comes rising demand, causing users to consume more. This incurs a higher cost to the ISP, since it needs to provision extra capacity. Being that an ISP's profit is the difference between revenue (i.e., the amount it charges customers) and cost, usage-based pricing is a way to help revenue catch up with the cost of supporting the rising demand.

2. We need government regulations that allow pricing practices that match cost.

   While there are many regulatory issues, allowing the monthly bill to be proportional to the amount of usage is among the least controversial.

By 2010, the migration to usage-based pricing had begun. Let's look at a timeline of events for AT&T and Verizon, the two major operators in the US.

- *April 2010*: AT&T announced its usage-based pricing for 3G data users.
- *March 2011*: Verizon Wireless followed suit, for its iPhone and iPad users, and later that year for all of its 3G data users.
- *March 2012*: AT&T announced that existing customers on unlimited cellular data plans would see their connection speeds throttled (*i.e.*, slowed down) significantly once the usage exceeded a certain amount, effectively eliminating the unlimited data plan. The 4G data plans from both AT&T and Verizon Wireless for the "new iPad" launched soon after no longer offered any type of unlimited data options.
- *June 2012*: Verizon Wireless updated their cellular pricing plans. A customer could have *unlimited* voice and text, in exchange for turning an unlimited data plan to *usage-based.* AT&T followed with a similar move one month later.

Similar measures have been pursued, or are being considered, in many other countries around the world for 3G, 4G, and even wired networks. This highlights just how much of an issue data demand has become: providers see it more important to have usage-based pricing for data than for voice. You could argue that the "primary" use of cell phones has now switched from voice to data.

So, what do usage-based plans "look like?" In Figure 5.7, we show a few of the plans that Verizon Wireless offered as of Spring 2012. In general, there are three main characteristics of such pricing plans.

- *Baseline*: There is a baseline under which the charge is still *flat-rate*,

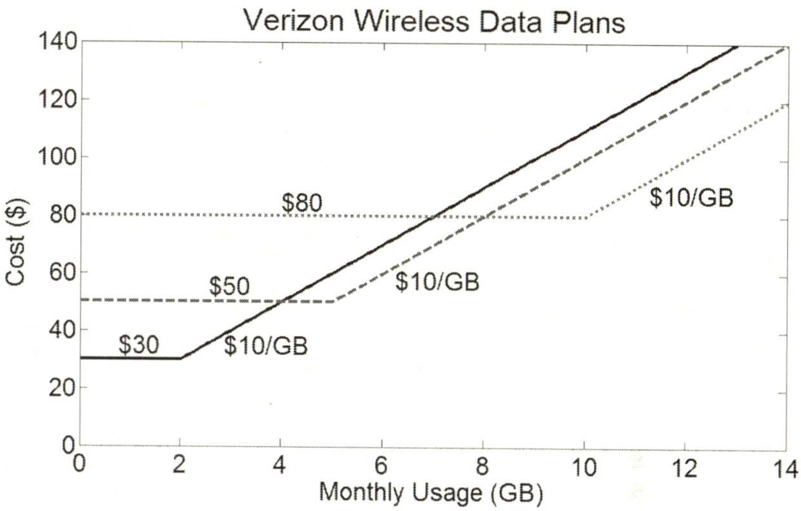

Figure 5.7 Verizon Wireless' data plan options in Spring 2012. The plans have a flat-rate component below a baseline, and then a usage-based component (e.g., $10 per GB) beyond that.

represented by the horizontal segments. For instance, the middle line will charge a customer $50 if his usage is anything below 5 GB.

- *Slope*: Above the baseline, the pricing curve becomes a straight line with a positive slope, $10/GB, as the usage grows. For the middle line, the customer begins to incur the $10/GB charge above 5 GB. So for 5 GB he would pay $50, for 6 GB $60, for 7 GB $70, and so on.
- *Overall charge*: The overall charge is based on the total monthly usage. It does *not* matter when you use it, where you use it, or what you use it for: only *how much* you used.

Even at a first glance, charging based on consumption probably sounds intuitive to you. Indeed, that's how most utilities and commodities are charged. But those who are used to flat-rate Internet connectivity will find it quite annoying. And if content consumption is suppressed as a result, usage pricing will influence the entire industry ecosystem, including consumers, network providers, content providers, app developers, device manufacturers, and advertisers. Yet we will see that there are several strong reasons, including those in the interests of consumers, that support usage-based pricing as a better alternative to flat-rate pricing.

Besides issues of sheer capacity, there are other reasons why ISPs

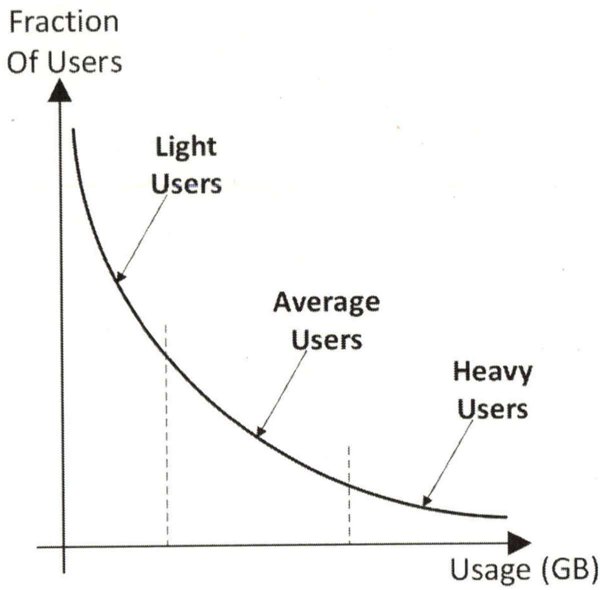

Figure 5.8 Distribution of users' capacity demand. Some are light users, some average, and a smaller amount heavy. These heavy users create a long tail, and dictate an ISP's cost structure. If the revenue model is based on the average user, the mismatch between cost and revenue will get larger and larger as the tail gets longer and longer.

needed to migrate to usage-based pricing. Here's another important one. Some users consume small amounts of data, and some consume large amounts. Say we can define three classes of users: light, average, and heavy, as illustrated in Figure 5.8. As we move to the right of the graph, we have less and less proportions of users demanding that much capacity (i.e., we have more people demanding 1 - 5 GB per month than we do 21 - 25). If we look at the distribution of capacity demand, the *tail* of that distribution (referring to the far right side) is often the dominant factor in an ISP's cost structure, and the heavy users are the constituents of the tail. This tail has always been long, meaning that there are some users consuming very large amounts of data, but it's getting longer and longer now. Traditionally, ISPs determined their fixed-price costs from the consumption trends of an average user. As the tail gets longer, the cost would be come increasingly larger than the revenue, unless a switch in pricing methodology was made.

Indeed, more users are consuming more GBs as months go by. This year's "heavy" users will become "medium" in just a couple of years'

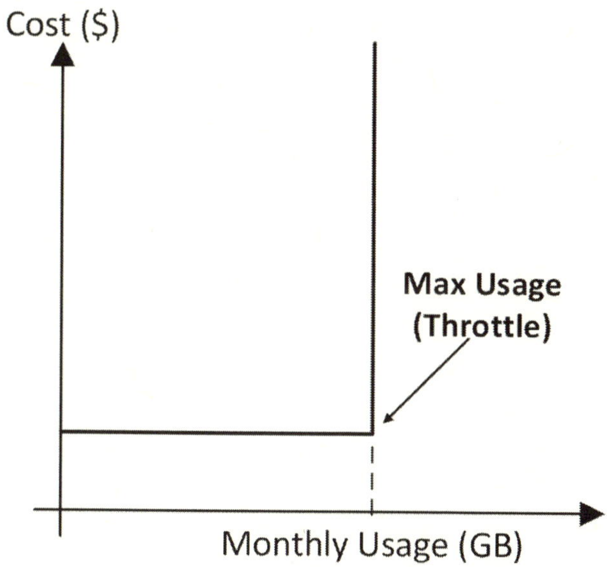

Figure 5.9 Illustration of throttling on a monthly pricing graph. Below a certain baseline, the user will pay a fixed cost for consumption. But the maximum quantity is capped at the baseline: after that, the price becomes "infinite". In practice, throttling is done either by prohibiting network use completely, or by throttling speeds so low that they cannot be used for practical purposes.

time. Further, with the 4G speed much higher than that of 3G, more and more of these GBs will be consumed on mobile devices.

Pricing: this way or that way?

One way or another, the cost of building and operating a network must be paid by someone. Usage pricing based on monthly consumption, however, is not the only way to tackle the above issues. ISPs have other choices, such as the following.

- *Increase the flat rate for everyone.*
  With a sufficiently high flat rate, the revenue collected will be adequate. But clearly this creates affordability and fairness issues.
- *Cap heavy users' traffic.*
  Once you exceed a threshold, you can no longer use the network. Or the speed will be **throttled** to the point that the quality of service becomes too low for practical use once the cap has been exceeded. This is actually a special case of usage pricing, in which the slope becomes infinite above the baseline, as shown in Figure 5.9.

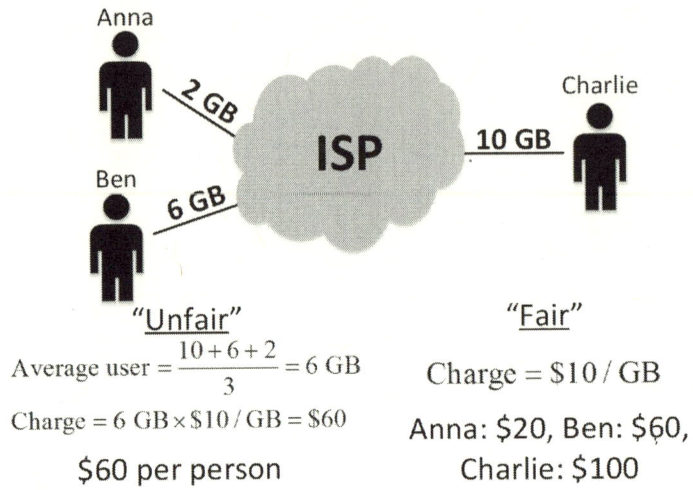

Figure 5.10 Here, we have three users, with different monthly consumption characteristics (light, medium, and heavy). If we charge based on the "average," then everyone will pay $60. That is certainly a great deal for Charlie, who is then paying $6/GB, but not so much for Anna, who pays $30/GB. Anna is subsidizing Charlie's lifestyle. On the other hand, with usage based pricing everyone pays $10/GB, and so Charlie is charged five times more than Anna.

- *Implement smarter versions of pricing.*
  We can charge users depending on the current congestion conditions in the network, or at different times of the day.

Most of the ISPs have realized the problem and started pursuing the usage-pricing solution. What are the criteria that we can use to compare alternative solutions to the exploding demand for mobile data? There are too many to list here, but the top ones include the following.

- *Economic viability*: As profit-seeking companies, ISPs need to first recover cost and then maximize profit. Their profit margins are in general declining, as many other transportation businesses have seen in the past. Mobile and wireless networks are bright spots that they need to seize.
- *Fairness*: Consumer A should not have to use her money to subsidize the lifestyle of consumer B. We show an example of this in Figure 5.10: under flat-rate pricing, light users will have to pay extra due to the presence of heavy users. More on this later.
- *Consumer choice*: Consumers should be able to choose among alter-

natives, e.g., more money to get premium services, or receive standard services with a cost savings.

Along all of the above lines, usage pricing makes more sense than fixed pricing, although it can be further enhanced with more intelligence. The key advantages of usage pricing are listed below and will be analyzed in detail in the next subsection.

- Usage pricing produces less "waste" and matches cost.
- Usage pricing does not force light users to subsidize heavy users.
- Usage pricing encourages customers to upgrade services when such upgrades are available.

5.1.2 Modeling utility and demand

In order to proceed further to understand Internet access pricing, we need to find a way to quantify consumer *demand.* How do we figure out how much data users will consume, given the network price? And how do service providers leverage such information?

Utility

Let's start with a motivating example. Suppose you are sitting down at a meeting, and your boss has brought pizza for everyone to eat (see Figure 5.11). This happens to be a type of pizza you like, and you're very hungry as well. Naturally, you want to get your hands on some

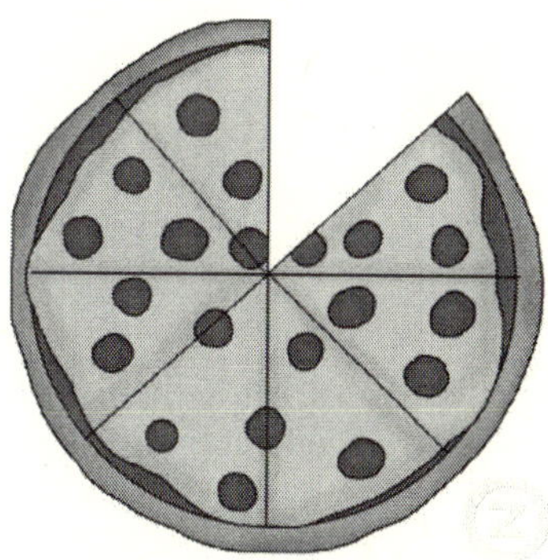

Figure 5.11 If you're really hungry (and you like pepperoni), eating more and more slices of pepperoni pizza will give you more and more "happiness." But the amount of happiness you gain with each additional slice is decreasing: the first is the best, the second is the second best, and so on. This illustrates the principles of utility and diminishing returns.

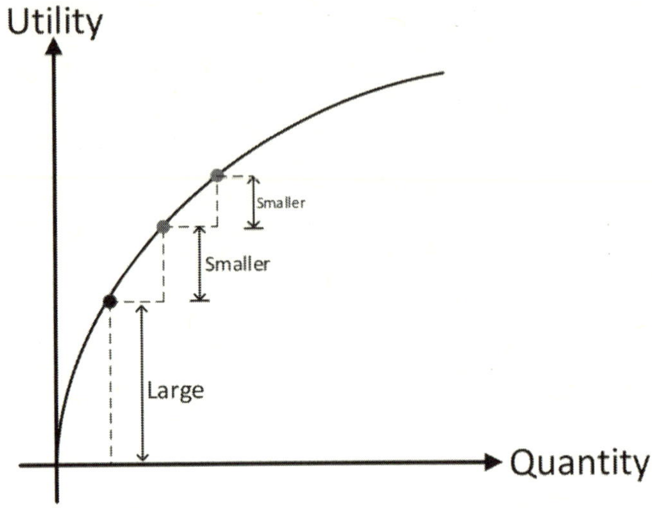

Figure 5.12 Typical utility-function shape, as the amount allocated changes. The utility is an increasing function, *i.e.*, increasing the amount will make the utility go up. But *gain* in welfare gets smaller and smaller as the amount gets larger and larger.

slices. So you take your first one. It tastes great, and it's everything you thought it could be. It certainly cut down your appetite some.

But you want more, so you take your second slice. Again, it tastes quite good, and is satisfying. But nothing could match the greatness that the first slice brought you: that was the one that started to satiate your appetite.

After your second slice, you're still kind of hungry, so you go ahead and eat a third. It still tastes good, and therefore is beneficial to you. But you're not as hungry anymore, so it's not doing you quite as much "good" as the first or second.

This process keeps going (depending on how much pizza you can stomach), until eventually, you are indifferent to eating another slice. Since it's free, you might go for it since it tastes good, but you certainly would not *pay* for another at this point.

We can generalize our findings here to the allocation of any resource, whether it be food, consumer electronics, or more cellular data. It is important to capture how "happy" a user will be if a certain amount is allocated, and we call this the user's **utility**. It's the *satisfaction* the user will have with the amount they obtain (*i.e.*, the amount of slices he got). We can make two general remarks about how utility will behave as we change the amout the user is given.

- *Increasing*: The utility needs to always *increase* as the quantity allocated increases. You would keep eating pizza as long as it was being given to you.
- *Diminishing returns*: Above a certain point, the utility needs to increase *slower and slower* as the quantity increases. The first slice or two are the best, and after that, the successive gains you are getting drop off. This is called the principle of **diminishing marginal returns**. Typically, the returns begins to diminish right away, *i.e.*, right after the first slice, but this does not always have to be the case.

A typical example of such a utility function is shown in Figure 5.12.

How do we model a specific user's "happiness?" One source is human subject experiments. For example, when you're on the phone, your utility will change depending on the perceived delay of the call, as well as the rate that is being allocated to your call. So researchers may run tests on various people with differing levels of delay and rate, and record their perceived "satisfaction." As another example, with a video, your satisfaction will vary depending on whether you are watching in high or standard definition, how the sound quality is, and other factors as well. If enough experiments are run, utility can be modeled accordingly.

Demand

Another way we can model utility is by actually observing consumer behavior. Whenever users purchase a certain amount of something, they are interested in maximizing their **net utility**. Net utility is the resulting payoff, or the difference between the customer's satisfaction and the total price paid. If the customer purchases an amount of a resource at a specific `Price/Unit` (*i.e.*, \$10/GB), the resulting payoff is given by:

$$\text{Payoff} = \text{Utility} - \frac{\text{Price}}{\text{Unit}} \times (\text{Quantity Purchased}). \qquad (5.1)$$

This is a generalization of what we saw in Chapter 2 when discussing auctions.

The quantity that a user purchases for a given price is given by the **demand**. The demand function captures the "desire" for a resource depending on the price offered. As you can probably guess, demand is a decreasing function of price: a higher price induces a lower demand.

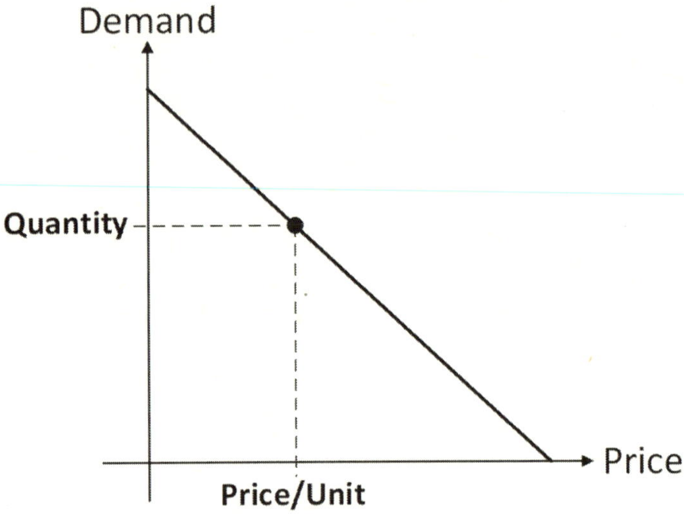

Figure 5.13 Depiction of a user's demand curve, which gives the amount of a resource he will obtain if the seller is offering it at a specific Price/Unit. The demand curve is a decreasing function, since a higher price naturally implies that a person will purchase less. We assume demand curves to be linear.

If the price of data doubled, you would consume a lot less. On the other hand, if it halved, you would use a lot more (within reason).

The demand functions we will consider here are linear, as shown in Figure 5.13. For the moment, we will assume that the price being charged is usage-based. The seller incurs a certain cost in producing the items, and the price charged to the customer is the equivalent cost per quantity that the seller needs to make back.

In this case, the user will consume *exactly* the amount given by his demand curve. It's relatively easy to find a user's utility given the demand curve and the price that is being charged. From this, we can also find the net utility.

- From the offered price, we can directly obtain the user's demand. This demand corresponds to the amount that the user will purchase to maximize his net utility.
- The price that the user is charged is simply the product of `Price/Unit` and the quantity purchased, as shown in Figure 5.14. Geometrically, this is the area of the rectangle `B` highlighted in the figure.
- Without going into mathematical detail, the user's utility can be obtained by finding the total area to the left of the demand curve and below the quantity purchased. In other words, it is the combined

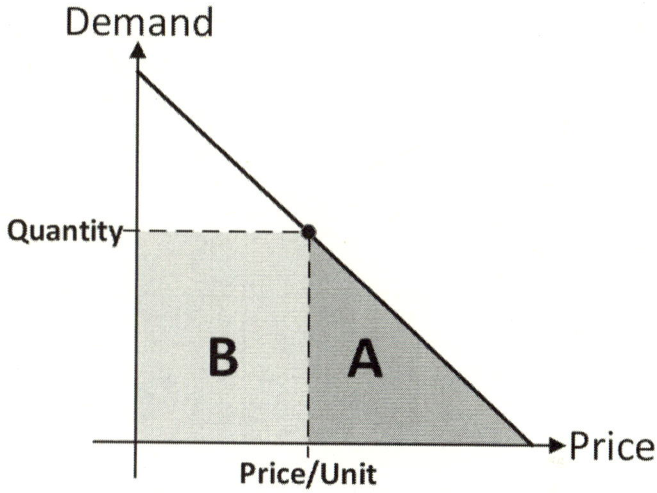

Figure 5.14 Geometric illustration of the process of obtaining utility and net utility from a user's demand curve (assuming a usage-based price). The utility is the area to the left of the demand curve and below Quantity, or `A + B`. The amount charged is given by the product of `Quantity` and `Price/Unit`, which is `B`. The net utility is the difference between these, or `A`.

area of the triangle `A` and the rectangle `B` shown in the figure: `A + B`.

- Finally, we can obtain the net utility from Equation (5.1) by subtracting these two areas in the graph: Payoff = (`A + B`) - `B` = `A`, the area of the triangle `A` in the figure.

This is great and all, but why is the quantity the user consumes given exactly as a point on the demand curve? Why would he not be incentivized to consume more, or to purchase less?

Well, first consider what happens if the user decreases the amount purchased to some amount `Quant Dec`, as shown in the left side of Figure 5.15. On the one hand, the price charged decreases by the area of the rectangle enclosed by `Quant`, `Quant Dec`, and `Price/Unit`, which is `B1`. But the utility decreases by the *entire* area to the left of the demand curve and between `Quant` and `Quant Dec`, which is `A1 + B1`. Hence, the net utility will *decrease* by `A1`. The lessened utility outweights the lessened cost.

On the other hand, consider what happens if the user *increases* his consumption to some amount `Quant Inc`, as shown in the right side of Figure 5.15. The price charged increases by the area of the entire

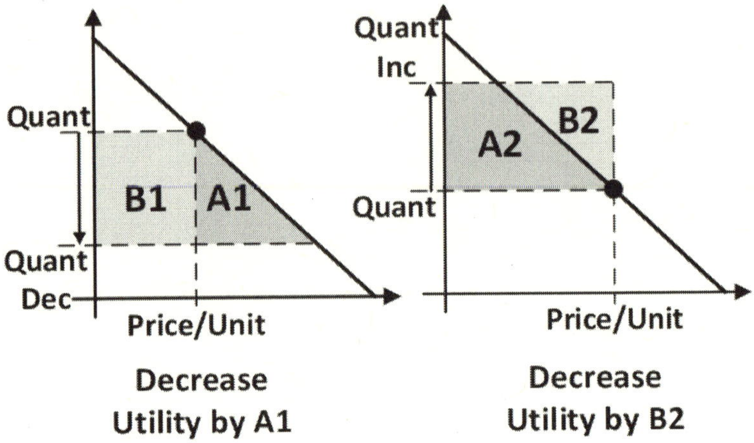

Figure 5.15 On the left hand side, we illustrate what happens if the user decides to consume less than his demand: the net utility will decrease by `A1`. On the right, we illustrate what happens if the user decides to consume more than his demand: the net utility decreases by `B2`.

rectangle `A2` + `B2`. On the other hand, the utility only increases by `A2`. So the net utility will *decrease* by the area `B2`. The added cost outweighs the added utility. From each of these cases, you should be convinced that the person should indeed consume exactly what is given by the demand curve.

5.1.3 Why usaged-based pricing?

Recall that flat-rate pricing charges a single, fixed amount irrespective of the amount consumed. On the other hand, usage-based pricing charges proportional to the quantity. We will now move to understand why usage-based pricing is generally preferred.

The tragedy of the commons

In Figure 5.15, we explained why a user would not want to stray from the demand curve under usage-based pricing: if she does, her payoff will decrease. But is the same true for a flat-rate?

The answer is no: it's in the best interest of the customer to grab as much as possible, until there is no more "utility" to be had. Think about it: if you pay \$20 each month for your data plan, what would dissuade you from streaming 100 videos as opposed to 10? Clearly not money. You would use as much data as you could possibly need each

Figure 5.16 We have a group of herdsmen (two shown here), who are raising livestock on a shared parcel of land. This will be used to illustrate "the tragedy of the commons."

month. In general, if a user is charged the same amount irrespective of the volume of his consumption, he will consume *as much as possible*, until his utility cannot increase anymore.

But why is this undersirable? To understand, we will appeal to a well-cited analogy in economics: "the tragedy of the commons", which was popularized by Garrett Hardin in 1968.

Consider a group of herdsmen, who are sharing a common pasture that they are each using to feed their own livestock. This land is open to each person, and each can readily add more cattle to graze on the land (see Figure 5.16).

Now, let's pretend you are one of these herdsmen, and you're trying to decide whether or not to put another animal in your herd. Since you are trying to maximize your gain, you ask the following question: "if I add one more animal, how will that affect my net utility?" This utility has two components to it.

- *Positive component*: If you add another, you will receive *all* of the revenue associated with the sale of that animal. View your gain here as +1.
- *Negative component*: On the other hand, adding another brings the land one step closer to becoming *overgrazed*. Once there are too many, the pasture will deplete, and all of the animals will perish. But this cost is spread evenly across *all* the herdsmen, so to you, it is only a *fraction* of -1. If there were ten, it would be something like $-1/10$ for you.

Since 1 is greater than a fraction thereof, it's in your best interest to

Figure 5.17 Rationally, it's in each herdsman's best interest to keep putting additional cattle on the land. But then the pasture will become overcrowded, and eventually deplete due to overgrazing. At this point, all the animals will perish, leaving the herdsmen with nothing.

add another animal to your herd. The revenue outweighs the cost, and your net utility is positive. And then you add another, and another, and so on, since the logic doesn't change. Every other herdsmen faces the same decision, and will do the same.

But what is the end result? The pasture is only finite, so as we add more and more livestock, it will become overcrowded (see Figure 5.17). Eventually, there will be no land left to graze. At this point, each animal will starve, and this is the worst scenario for everyone. Driven by self-interest, you and the other herdsmen have collectively driven the pasture to ruin, and therein lies the tragedy.

What do we make of this? Think of why each herdsman wants to keep increasing her livestock. The cost she incurs for doing so is just a tiny fraction of the gain. Back to mobile data pricing, that's analogous to a *flat-rate* scenario, where you incur no additional charge each time you draw from the network. From your perspective, the only "cost" is the miniscule congestion you are adding, so it's in *your* best interest to keep requesting as much as you can. But this could eventually lead to a "tragedy" where the network buckles under too much *collective* demand from everyone. Eventually, you will indirectly feel the effect of "overgrazing," through an ever-increasing price that the ISP is charging you to keep up with the rising demand.

This is another example of the negative network effect, like we discussed in Chapter 2 in the context of Wikipedia consensus formation. The more users we have chasing their own self-interests, the more effect this has on the whole, and eventually we reach an undesirable scenario. How do we mitigate this? We have to change the situation so that it's *not* in everyone's best interest to consume as much as possible.

The problem lies in the fact that we don't have an efficient pricing signal. Additional data translates to more utility, and there's no extra cost. But every data query is creating a *negative externality* on the network as a whole, just like in Chapter 1 where we saw that your cell phone creates interference for everyone else. To fix this, we have to force everyone to *internalize* their negative externalities. And the way we do that is by making them pay for it. In this case, that becomes monetary: we have you pay for each piece of data you request, or for each additional animal you want to add to your herd. With higher price comes lower demand, and then we can avoid the tragedy in the first place.

Over the years, there have been many studies comparing flat-rate and usage-based pricing. The tragedy of the commons is one useful analogy for doing so. We will now return to our price/demand relationships, as the main results in comparing the two can be summarized in a few graphs.

Flat-rate creates waste

In Figure 5.18, we represent the difference in pricing signals between flat-rate and usage-based graphically. On the left is usage-based: the provider incurs a certain cost per GB consumed, and charges accordingly. To maximize payoff, the user will consume the corresponding number of GB on the demand curve.

On the right, we have a flat-rate pricing scenario. The provider incurs the same cost per GB, but does not charge users accordingly: it has to determine what the best monthly price to offer is. The user, being charged whatever this flat rate is, is incentivized to increase the number of GB consumed *to the top* of the demand curve.

To look at the implications here, we define the **user surplus** as the difference between utility achieved by the customer, and the cost incurred by the provider. Clearly, larger surplus is better.

- For usage-based, the consumer's utility is `A` + `B`, and the cost incurred by the provider is `B`. Hence, the user surplus is (`A` + `B`) - `B` = `A`.
- For flat-rate, the total utility is the *entire* area of the triangle: `A` + `B` + `C`. On the other hand, the price incurred by the provider is the area of the rectangle formed by $/GB and GB: `B` + `C` + `D`. Therefore,

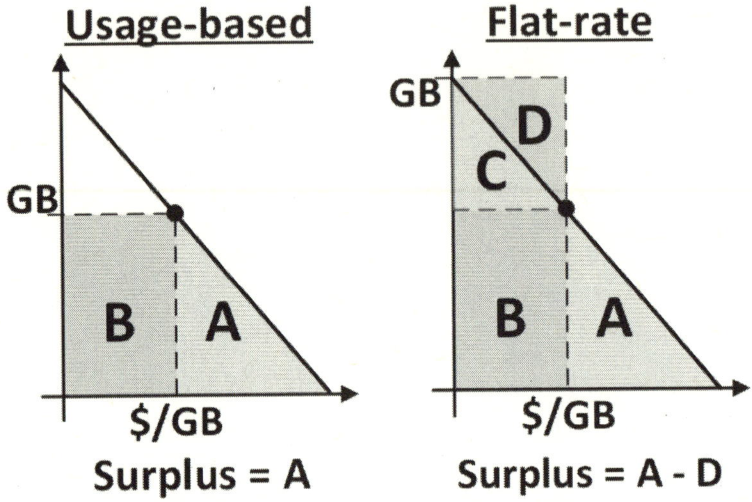

Figure 5.18 Comparison between the user surplus under usage-based (left) and flat-rate (right) pricing schemes. On the left, the surplus is A, whereas on the right, it is A - D. The area D creates negative surplus, and is nothing but waste: it is the extra cost incured that is not counteracted by any change of utility.

the surplus is (A + B + C) - (B + C + D) = A - D. This surplus is less than that of usage-based pricing by the area D.

The area D in Figure 5.18 is nothing but "waste:" it's added cost is not counteracted by any gain in utility. Hence, it detracts from the user surplus. Just as with the tragedy of the commons, this highlights that usage-based pricing is more efficient because it sends the right signals to the consumers: don't take more than your demand calls for. No network service is actually free, so "free" is not a particularly efficient pricing signal.

Flat-rate favors heavy users

Let's consider three different types of users: a light user, who has relatively small need for data, an average user, who has typical needs, and a heavy user, who consumes a lot. The demand characteristics of each of these three types of users are shown in Figure 5.19: the light user is the lowest one, and the heavy user is the highest one, because for a given price, the former will demand the least data, while the latter will demand the most.

Clearly, we have large diversity in the consumption characteristics of

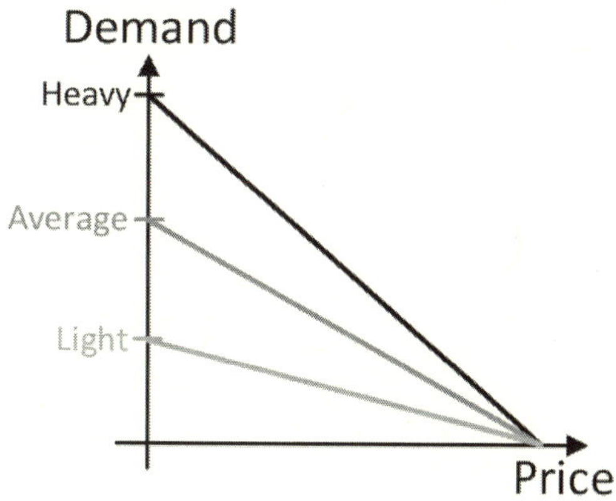

Figure 5.19 Demand curves of three different types of users: light, who have the lowest demand for any given price; average, who are in-between; and heavy, who have the highest.

each consumer. Usage-based pricing has no problem dealing with this, because by charging per GB of data used, light users will be charged the least, and heavy users the most. This is straightforward, intuitive, and fair.

On the other hand, for flat-rate pricing, it's a different story. How does the provider determine what price to charge? Well, it needs to make sure its revenue is high enough. If it knows the total number of customers it has, it can estimate the total capacity cost it will incur based on how much everyone will consume in total. Now, everyone needs to be charged the same price. The only way to do this while recovering expenses is to *average* the total cost over all of the users.

As a result, everyone will get charged based on the consumption of the *average user*. We illustrate this in Figure 5.20. Based on the amount (in GB) consumed *on average*, the provider sets its own cost incurred (in \$/GB) to guarantee that the resulting *revenue* obtained from each user will suffice when added up. The amount it charges each user is hence given by the area R. We make two observations:

1. *Light user*: On the left side, we highlight the case at hand for light users. Under a flat-rate scheme, the user's utility is given by the area of the entire triangle, `U1`. Now, compare this with the amount that it will be charged, given by the rectangular area `R`. Clearly, the utility is *less* than the revenue, and hence, we have a *negative* surplus for light users.
2. *Heavy user*: On the right side, we instead show the case for heavy

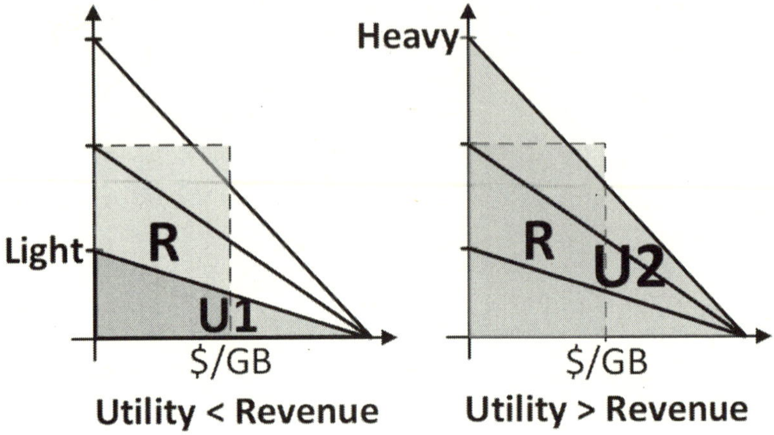

Figure 5.20 Under a flat-rate scheme, users are charged based on the consumption amount of an average user. The provider must set this price high enough to recover its capacity costs. As a result, on the left, we show that light users can have a very negative surplus, while on the right, we show that heavy users can have a very high surplus. Light users subsidize the lifestyles of heavy users under flat-rate pricing.

users. The utility is given by the triangular area `U2`. And the amount charged is the *same* as it was for light users. Clearly, the utility is *greater* than the revenue, and hence, we have a *positive* surplus for heavy users.

The result of this is that heavy users can obtain very high payoffs, while light users may have very negative payoffs. By charging a flat-rate price, light users are effectively subsidizing the lifestyles of heavy users, who incur no additional charges for their voluminous consumption.

We can see now that there are a different benefits of usage-based pricing, especially in the wake of enormous demand increases that ISPs are facing today. The key idea is that it sends better pricing signals to the consumers, and forces them to internalize the negative externality of added congestion that they impose on the network.

Summary

We've just walked through some of the intellectual foundation behind network pricing. We saw why providers like Verizon and AT&T have been largely switching from flat-rate to usage-based schemes; the former sends a "free" pricing signal to consumers, where each extra GB incurs no additional cost, resulting in the tragedy of the commons which is not sustainable given the explosion of demand. Charging based on usage, on the other hand, forces users to internalize the negative externality they impose on the network.

6 Network Of Networks

We have mentioned the Internet many times so far, and all the previous chapters rely on its existence. It is about time to get into its nuts and bolts. As you will see, it's indeed a network composed of many networks.

6.1 Packet switching

There are three fundamental concepts behind the Internet: packet switching, distributed hierarchy, and layering. We will take this section to look at the first two, and will return to the third in the next chapter.

6.1.1 Sharing revisited

All the way back in Chapter 1, we discussed different multiple access techniques that allow users to share network resources: either transmit at different frequencies, different times, or with different codes. Regardless of the specific technology, what's key here is that we have a *dedicated* resource for each user, which is how they are distinguished along their paths from beginning to end. It gives us a way of allowing them to use dedicated network resources. Such networks that assign fixed portions of the resources to each user are termed **circuit-switched**.

But there was an underlying negative to circuit-switching: not *all* resources will be used constantly, since not everyone is constantly using the network due to the bursty nature of data applications like web, email, and file transfers. So, when typical users don't really require a *dedicated* resource, why not allow them to *share* these resources, in addition to sharing the paths?

The word "user" here is used interchangeably with "session." The logical unit is an application session rather than a physical user or device. For now, assume a session has just one source (i.e., the sender) and one destination (i.e., the receiver) which is known as a **unicast** session.

When a session is established on the Internet, resources lie along the entire path from one end of the communication path, the sender, to the other end, the receiver. How can we distinguish such sessions

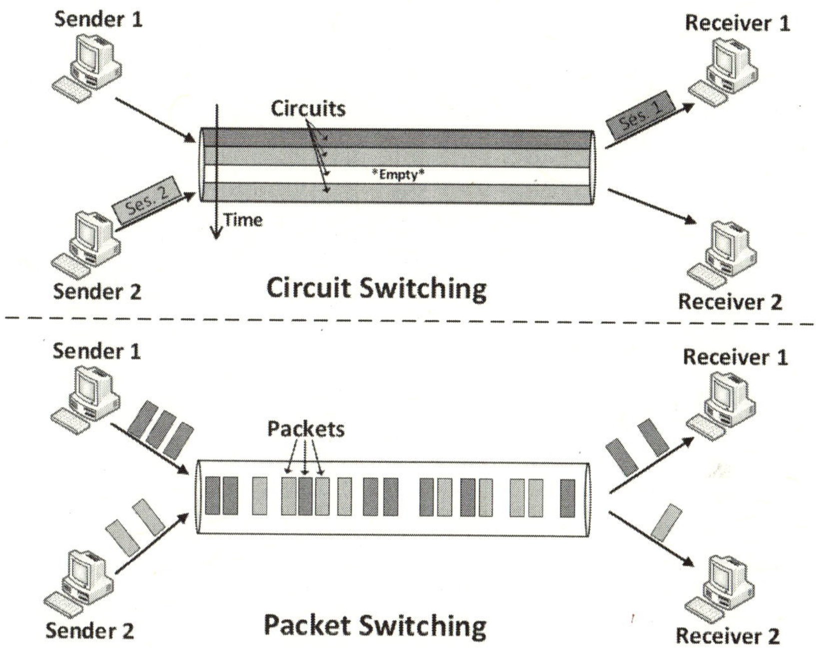

Figure 6.1 A simple network with two sessions (sender–receiver pair) along one link. At the top, we show circuit switching, in which each session has a *dedicated* resource along the link. Specifically illustrated is TDMA, in which each session is allocated a different timeslot (we show the pipe divided vertically into timeslots). Session 1 transmits, then 2, then 1, then 2, ... and so on. On the other hand, at the bottom, we show packet switching, in which each session divides its messages into packets and sends along the link, whenever there are packets to send. The path is shared along all timeslots and frequency bands.

from one another? Well, one way is through circuit-switching: dedicate a fixed portion of the resources along the path to each session. But alternatively, we could mix and match pieces of information belonging to different sessions, and still share all the paths. This is the essence of a **packet-switched** network: through smaller chunks of information, let everyone share everything and come up with other ways of managing the sessions.

A simple illustration of packet vs. circuit switching is shown in Figure 6.1. Each circuit in circuit switching may occupy either a particular frequency band or a dedicated portion of timeslots. In contrast, in packet switching, there is no dedicated circuit for each session. Each link is shared by multiple sessions, and each session may use multiple paths.

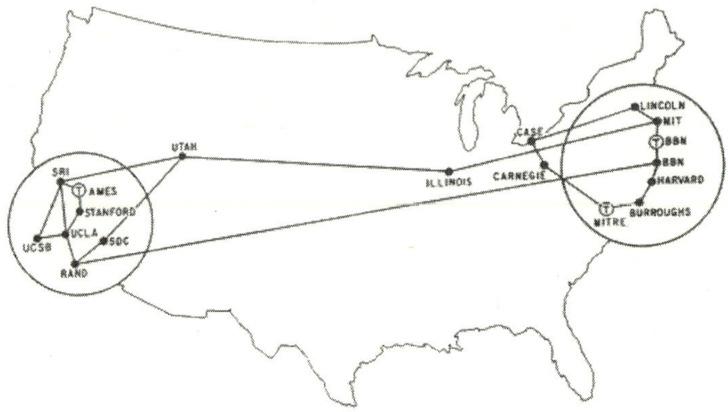

Figure 6.2 Map of the ARPANET in September 1971, with the first 18 hosts. It stretched from West to East Coast of the US, with some points in between. Some of these locations were universities and others were technology centers (e.g., BBN). The ARPANET grew into the modern Internet.

As we highlighted in Chapter 1, before the 1960s networking was mostly about connecting phone calls in *circuit-switched* public telephone networks. There continued to be active research in this area all the way into the early 2000s. A revolution, which came to be known as the Internet, started during the 1960s-1970s, with a shift to *packet switching* as the fundamental paradigm of networking. In the mid-1960s, researchers formally developed the idea of chopping up a session's messages into small packets and sending them along possibly different paths, with each path shared by other sessions. In fact, many of the concepts behind computer networking that were formulated then permeate through into the present day. Let's walk through some history now, to see how the Internet and packet switching came to be.

From ARPANET to Internet

The physical building blocks of the Internet spawned from the US Department of Defense. In the mid 1960s, researchers in the Advanced Research Project Agency (ARPA) expressed continued interest in creating a large-scale computer communications network based on packet switching. By mid-1968, they had unknowingly prepared a plan for a computer network that would ultimately change our lives forever.

In 1969, ARPA awarded a contract to BBN Technologies to put together the first computers that could support a network based largely on the initial proposal. Equipped with these machines, four institutions

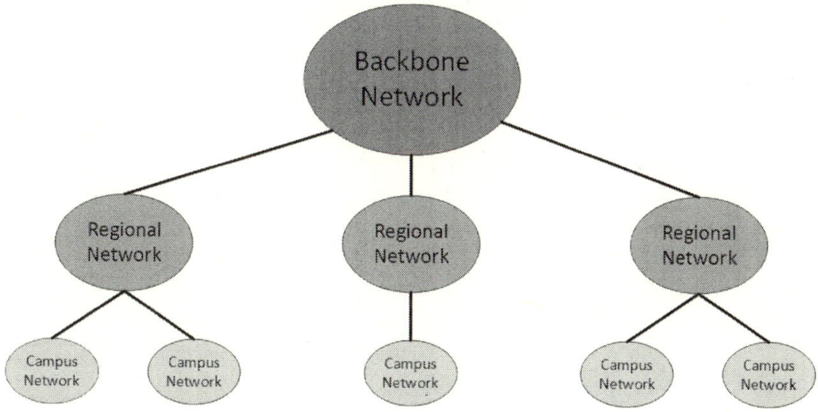

Figure 6.3 The idea of the NSFNET was to build a three-tiered (*i.e.* leveled) "network of networks." Campus networks are at the lowest level. Groups of campus networks close to one another would be connected by a regional network. And all the regional networks were connected together through the national backbone.

– UCLA, Stanford, UCSB, and the University of Utah – put together the first prototype of a packet-switched network, which came to be known as the **ARPANET**.

After its inception, the ARPANET started to grow. In March 1970, it reached the East Coast of the US, at Cambridge. 9 institutions were interconnected by June, 13 by December, and by September of the following year there were a total of 18 sites comprising the network. The map of the ARPANET at that time is shown in Figure 6.2.

In the fall of 1972, members of the ARPANET team connected 20 different computers at an international conference to demonstrate their achievements. Present there was one of the driving forces behind the ARPANET, Robert Kahn, who later remarked that this was the time when others began to realize that packet switching was a "real technology." In 1974, he along with Vinton Cerf published a landmark paper detailing their developement of a **protocol** for packet-switched networks. A protocol is simply a set of rules that devices use to communicate with one another, and it specifies the common "language" they speak. This one became known as the Transmission Control Protocol/Internet Protocol (**TCP/IP**), and gave a *scalable* way to connect the hosts in the ARPANET through end to end control over a *network of packet-switched networks* that allow for many variations. In other words, TCP/IP didn't need to be modified when large numbers of de-

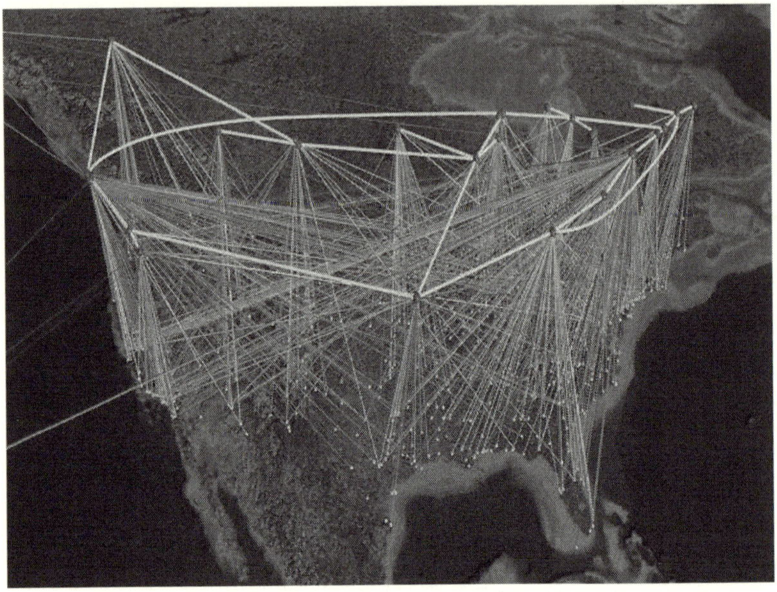

Figure 6.4 NSFNET in 1991, shown throughout the US. The backbone nodes are shown at the top, and the regional networks are indicated below. You can see how much the Internet had grown by this time, from its inception back in the late 1960s.

vices were added to (or removed from) the network. It wasn't until 1983, however, that TCP/IP would replace ARPANET's original protocol. And by this time, the ARPANET had become one subnetwork of the early Internet.

Being tightly contolled by ARPA, there tended to be funding and authorization issues that prevented groups from connecting to the ARPANET. Largely in response to this, the Computer Science Network (CSNET) was deployed in 1981 to provide Internet services to academic computer science departments first in the US, and later around the world.

Following this, the US National Science Foundation (NSF) largely took over Internet development. Their aim was to create an academic research network, allowing researchers to have access to large computing centers throughout the US. From 1985 to 1995, they sponsored the creation and operation of an ever-increasing *network of networks*, which became known as the **NSFNET**.

The idea was to have a three-tiered structure, as shown in Figure 6.3. Certain campus networks were connected together through a regional network. And all the regional networks were combined through the NSFNET's backbone, which spread across the US, and later, to international locations. The idea is that as you go "up" the tree, the geographical area that the nodes cover gets larger and larger.

The NSFNET grew as more and more network providers joined. To satisfy the increasing demand, the backbone, which held everything together, had to be improved at different times. In 1988 it had 6 nodes, and by 1991, 14. Figure 6.4 shows just how dense and rich the connectivity had become by this time.

The NSFNET was appropriated by "research and education," for "science and engineering." This meant that, strictly speaking, commercial activities were not allowed. The Acceptable Use Policy in 1990 stated that commercial enterprises could use the network, but only as it was related to research, not to promote marketing and sales. This may sound counterintuitive now, because most of what we use the Internet for today is commerce and other, non-research related activities.

With demand on the rise, in the early 1990s, a number of Internet Service Providers (ISPs) emerged to extend the Internet for public use. Commercial interests and entrepreneurial activities dramatically expanded this inter-connected network of networks. By 1994, the World Wide Web and web browser for the standard user had matured, and the world quickly began developing commercial applications on top of this network: email, file sharing, web surfing, you name it. In 1995, the NSFNET was officially decommissioned and was superseded by the commercial "Internet."

Today, the Internet has blossomed into an essential part of how people live, work, play, talk, and think. There are now more Internet-connected devices than people in the world, and it is projected that by 2020 there will be more than six times as many connected devices as people (see Figure 6.5). It has been a truly amazing five decades of technological development.

Packet switching or circuit switching?

We just briefly explored some of the history behind how the Internet came to fruition. But we are still left with this question: why base it on packet switching rather than circuit switching?

The debate between dedicated and shared resource allocation runs far and deep. We have seen such a design decision previously: TDMA vs. CDMA in Chapter 1. We will also see two more later: client-server vs. peer-to-peer and local storage vs. cloud services.

There is one big advantage of circuit switching, or dedicated resource allocation in general: *guarantee of quality*. Each session gets a circuit

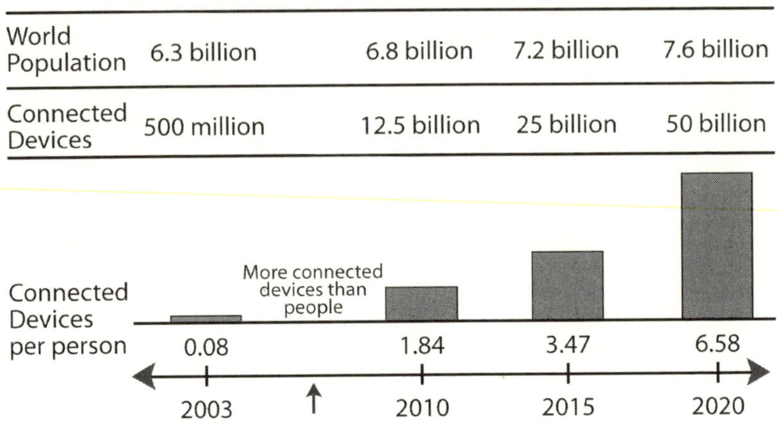

Figure 6.5 In recent years, the number of Internet-connected devices has caught up with the world's population. In fact, by 2010, it was almost double. By 2020, it is estimated that there will be more than six connected devices for every person on the Earth.

devoted to it. Throughput performance, *i.e.* the rate of successful message delivery, and delay performance, *i.e.* how long it takes to deliver a message, are guaranteed and will not change.

In contrast, packet switching can allow a session's traffic to be split across different paths, each of which is shared with other sessions. Pieces of a message can arrive out of order (the receiver can re-order them). Links may get congested. Throughput and delay performance become uncertain.

In the face of such uncertainty, the Internet claims to offer **best-effort** service, meaning that it will do its best to transfer your message with high performance, but there's no guarantee. This is perhaps more accurately described as *no effort* to guarantee performance.

On the other hand, there are two big advantages of packet switching: (1) *ease of connectivity* and (2) *scalability*.

1. Ease of connectivity is easy to see: there is no need to search for, establish, and maintain end-to-end resources for each session. The network doesn't have to make sure that your resources are saved for you, and at the same time, you don't have to wait until they are. You can transmit at will, as long as you follow the Internet's protocols (we will discuss these later).
2. Scalability here refers to the ability to take on a *large number* of *diverse* sessions. Some may be short in duration, and others long.

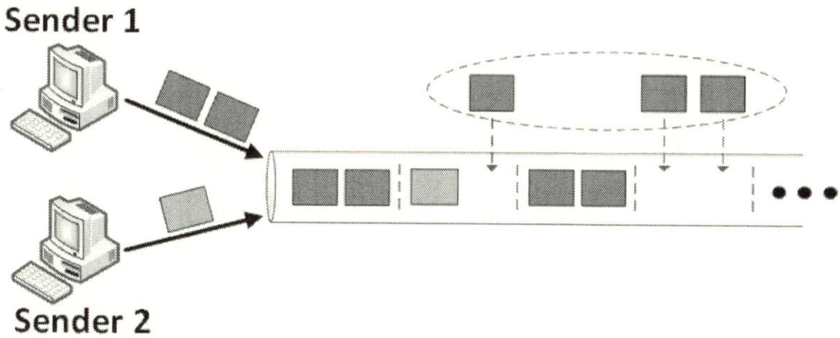

Figure 6.6 Here, we are illustrating the benefit of statistical multiplexing, where multiple sessions can use the same resources. Sender 2 has much less demand for the link than Sender 1. Under a circuit-switched system, portions of Sender 2's timeslots would be wasted. On the other hand, packet switching would allow Sender 1 to "fill up" this available supply, since these timeslots aren't dedicated to anyone.

Such scalability is obtained by achieving high *efficiency* in network resource usage.

Why is packet switching highly efficient? There are two main reasons. The first is that many sessions can share one path (and its resources), and the second is the complementary side: one session can use many paths. These features are known as **statistical multiplexing** and **resource pooling**.

- **Statistical multiplexing**

  There's not just one session on each path, occupying each resource on that path. Since sessions don't have dedicated resources, they are not wasting anything when they're idle. Other sessions with higher demands can make use of this.

  As an example, take a look at Figure 6.6. Sender 1 has much higher demand than Sender 2. Under circuit switching, each are allocated different timeslots. As a result, many portions of Sender 2's timeslots are wasted. On the other hand, packet switching allows Sender 1 to "fill up" this available time, since the resources are shared, not dedicated.

  As another example, consider our cocktail party analogy from Chapter 1. If everyone is taking turns speaking in the same room (*i.e.* one path), what happens when one group (*i.e.* session) is scheduled but they have nothing to say? Under circuit-switching, nothing: we would have dead time where others were dying to speak. Packet

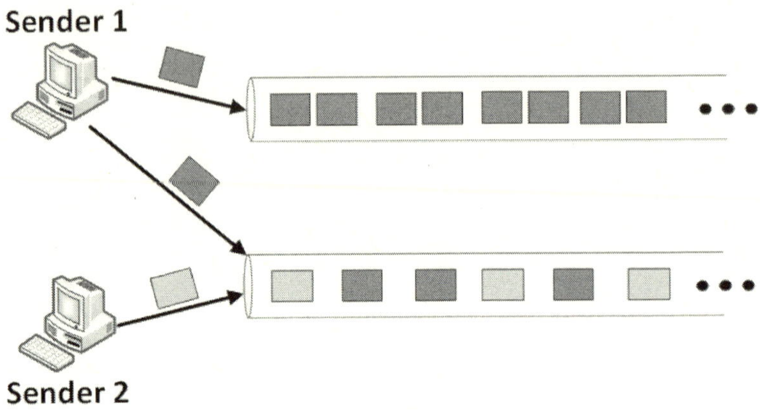

Figure 6.7 Here, we illustrate the benefit of research pooling in packet switched networks. We have two senders and two links. Sender 1 has more demand than the top link can handle. On the other hand, Sender 2 has much less demand than the capacity of the bottom link. Viewing the two links as a single, "pooled" resource, Sender 1 can transmit some of his message over the bottom link, increasing efficiency.

switching avoids this: available supply is used as long as there's unserviced demand.

- **Resource pooling**

  One session can use many paths to transfer information. Demonstrating the efficiency of this can get pretty complicated, but the basic idea is straightforward. Instead of having two sets of resources, say two isolated links, we can combine them and utilize them as one large resource.

  Consider the situation in Figure 6.7. Here, we have two sessions and two links. Sender 1 has high demand for the system, and has used up the available capacity on the top link. On the other hand, Sender 2 has low demand, and as a result, his link has low utilization. Under packet switching, paths become *resources* rather than dedicated circuits. In effect, by aggregating both links into a single, larger "pool," Sender 1's additional utilization can be met.

  Now, back to our cocktail party. Suppose there are two rooms (*i.e.* two links), A and B, each with ten pairs of people talking. Assume these are small rooms, so that's quite uncomfortable and congested. Then, in room A, six pairs decide to leave, leaving four in A and ten in B. Packet switching would look to alleviate congestion by

Property	Circuit Switching	Packet Switching
Guarantee of Quality	✓	X
Ease of Connectivity	X	✓
Scalability	X	✓

Figure 6.8 Summary of the different properties of circuit switching and packet switching networks. In the end, packet switching won the day for the Internet.

equalizing demand in each room, by telling three sessions in B to move to room A. On the other hand, circuit switching looks at these rooms as two separate resources and would not attempt to shift sessions from one to the other.

In the end, the abilities to easily provide connectivity and to scale up with many diverse users won the day, although that was not clear until the early 2000s. In contrast to quality guarantee, which is certainly nice to have, these properties are *essential* for a large and dynamic network like the Internet. Once the network has grown in an easy and scalable way, we can search for other solutions to take care of quality variation. Figure 6.8 summarizes the three key differences between what each packet switching and circuit switching offer to a network.

6.1.2 Distributed hierarchy

Now, you're convinced of how great packet switching is in facing dynamic, bursty traffic. But managing such a network is actually quite complicated. For instance, we discussed how *big* the Internet has become, in terms of the number of people using it.

At the same time, the Internet has extended to almost every corner of the globe. Many different ISPs have emerged throughout the world, each owning a different portion of the network. As a result, any given end-to-end session may traverse links across *multiple* ISPs. For instance, a YouTube session from Google to your iPhone will go through a wireless interface, then probably a few links in the core of the cellular network, followed by a sequence of even more links across many different ISPs.

Each ISP sits at a different level of an overall *hierarchy*, as shown in Figure 6.9. The levels are divided into tier-1, tier-2, and tier-3. The exact definition of each tier is debatable, and the lines between them are somewhat blurred, but we will point out the basic ideas here.

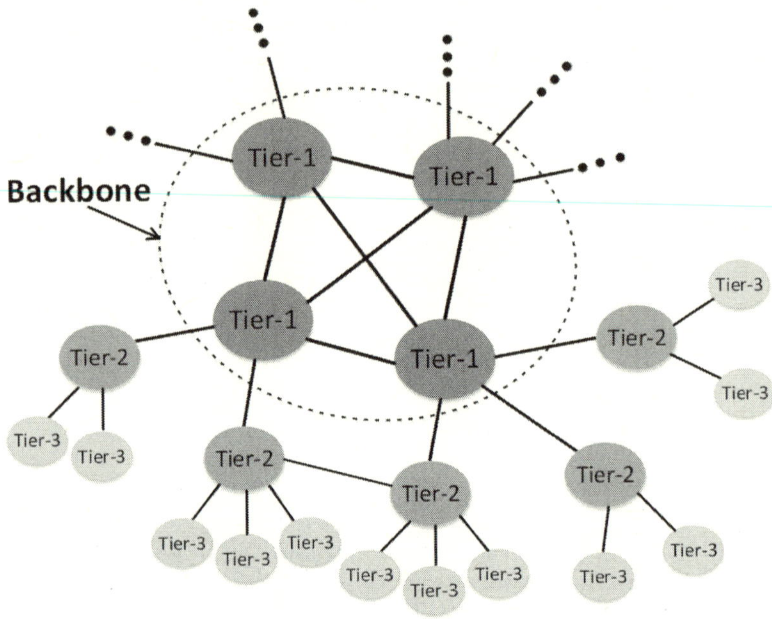

Figure 6.9 The Internet separates ISPs into different levels. The tier-1 ISPs are all connected to one another, and form the backbone of the Internet. Each of them is connected to a number of tier-2 ISPs, and these in turn connect to tier-3's. An ISP of a lower level becomes a *customer* of the *provider* ISP one tier above, and payments are made accordingly. Peer ISPs of the *same level*, like the tier-1s that form the backbone, do not pay for passing traffic through one another. Note that ISPs at any level may connect directly to customers, which are not shown here.

- A few very large ISPs become **tier-1 ISPs**. Each tier-1 has a global footprint and can reach any other tier-1 without having to go through a *lower level*. This means they form a *full mesh* **peering** relationship, and can pass traffic through each other for free. The full mesh of tier-1 ISPs is sometimes called the Internet *backbone*. Examples include AT&T, Verizon, BT, Level 3 Communications, and Sprint.
- Many more ISPs with regional footprints become **tier-2 ISPs**. These may also form peering relationships among one another, but cannot physically reach the Internet backbone without passing through their tier-1 provider(s). When a tier-2 and tier-1 ISP are connected, they form a **customer–provider** relationship, where the former has to pay a fee to the latter in order to pass traffic through.
- Each of these tier-2 ISPs provide connectivity to many **tier-3 ISPs**, forming another type of customer–provider relationship. Tier-3 ISPs

take traffic only to or from their customers, and not from any other ISPs. Typically, campus, corporate, and residential ISPs in rural areas belong to this group.

Another useful concept here is that of a **domain**. Each business entity forms a domain called an **Autonomous System** (AS). There are more than 30,000 ASs throughout the Internet. Dealing with Internet traffic that stays *within* an AS, or intra-AS, is much different from dealing with traffic that *traverses* two or more ASs, or inter-AS. We will focus exclusively on intra-AS traffic in this book.

We have now seen two fundamental concepts behind the Internet: packet switching and distributed hierarchy. But we have only just started. There are many tasks involved in managing the ever-expanding network of networks: we have to figure out how to get from point A to point B, manage congestion along links, and so on. We will turn to routing in the next section before going into detail on layering – the third fundamental concept – in the next chapter.

6.2 Routing traffic

How do we get traffic through the Internet, from source to destination? In a nutshell, the Internet contains devices called **routers** that direct packets of data where they need to go. This is done hop-by-hop, e.g., one step at a time.

The process of selecting paths in a network along which to send traffic is refered to as **routing**, a word you've probably heard used in everyday speech. Actually, it's very similar to the "route" you would follow when driving from one place to another, and major roadways themselves are commonly refered to as "routes." Before we dive into an algorithm that can be used to figure out these paths, we must point out some of the main ideas behind routing.

6.2.1 The "postal mail service" of the Internet

Transportation networks often offer interesting analogies for communication and social networks. In the case of routing, we can draw a useful analogy from the postal mail service. In order to route a letter from sender to receiver, we need three main features: addressing, routing, and forwarding.

Figure 6.10 When you send a letter to your friend through the postal service, you have to put his address on the envelope. This address serves as a unique identifier for him, so the mail system knows where to go with the letter. This is analogous to addressing in the Internet: every device that we want to send/receive information to/from has a unique identifier that the network will use in routing.

Addressing

When you send a letter to your friend through the mail, you have to put your friend's address on the envelope (see Figure 6.10). Otherwise, the postal service would have no idea what to do with the letter! You put their house address, town, state, country, and zip code on the front, which tells the post office where it should have the letter taken to.

Your friend's postal address gives him a *unique label* that nobody else in the world shares (except people living in the same house), so that there's no ambiguity as to where you want the letter to go. This is precisely the way addressing works within in the Internet. Before anything, we need to attach a unique label to each node in the network. Otherwise, we cannot even identify sources or destinations. We do this by assigning each network device an Internet Protocol address, or **IP address**, typically represented as four decimal numbers separated by dots. You've likely come across one before: for instance, 127.12.5.88 is one. Each of the four decimal numbers ranges from 0 to 255, which gives us over four billion possibilities. We show a few common Internet devices in a home, and which ones are often assigned IP addresses, in Figure 6.11. Some devices, like modems, don't use the Internet Protocol to communicate, and aren't assigned IP addresses (more on this in the next chapter).

Typically, we refer to the length of an IP address as the number of bits that we use to represent it. If you remember from Chapter 1, a bit is simply a 0 or a 1. Every decimal number can be represented using a sequence of bits, forming a **binary number**, and vice versa. Each of

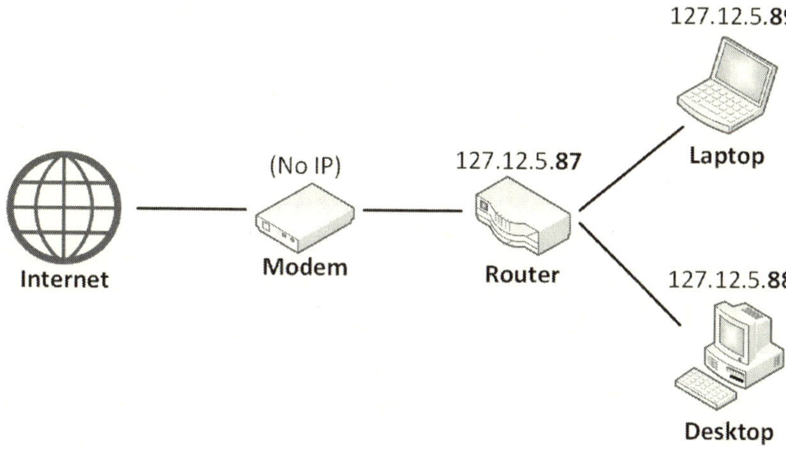

Figure 6.11 Each network device has a unique identifier called an IP address. They are represented as four decimal numbers, separated by dots. Just as the postal service uses the addresses on envelopes to tell where letters are destined, the Internet uses IP addresses to identify recipients. Here, we show four common Internet devices in a home, three of which have IP addresses. Your home router and computers each will have one. Your modem, on the other hand, doesn't have one because it's not an IP device.

the four numbers making up an IP address consists of eight bits, for a total of 32.

We should mention that there are two versions of IP: version 4 and version 6. IPv4 uses 32 bits for addresses, as in our example above. However, this ran out as of early 2011. IPv6 uses four times as many bits, 128, which gives us 4 billion × 4 billion × 4 billion × 4 billion addresses. That might sound like a lot, but with the proliferation of Internet-connected devices, we are well on our way to using many of these addresses as well.

The IP address in the Internet is analogous to the postal address in the mail system. Every element in the Internet that uses IP to communicate is stamped with a different one. This is how your computer, your home router, your friend's computer, and all the routers in between are identified.

Let's dig a litte deeper. Just as the zip code of a postal address gets our letter to the specific town or city, the **prefix** of an IP address gets us to the destination router. For our standard, 32 bit address, the leftmost three decimal numbers (or 24 bits) might indicate the prefix. This is denoted with a forward slash: 127.12.5.0/24 means we are referring

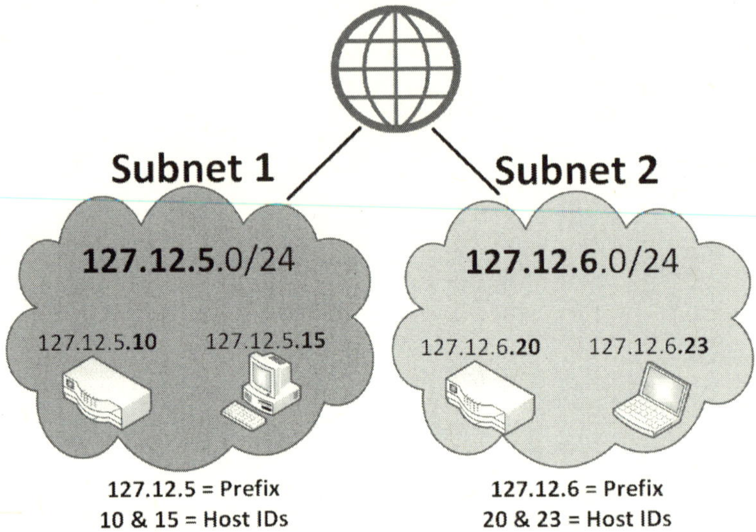

Figure 6.12 An IP address can be divided into two parts: the prefix, which indicates the subnet, and the host identifier, indicating the specific machine within the subnet. Here, we have two subnets, with prefixes 127.12.5 and 127.12.6, respectively. Within the first subnet, we show two machines with host IDs 10 and 15, and in the second, we show two with IDs 20 and 23. In the postal system, the prefix is analogous to the zip code, and the host ID to the house address.

to prefix 127.12.5. Using this, the router will bring us to the proper **subnet**, which is the group of devices belonging to this prefix. Any host within the subnet must have an address starting with 127.12.5, and will end with another decimal number (or 8 bits) that identifies the specific host. The last decimal number is then referred to as the **host identifier**. So your computer might be 127.12.5.88, and your router 127.12.5.1, indicating host IDs of 88 and 1, respectively. Similarly, your zip code gives the town or city you live in, and your street address identifies your particular house. We illustrate these notions in Figure 6.12.

Of course, the prefix doesn't have to be 24 bits long. The number after the slash will always tell us how many bits make up the subnet. For instance, with 127.12.0.0/16, the subnet is 127.12, and for 127.0.0.0/8, it is 127. Note that when we are specifying the subnet, we have filled digits that would make up the host ID with 0's.

An end-user device typically won't have a fixed, static IP address. Rather, one is automatically assigned and leased for a given amount

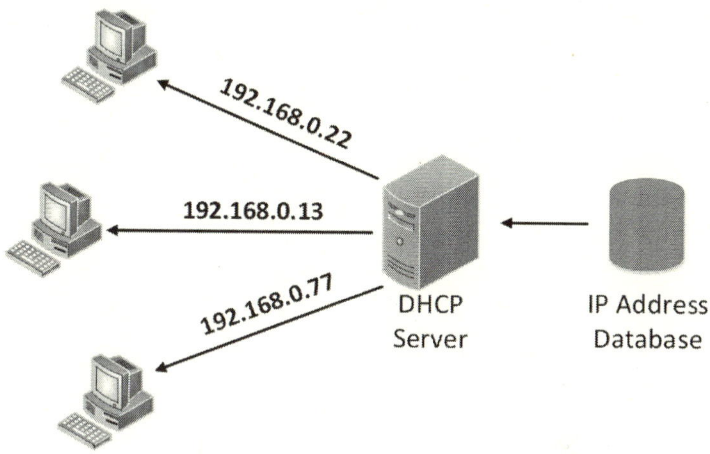

Figure 6.13 A DHCP server is responsible for giving your device an IP address when you want to be connected to the Internet. To do so, it keeps track of the available addresses that haven't been assigned yet in a local pool, using a database. Once it determines you aren't using your address anymore, the DHCP server will put it back in the database so it can be assigned to someone else.

of time by a server, often in your local network. This service is offered by a Dynamic Host Configuration Protocol (**DHCP**) server, providing pertinent IP addressing information to the end-user. The DHCP server keeps track of what IP addresses are free for you to take. Your device contacts the DHCP server and obtains a currently unused IP address for a finite "lease" time. When the lease expires, it can either be renewed, or the server will return it to the address pool so another device can obtain it. DHCP is illustrated in Figure 6.13.

And sometimes, the address given to your device within your *local network* (e.g., on your college campus) is different from the one seen by the *outside world* (e.g., the rest of the Internet). A Network Address Translation (**NAT**) router is in charge of "translating" back and forth, so people from outside your local area can address you based on what they see, as is illustrated in Figure 6.14. As an analogy to NAT, you can think of a mailroom in a company. When a package is sent to an individual in the company, the postage system only uses the building address (e.g., the public IP) to get it there. Then, a sorter (e.g., the NAT router) in the mailroom uses the person's name (e.g., the private IP) to determine whose mailbox to place it in, or which desk to deliver it to.

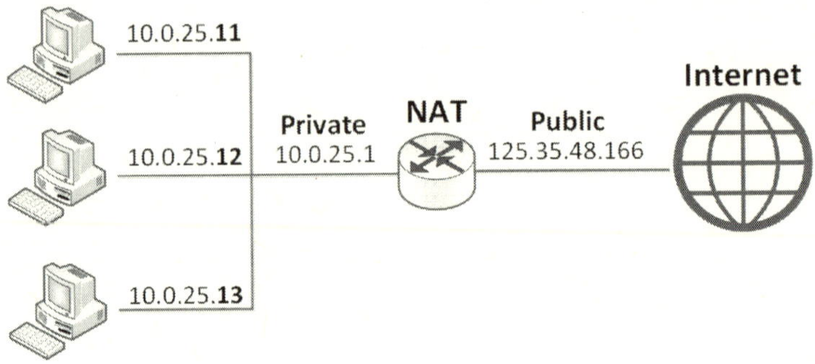

Figure 6.14 Sometimes, your public IP address (e.g., the one seen by the world) might be different from your private IP address (e.g., to everyone on your local network). A NAT router is in charge of converting back and forth: when a packet comes to you based on your *public* address, it translates to your *private* address and passes it along to you. Similarly, when you send something to the Internet, it converts from your private to public address for everyone to see.

There are multiple types of NAT, which are beyond the scope of this book. However, it should be noted that NAT has been a key factor in conserving IPv4 address space, which we noted was an issue before.

Routing

After addressing your destination, you have to decide the *path* that a session will take to get there. Let's assume that each session uses only one path (single-path routing), though this doesn't have to be the case (like in multipath routing).

In the postal system, routing decides ahead of time which intermediate cities the mail will go through in order to reach, say, Miami, Florida from Princeton, New Jersey. From your post office, your mail will be transferred to a large, regional office somewhere in NJ, then down to a large office somewhere in FL, then to the recipient's local post office, and finally to the recipient's door.

In the Internet, the paths our messages take will vary depending on the routing method we are using. There are two broad classes: *metric*-based, and *policy*-based. Inside an AS, we use metric-based routing, usually trying to find the shortest or least-congested path to the destination. In between ASs, however, routing is based on policies. For example, AS 1 might suspect that AS 2 has a lot of hackers, and therefore will want to avoid routing packets along any path that goes through AS 2.

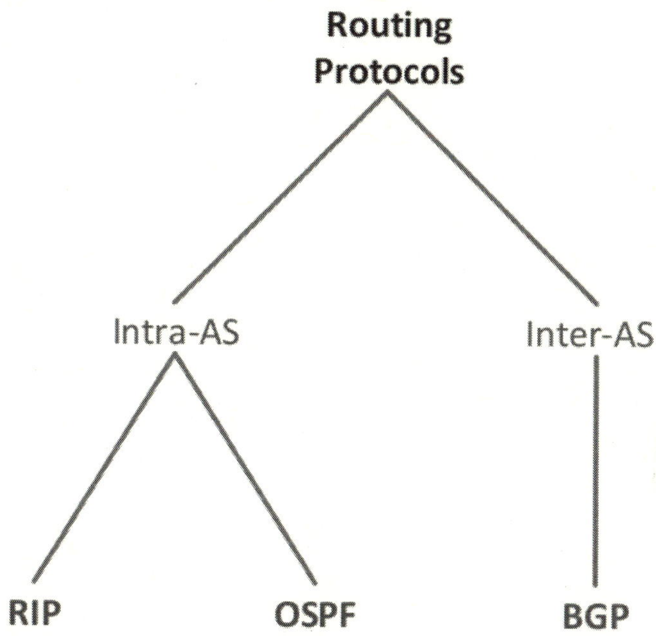

Figure 6.15 Classification of the three major routing protocols in use today. RIP and OSPF are metric-based, used for routing *within* an AS (e.g., intra-AS). On the other hand, BGP is the dominant policy-based protocol, used for routing *between* ASs (e.g., inter-AS).

As we have mentioned, inter-AS routing is very different from intra-AS routing. Border Gateway Protocol (**BGP**) is the dominant protocol for routing between different ASs. It "glues" the Internet together. However, it's quite complicated and messy, as most policy-based protocols are.

Within an AS, there are two main flavors of routing protocols: Routing Information Protocol (**RIP**), where each node collects information about the distances between itself and other nodes, and Open Shortest Path First (**OSPF**), where each node tries to figure out what the whole network looks like, rather than focusing on its neighbors. The classification of these three protocols is illustrated in Figure 6.15. The details of each are too complicated for us to get into. But in the next subsection, we will illustrate the main ideas behind RIP.

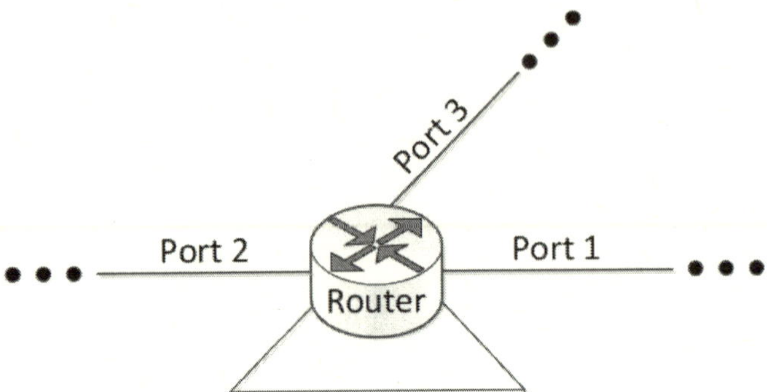

Routing Table	
Dest. IP	**Port**
10.1.1.0	1
10.1.2.0	3
10.8.4.0	2

Figure 6.16 A router has a number of ports, which are where packets arrive and depart. When a packet of data arrives at an input port, the router will look at the destination IP address and use its routing table to determine which port to send off to.

Forwarding

Forwarding is the method that actually implements the routing policy. It is done one **hop** at a time. Note that a hop, in general, is one portion of the network path between source and destination.

Back to our NJ to FL example. The post office in NJ will look at the zip code on the envelope, and see it is going to FL. It probably doesn't care about the specifics, like what town or city in FL, just that it has to get to some regional office in the state. So the decision it makes is to put the letter on the plane destined for this location. Then, once it arrives in FL, someone will look at the zip code again and know they have to get it to Miami. So they will put it in the mail truck that is headed over there. Finally, once we are within Miami's zip code, the local post office will look at the home address and deliver it to the recipient.

In the Internet, the action of forwarding happens each time a packet of data is received at a router. When one comes in, the router figures out where it has to go, and sends it off to the right neighbor. Then, this neighbor receives the packet, looks up the destination, forwards it along, and so on. Depending on how far away your destination is, the first routers will probably only care about the prefix. In other words, they first forward towards the destination subnet, just as the mail system first forwards based on the destination zip code. Finally, once the packet reaches the destination subnet, the routers will look to identify the specific recipient, taking into account the last digit of the IP address.

A packet arrives at one of the router's *input* ports and is sent off one of its *output* ports. In between is hardware that physically moves the packet from one to the other. So, how does a router determine which output port is the "right" one? This information is contained in what is called a forwarding table, which maps destination IP addresses to output ports. Basically, a router looks at the address, finds its entry in the table, and chooses the corresponding output port. This process is similar in function to a phonebook, except we look up ports from addresses rather than numbers from names. We illustrate this in Figure 6.16.

We are now going to dive more specifically into how routers determine which neighbor to forward to, based on the destination.

6.2.2 The shortest path problem

So what is the objective behind routing? Well, as we said, we have to figure out how to get from one point to another. The intermediate "nodes" that we pass messages to are the routers.

We have looked at graphs many times in the past chapters. To discuss routing decisions, we have to introduce yet another one: the graph of *routers*, where links between them represent physical connections. Such a graph is shown in Figure 6.17. The source's job is to determine which of its neighbors, A – C, to forward the message to. Then they will forward to other, intermediate routers, and so on, until it arrives at the destination.

Not all links in routing graphs are the same. Each has a *cost* associated with it. Often, this cost is directly related to the distance between the two endpoints on the link, i.e., the one between two routers in the same room might be less than the one between two in different buildings. These costs are indicated in the graph, on top of the links: for instance, sending from the Source to A costs 5, while sending from the Source to B costs 2. In general, we call a graph that has costs, or weights, on the links a **weighted graph**.

While we want to get between two routers, we also want to do so with the *minimum cost possible*, where the cost is the total over all the links used. This is known as the **shortest path problem**, the name

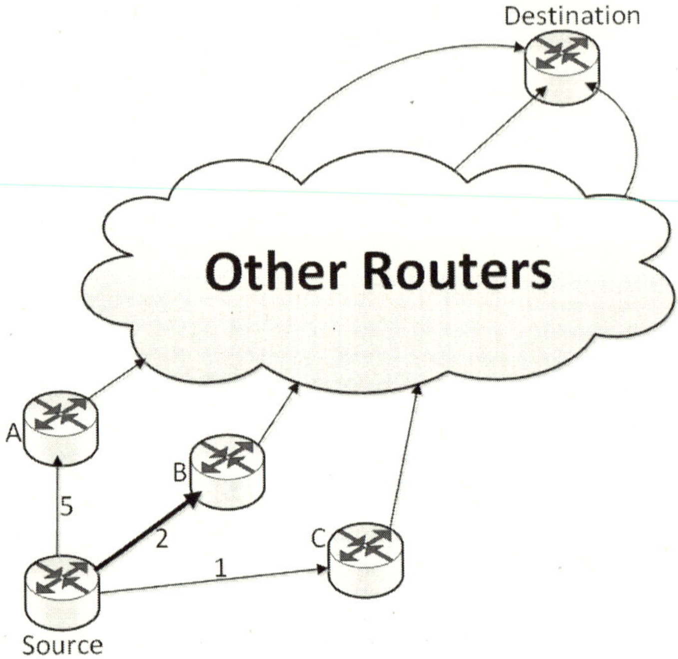

Figure 6.17 Illustration of the basic principle behind routing. We have some source router that wants to transfer traffic to some destination router. The source needs to know which of its neighbors (A, B, and C) is best to *forward* the traffic to. The neighbor that lies on the *lowest cost* path will be the one that is selected. Then, this neighbor will forward to the next router along the minimum cost path, and so on, until the message reaches the destination.

of which is perhaps misleading, because we are not concerned with taking the minimum number of hops possible, but rather with finding the minimum cost. "Minimum cost" would become "shortest path" if we assume that all of the link costs were 1: then, one link would cost 1, two would cost 2, and so on. Even so, "shortest path problem" has stuck universally, so we will stick to this name.

A simple example of the shortest path problem is shown in Figure 6.18. Here, we have four routers, A – D, and corresponding link costs. For A to send to D, it can either forward to B or to C. The cost along path `(A, B, D)` is $2+4=6$, and that along path `(A, C, D)` is $3+5=8$. The shortest path problem would tell us to choose `(A, B, D)`.

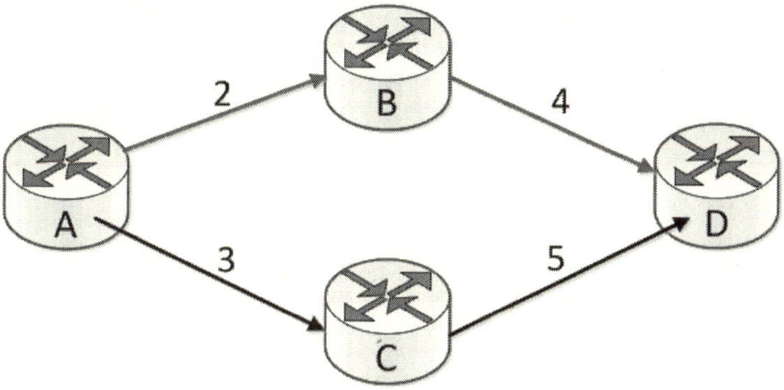

Figure 6.18 Simple example illustrating the minimum cost path between routers A and D. Path `(A, B, D)` costs 6, while path `(A, C, D)` costs 8. If A wants to send a message to D, it will forward to B, which will in turn forward to D.

The Bellman-Ford algorithm

So finding the shortest path is easy, right? Don't let this example mislead you. The Internet contains an enormous amount of routers, and solving this problem by inspection would be just as convoluted as doing so for PageRank, which we saw in Chapter 2. In fact, the shortest path problem has been studied extensively since the early 1950s, and there are several famous approaches: the Bellman-Ford algorithm and Dijkstra's Algorithm, among others. We will focus on the **Bellman-Ford Algorithm**, since it's simple and illustrates the fundamental principles behind routing algorithms. It also leads to implementations used in some famous routing protocols, such as the original ARPANET.

The Bellman-Ford algorithm solves the shortest-path problem by finding the minimum-cost paths for different numbers of hops. And they all build on each other, in an *iterative* manner, similar to the iterative algorithm of Distributed Power Control in Chapter 1. To get started, we first ask (rather trivially), what is the minimum cost path using no more than zero hops? Then, what is it using no more than one hop? We continue with two, three, and so forth, until we know the algorithm has terminated. Allowing the nodes to use more and more hops adds more and more possibilities, so the cost can only decrease over time. We will see how this is done through the following example.

Consider the router graph in Figure 6.19. There are six routers, A – F, and the link costs are as indicated. The objective is for each router A – E to figure out the minimum cost path to get to F.

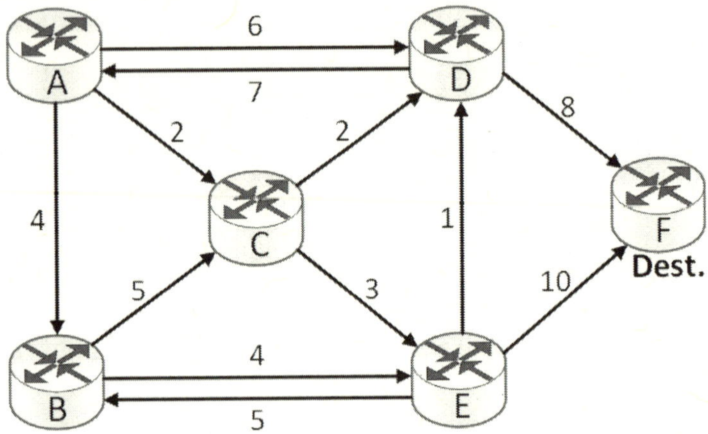

Figure 6.19 Example router graph that is used to illustrate the principles behind the Bellman-Ford algorithm. There are six nodes (A – F), with link costs as indicated. The objective is for each node to figure out the minimum cost path to get to F.

- *Zero hops*: In zero hops, it's not possible to get from any of the source nodes to F. Whenever the path is not possible, we say the cost is infinity, or ∞. On the other hand, it's obviously possible to get from F to itself in zero hops, and this cost is 0. Letting "Cost A" be the cost from A to the destination, in zero hops we have Cost A = Cost B = Cost C = Cost D = Cost E = ∞, and Cost F = 0.
- *One hop*: How about in one hop? Well, A, B, and C still can't get to F, because it is not any of their immediate neighbors. Therefore, Cost A = Cost B = Cost C = ∞. But D and E can: their direct links cost 8 and 10, respectively. So, Cost D = 8, and Cost E = 10. No need to update F.

We are at a position now to make a generalization about the cost updates. We have just determined how each router will get to the destination using no more than one hop. How can we use this information to determine the costs using no more than *two* hops? Quite intuitively actually: each router can look at its *neighbors*, to which it takes one hop to get to. For each neighbor, it can see the following:

- The cost that it takes for the router to get to the neighbor.
- The cost that it takes the neighbor to get to the destination in no more than one hop.

Adding these two costs, we get the total incurred over *two* hops by for-

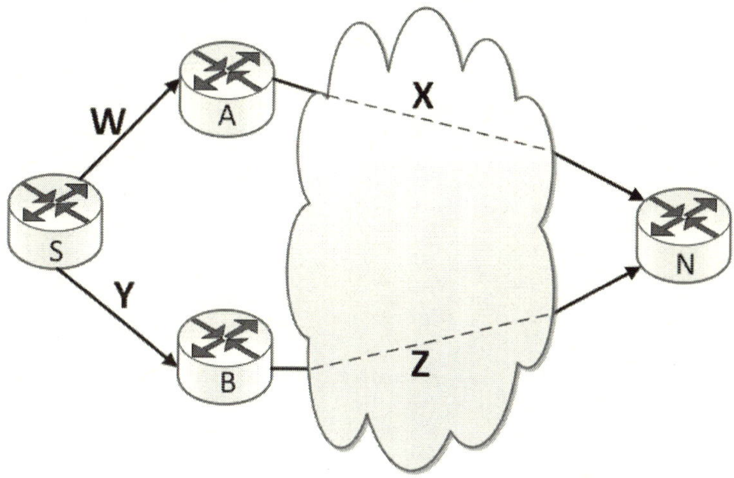

Figure 6.20 In the Bellman-Ford algorithm, a router (S) needs to determine the best neighbor to forward to, to minimize total cost to the destination (N). In doing this, it must know two things: the cost to get to its neighbors (W and Y) and the total cost from each neighbor to the destination (X and Z) in a certain number of hops. By adding these two, the router knows how much it will take to get to the destination with one additional hop ($W + X$ and $Y + Z$), and will choose whichever is smaller.

warding to this neighbor, since it takes one hop to get to this neighbor in the first place.

Basically, a router looks at a neighbor and says, "Well, if it costs me W to forward to you, and it costs you X to get to the destination, then it must cost me $W + X$ to get to the destination in one more hop." By considering all neighbors, the router can determine the best place to forward by finding the minimum of these costs. We illustrate this in Figure 6.20. Here, Router S has two neighbors, A and B. If A can reach the destination (N) in, say, three hops with a cost of X, and the cost of link `(S, A)` is W, then S can reach the destination through A in *four* hops, with a cost of $W + X$. Similarly for B, the cost will be $Y + Z$. Whichever of $W + X$ and $Y + Z$ is smaller will be the path selected.

We return now to our example. Let's recap: in one hop, D and E can reach F with costs of 8 and 10, respectively. On the other hand, A, B, and C are more than a single hop away, so their costs are ∞. Naturally, F can reach itself, with a cost of 0. This is summarized in Figure 6.21, where we have highlighted the links currently in use and indicated the costs to F in parentheses.

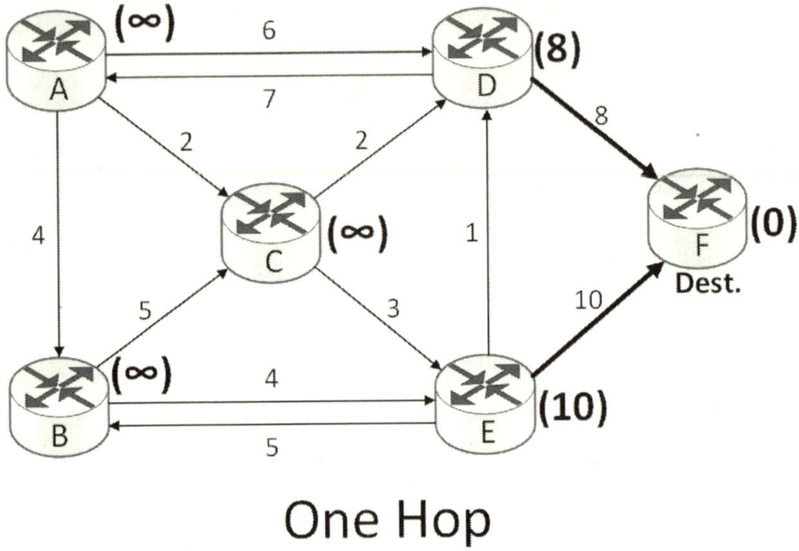

Figure 6.21 Shortest path costs and paths in our example, for one hop. The costs from each node to F are indicated in parentheses. In one hop, D can reach F with a cost of 8, and E can with 10. A, B, and C cannot yet reach F, and so their costs are ∞. Naturally, F can reach itself with a cost of 0, and we have indicated this as well.

Let's move now to two hops. We will analyze each router separately.

- *Router A*: A has three neighbors: B, C, and D. B and C cannot reach F in one hop, so we won't consider them at this point: the total cost would be ∞. But D can reach F with a cost of 8, and it costs 6 to reach D. Therefore, by forwarding to D, A will reach F in two hops, with a total cost of $6 + 8 = 14$.
- *Router B*: B has two neighbors: C and E (B cannot forward to A because the link goes from A to B, not the other way around). C cannot reach F in one hop. But E can get there, with a cost of 10, and B's link to E costs 4. So by forwarding to E, B can reach F in two hops, with a total cost of $4 + 10 = 14$.
- *Router C*: C has two neighbors: D and E. They can both get to F in one hop.
  - The cost from C to D is 2, and that from D to F is 8, so the total cost through D is $2 + 8 = 10$.
  - The cost from C to E is 3, and that from E to F is 10, so the total cost through E is $3 + 10 = 13$.

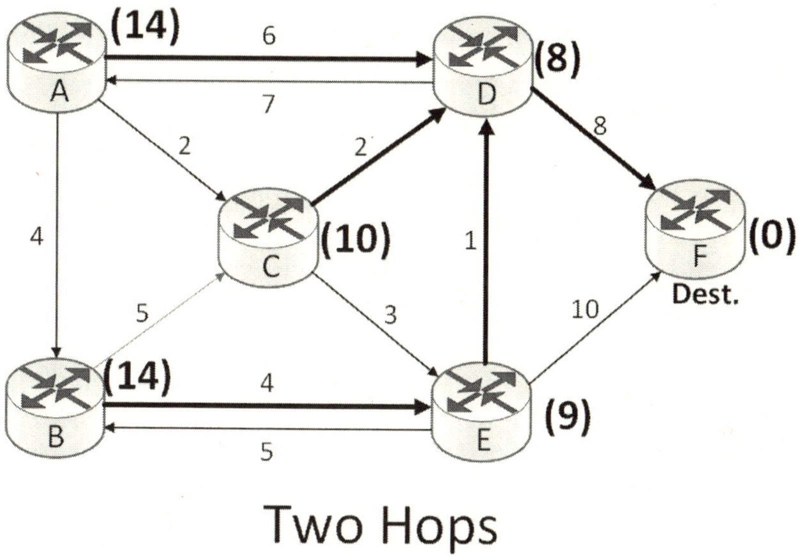

Figure 6.22 Shortest path costs and paths in our example, for two hops. Unlike in Figure 6.21, we see that A, B, and C can now get to F. Also, E has decided to route to D rather than directly to F to lower total cost.

Since $10 < 13$, C will choose D to get to F, with a total cost of 10.

- *Router D*: D has two neighbors: A and F. In one hop, A cannot get to F, so its only choice is to route directly to F, with a cost of 8.

  Two points here. First, you're probably wondering why we chose a path with one hop, when we said we were looking for two. Remember, we are not concerned with finding the minimum cost path using *exactly* two hops, but rather, the minimum cost path using *at most* two hops.

  Second, why did we even consider A? Would a router ever want to send somewhere else, if it was linked directly to the destination? Well, it's possible for the direct path cost to be high, and for there to be intermediate links with much lower total cost. Remember, we are considering minimum *cost*, not minimum *distance*. More on this next.

- *Router E*: E has two neighbors: D and F. Going directly to F costs 10. On the other hand, D can get to F with a cost of 8, and the link to D costs 1, for a total of $8 + 1 = 9$. Since $9 < 10$, E will route *to D*, with a total cost of 9.

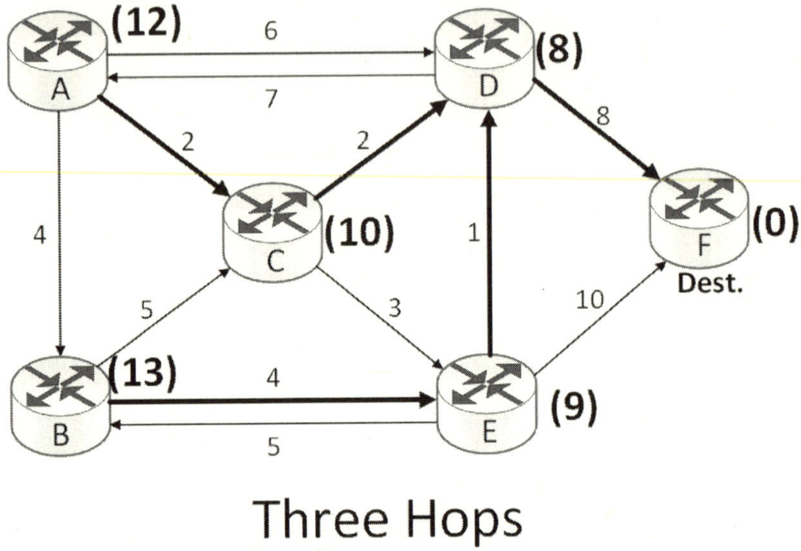

Figure 6.23 Results of applying the Bellman-Ford algorithm to our example graph with three hops. Now, A has decided to forward to C, rather than B, to minimize total cost. At this point, we don't have to consider a larger number of hops, since nothing will change; you can easily verify this.

The net result for two hops is shown in Figure 6.22. Now, A, B, and C can reach the destination. Further, the path chosen by E has changed.

Let's continue now for three hops. The procedure is the same as we just saw, except now we will be using the results for two hops. Each node can now reach the destination.

- *Router A*: A's three neighbors, B, C, and D, are advertising that they can reach the destination in (at most) two hops with costs of 14, 10, and 8, respectively.
  - The cost from A to B is 4, so the total cost through B is $4+14=18$.
  - The cost from A to C is 2, giving a total of $2+10=12$.
  - The cost from A to D is 6, and hence, the total cost is $6+8=14$.

  Since 12 is the smallest, A will forward to C, with a path cost of 12. Again, we see that though this path has more total hops, it has a lower cost, and is thus selected by the Bellman-Ford algorithm.
- *Router B*: B can forward to either C, with a total cost of $5+10=15$, or to E, costing $4+9=13$. As before, B will decide to go through E.

- *Router C*: As before, C has two choices: forward to D, costing a total of $2 + 8 = 10$, or forward to E, costing a total of $3 + 9 = 12$. C will forward to D.
- *Router D*: D will again choose to go directly to F, with no change: its alternative is to forward to A, with a total cost of $7 + 14 = 21$, which is larger than 8.
- *Router E*: E will again forward to D, with no change.

Figure 6.23 shows the result for three hops. So, when attempting to send a message to F, A will forward to C, B to E, C to D, D to F, and E to D. A summary of our results is shown in Table 6.1.

Router	Destination	Next Hop	Cost
A	F	C	12
B	F	E	13
C	F	D	10
D	F	F	8
E	F	D	9

Table 6.1 Summary of routing decisions to get to the destination. For each router, we have indicated the next hop as well as the total cost.

There's actually no need to continue to four, five, or more hops, because the algorithm has already determined the minimum costs. You should note that in general, if we have a number of nodes `Num` trying to get to a common destination (not counting the destination), we are *guaranteed* to have found the best solution after `Num` hops. The reason is that any path traversing `Num`+1 or more hops will have to go around a cycle. A cycle can only add to the length, and thus is never optimal. Here, we have five nodes A – E trying to get to destination F, meaning we would theoretically have to consider up to 6 hops in a general scenario. Even so, in a real implementation of routing, we do not know how many nodes there are, so we would have to rely on the lack of *new* messages to determine when it's safe to stop the algorithm.

RIP and mesage passing

We just illustrated the Bellman-Ford algorithm in detail, and showed how we can find the minimum cost paths from a given source to destination. But how this is done in practice is a bit more involved than the repeated application of the equation, as we just did.

Imagine you are one of the nodes trying to figure out how to reach the other nodes in the network. How do you know the cost to reach different nodes, and how do you even start?

What we haven't shown here is explicit *message passing.* Routers have to pass the path-cost information to their neighbors, and continually update based on the messages they are getting. When we applied the Bellman-Ford algorithm, we assumed everything was done by some central controller which knew all the path-cost information for the previous number of hops.

Without going into detail, we will point out several key features of message passing.

- *Short messages*: All the detailed network information (about who connects to whom and what the link costs are) can be *summarized* into these messages that they send to one another.
- *Local interaction*: Neighbor-to-neighbor message passing is enough to propagate this summary so that end-to-end shortest paths can be discovered. There is no need for broadcasting all the summaries to all the nodes.
- *Local view*: Even when these optimal paths are obtained, each node still has only a local view: it knows only to which neighbor to send a packet with a given destination address, but has no idea what the actual end-to-end path looks like. As far as forwarding packets using the knowledge of their destination IP addresses is concerned, it does not need to know, because the routers will forward packets hop by hop.

A combination of this message passing with an application of the Bellman-Ford algorithm is how RIP is implemented.

RIP is one of the oldest routing protocols, and is still in use today. The one-hop message passing between routers makes it simple to implement. But another protocol, OSPF, has gained much popularity over the past decade. There are several differences between RIP and OSPF; the main one is that while RIP only keeps information about its neighbors, under OSPF, each router actually tries to construct a local view of the network, including the state (e.g., cost) of each link. Hence, OSPF is a type of **link state** routing, whereas RIP is **distance-vector** routing. OSPF is preferred especially in larger networks, where link conditions may change quickly.

Still, RIP and OSPF are both intra-AS protocols. For inter-AS routing, the decisions become much more complex. They must factor in the price that each service provider charges to route traffic through their network. As we said, the most common inter-AS protocol is BGP.

Summary

In this chapter, we introduced some of the principles that govern the Internet and its operation. We started with two fundamental concepts: packet switching, where we saw the advantages of shared rather than dedicated resource allocation, and distributed hierarchy. Subsequently, we saw how traffic gets through the Internet by employing three main features of routing: addressing, forwarding, and determination of the next hop through an appropriate routing algorithm.

With these topics in our back pocket, we can move on to discuss some more of the tasks the Internet performs: congestion control and multicasting. We will segue into this with a discussion of layering, the third fundamental concept behind the Internet.

7 Layers On Layers

7.1 Divide and conquer

By now, you should be convinced that the Internet is a highly complex system. We saw how it's able to remain scalable and efficient through packet switching in the face of an ever-increasing number of users, as well as how it manages its physical span by establishing a spatial hierarchy of ISPs.

But the Internet's complexity is not just caused by its enormity. It is also due to the number of *tasks* it has to perform, routing being one of them. It is a natural practice when engineering such a complex system to *modularize* it, which means breaking it down into smaller pieces. For the Internet, we have different *layers* that can be controlled one at a time. Divide, and then conquer.

This process of **modularization** created the **layered protocol stack** for the Internet. In other words, each layer is controlled by its own protocols. Let's explore that now.

Protocol stack

A typical layered protocol stack is shown in Figure 7.1. We have five "basic" layers, typically referred to from bottom to top as 1 through 5.

- At the bottom lie the physical and link layers, 1 and 2, respectively. The **physical layer** is the medium over which we are sending signals. This could be copper cables, fiber cables, a wireless interface, and so on.
- In the middle, we have the network and transport layers, 3 and 4, respectively.
  - The **network layer** is where routing is handled, which we looked at in detail in Section 6.2. The protocol here is IP.
  - The **transport layer** focuses on end-to-end session management over the paths, with TCP as the dominant protocol. We will look at this in detail in the next section.
- At the top, we have the **application layer**, 5. This is the most "apparent" layer to the end user. Since the 1990s, one common protocol here is the Hypertext Transfer Protocol (**HTTP**), which is the basis

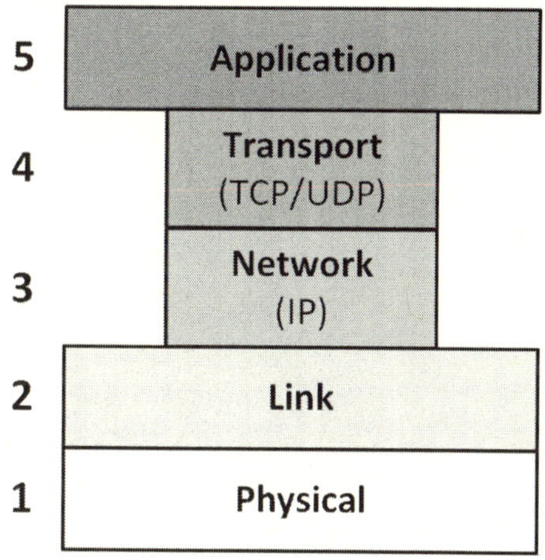

Figure 7.1 Modularization in networking: a typical model of a layered protocol stack. Each layer is in charge of a particular set of tasks, using the service provided by the layer from below and in turn providing a service to the layer above. The horizontal lines that separate the layers represent some kind of limitation of what each layer can see and can do. The two middle layers, dominated by TCP/IP, have remained largely unchanged over the years. They are the "thin waist" of the "hour-glass" model of the protocol stack. Note that the layer numbering starts at the bottom.

for the World Wide Web. We interact with applications every day: the web, email, Facebook, etc.

Over the short span of Internet evolution, the physical medium's transmission speed has gone up more than 30,000 times, evolving from 32 kbps dial-up modem to 10 Gbps optical fiber and 100 Mbps WiFi. At the same time, the applications have gone from command-line-based file transfer to Netflix and Twitter. Yet the Internet itself has continued to work, thanks in large part to the "thin waist" of TCP/IP that stayed mostly the same as the applications and the communication media kept changing.

Each layer provides a service to the layer above, and uses a service from the layer below. For example, the transport layer provides an end-to-end connection, running the services of session establishment, packet re-ordering, and congestion control, to the application layer above it that runs applications such as the web, email, and content sharing. In turn, the transport layer takes the service from the network layer below it, including the connections established through routing. The link layer is charged with controlling the access to the communication medium,

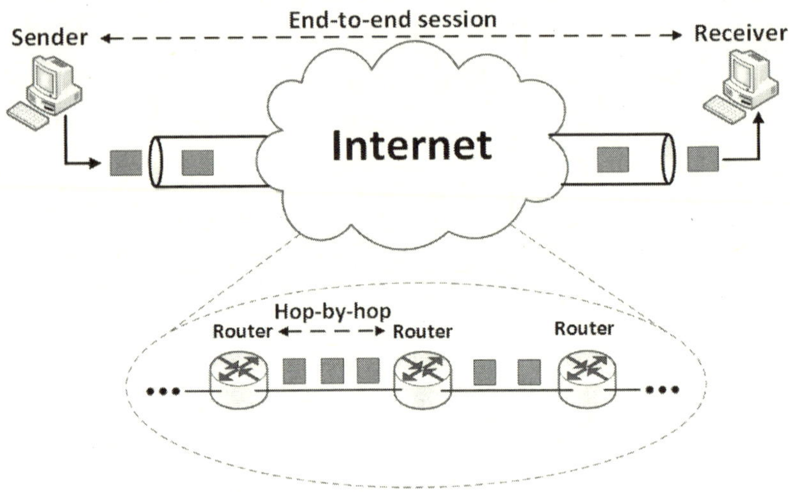

Figure 7.2 Here, we show the differences between the end-to-end nature of the transport layer and the hop-by-hop nature of the network layer. At the top, we have abstracted much of the Internet, focusing on the connection between the two end-hosts and transmitted/received data. At the bottom, we expand the "Internet," showing routers which are the most common type of layer 3 device. Here, we are concerned with connection only on a hop-by-hop (*i.e.*, between neighboring routers) basis.

and the physical layer controls the actual transmission of information on the physical medium.

Two points here are worth mentioning. First, the horizontal lines in Figure 7.1, denoting the boundaries between the protocol layers, are actually very complicated objects. They represent limitations as to what each layer can *do* and *see*. Second, the layering is not "clear-cut." There are functional overlaps between them. For example, each layer does some amount of error control, to try to detect and deal with mistakes in transmissions. And these overlaps were not by accident: they are there by design, to ensure robustness while allowing evolution of the network through layered structure.

Connection-oriented vs. Connectionless

Just between the transport and network layers, there are already quite a few interesting architectural decisions made in TCP/IP (the dominant but special case of the layers 4/3 protocol). First, the transport layer, in charge of end-to-end management, is **connection-oriented** in TCP, whereas the network layer, in charge of hop-by-hop management, is

Figure 7.3 A useful analogy to understand the difference between connection-oriented and connectionless services. When you make a phone call, the person on the other end has to pick up before you can have your conversation: this is a connection-oriented session. On the other hand, you can send someone a letter in the mail without them knowing it's on the way: this is a connectionless session.

connectionless in IP. A hop in a graph is a jump from a node to a neighbor, which is (likely) only one "piece" of the end-to-end path. This distinction is illustrated in Figure 7.2.

Let's consider two analogies (see Figure 7.3). First, suppose you are calling someone on the phone. Before you begin talking with them, they have to be alerted that you are calling (the phone rings) and pick up the receiver. Here, you are *establishing a connection* with them prior to your conversation. This is an example of a connection-oriented session.

On the other hand, let's think about the postal system again. Suppose you send someone a letter. First, the mailman takes it to the post office. Then, one post office transports it to another, closer to the destination. Then possibly another, and so on, one at a time. Eventually, a mailman in the destination's town will bring it to him. Along the way, the letter traverses a series of hops. All the intermediate points only care about where the letter is going *next*, not where it came from or where it's going. It only uses the final destination to determine the next hop on the path. This is an example of a connectionless session.

This also implies that the network layer doesn't care about congestion, or load conditions on the links. This is left to the transport layer, TCP in this case, which takes care of managing demand at the end-hosts. The Internet has run on this principle largely since its inception.

Encapsulation

We've mentioned the term "message" many times this chapter, meaning something you send, like an email, instant message, and so on. Its

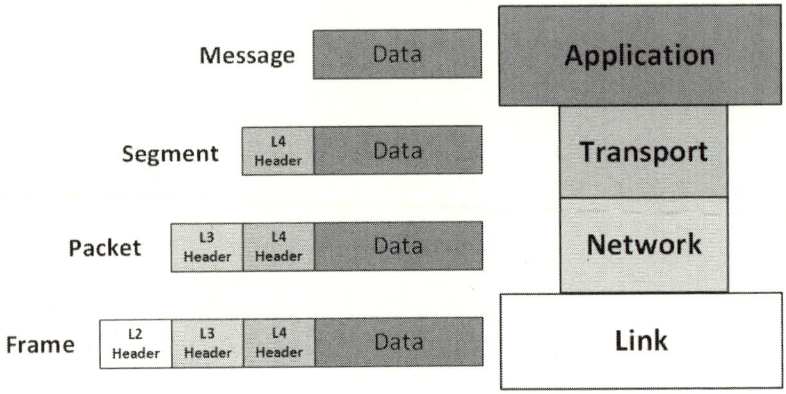

Figure 7.4 Basic principle of encapsulation. Starting with a message at the application, each layer down to the link layer adds its own header for transmission. Different elements in the Internet will make use of the different layers, depending on their functionality. For instance, routers, at layer 3, will use (and modify) the packet header. This process can get much more complicated when we need to divide each level: for instance, the network layer may break a segment into multiple pieces and form multiple packets.

technical meaning is similar. Let's take a look at that now, starting from the top at layer 5, and making our way down to layer 2, to see how these messages are **encapsulated** for transmission over the Internet.

5. When you go to send something on your Internet-connected device, you are telling the application layer to generate a sequence of **messages** to be transmitted. Messages contain the actual information that you want your destination to get.
4. Then, each message is divided into **segments** at the transport layer. Each of these segments contains two things: the **payload**, which is (some of) the actual content from the message, and a layer 4 **header**, added in front of the payload. These headers contain useful identification and control information about the segment.
3. Each of these segments are then passed to the network layer, which divides and encapsulates them as datagrams, or **packets**. Each of these packets have a layer 3 header in the front.
2. Each packet is further passed to the link layer, which adds a layer 2 header to form a **frame**.
1. Finally, the frames are all passed onto the physical layer for transmission as bits.

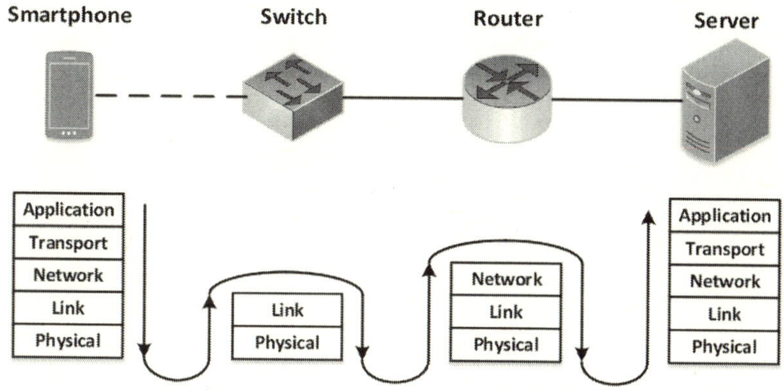

Figure 7.5 Different network elements process up to different layers in the protocol stack. The end-hosts process all the way up to the application layer. Switches that forward frames process up to the link layer. Routers that move packets across the network process up to the network layer.

This process of header-adding at each layer is illustrated in Figure 7.4. Doesn't it seem kind of redundant? Why does each layer add its own header? This actually creates a lot of **overhead**, meaning that we have a lot to send over the network that is not *actual* content. But such information is typically *necessary* for transmission.

On this point, we can return to our discussion on packet switching, which we said is *efficient* because everyone can *share* the same network resources. But how, then, can we distinguish between different messages? We need some way of knowing where they came from, where they are going, how long they have been in transit, etc. This is what the headers are for. As an example, the layer 3 header contains the source and destination IP addresses, which specify the next hop to take in the network. This is essential for routing, as we saw in Section 6.2.

Finally, note that network elements (e.g., computers, routers, modems, servers, etc.) run different subsets of the layered protocol stack. Each will decode and read only the header information associated with its subset. This is illustrated in Figure 7.5.

- Computers and servers, being end hosts, process all 5 layers.
- Routers, as network-level devices, go up to the third layer. This requires them to have, as well as process, IP addresses.
- Switches, as link layer devices, only process up to layer 2. They *do not* have, or process, IP addresses.

Equipped now with an understanding of the third fundamental concept behind the Internet – modularization – we will move on to discuss some of the specific functions of different layers in more detail. We already studied routing at the network layer in Section 6.2. In Section 7.2, we will look at congestion control at the transport layer, and the topic in Section 7.3 will be multicasting at the application layer.

7.2 Controlling congestion

In Section 6.2, we saw how to discover routes from source to destination. But we never said anything about *how much* these sources were sending. The cables that the Internet uses to transfer data across can only *handle* so much: it's left to the transport layer to make sure nobody is *asking* for too much.

We can look at the amount a user wants to send over the network as his demand and the amount that the network can handle as its supply. As discussed in Chapter 5, we measure these quantities as *rates*, *i.e.*, per second. Each link has a fixed capacity, which is a measure of the number of bits it can handle each second, or **bps**. With current technology, this is on the order of millions or billions, and as such, we typically speak of Mbps or Gbps. At the same time, we as users want to send a certain

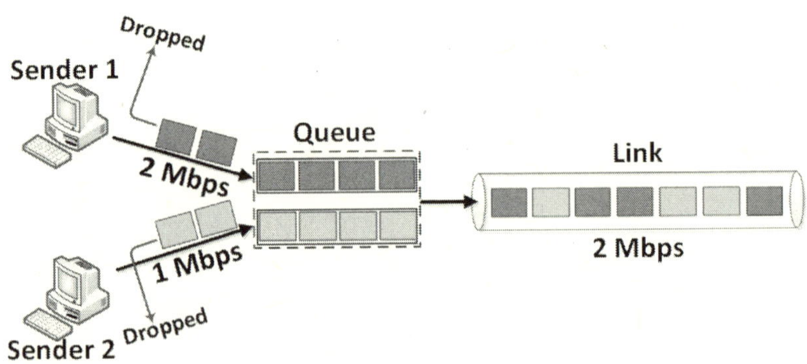

Figure 7.6 An illustration of congestion at one end of a link. Two senders, 1 and 2, are transmitting at 2 Mbps and 1 Mbps, respectively, making an aggregate demand of 3 Mbps. However, the link can only handle a maximum of 2 Mbps. Demand exceeds supply. As a result, packets will accumulate in the buffer at the front side of the link. Eventually, the buffer will be full, and packets will start to get dropped. Which specific packets get dropped depend on the details of the queue-management protocols.

Figure 7.7 Congestion in the Internet is similar to what you experience in a traffic jam. The amount of cars on the road is too much for it to handle: supply exceeds demand. As a result, everyone gets held up and delayed. This "hold up" is analogous to the time a packet spends in a queue. And as more and more cars pile on, the delay gets longer and longer, exacerbating the problem.

number of bits per second, maybe 1 Mbps depending on what we are doing.

It's the transport layer's job to make sure the demand for the Internet doesn't exceed its supply. And when it does, **congestion control** is essential in order to bring the demand back down to a tolerable level.

7.2.1 Principles of congestion control

To illustrate the notion of congestion, consider the example of a single link, as shown in Figure 7.6. The link has a capacity of 2 Mbps, which means it can't handle any more than 2 million bits each second. There are two senders trying to use this link, one of which is demanding 1 Mbps and the other 2 Mbps, for a total of 3 Mbps. You can see the problem: 3 is more than 2. Demand exceeds supply.

So what happens? Well, packets start to accumulate at the front side of the link, in what is known as a buffer. They are stored here, and wait until they can be transmitted. So, the first two million bits transfer in one second, the second two million in the next, and so on.

This is similar to sitting in a traffic jam, as shown in Figure 7.7. There are too many cars on the road, and as a result, you all can't drive on through at a normal speed. Take any car in this situation. It slows down with the other cars, and inches slowly on. Eventually, it will exit the jam and continue on at a normal speed. The time it was

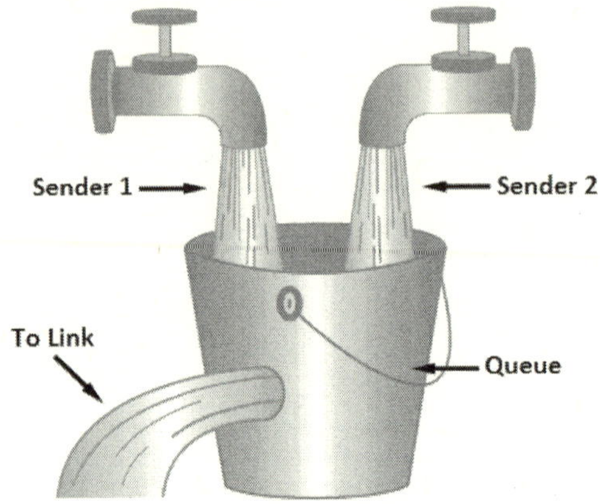

Figure 7.8 Dropping packets is similar to what happens when you fill up a leaky bucket too quickly. If you poor water in faster than the exit hole of the bucket can handle, the water level in the bucket will rise, and eventually, it will start spilling over the top. Similarly, if you transmit packets at a higher rate than the link's capacity, the buffer will fill up and eventually packets will be dropped.

in the traffic jam is analogous to the time a packet waits in the queue before transmission.

But think about what's happening behind this car. More cars are piling up behind, and more behind them, and so on, making the traffic jam even longer for the cars that get there later. Similarly, as more and more packets come in at the same rate, the queue gets larger and larger, which creates even more congestion. Eventually, the buffer will overflow and packets will get dropped, meaning the Internet will not even try to send them out.

The dropping of packets here is analogous to what happens when you fill up a bucket that has a hole in it, as shown in Figure 7.8. Think of the bucket as the queue and the hole as the exit point to the link (normally we think of a "leaky" bucket as a bad thing, but not in this analogy). As you fill the bucket, the water will start exiting through the hole. But if you fill it up faster than the hole can handle, the water level in the bucket will begin to rise. Eventually, water will overflow and spill out the top. This spilling is exactly analogous to the dropping of packets in an over-filled buffer.

Clearly, we have a problem already. But it's exacerbated even further. In order to promote reliability, the most common of the transport layer's protocols, TCP, has an acknowledgement system. What this means is that when your computer sends a packet to a destination machine, that machine replies with an acknowledgement (ACK) packet once it gets there. Then you as the sender will know that everything is good, and that your packet was properly received. Otherwise, after a certain amount of time, you will try to resend your packet and wait again for an ACK.

So what happens when the packets never reach the destination in the first place? Well, the sender will resend the unacknowledged packets. Then these packets are in the buffer. And then these don't get acknowledged either, so he tries to send again, and so on. This leads to a vicious cycle: congestion persists as the same set of senders that caused the congestion in the first place keeps resending the dropped packets. This pattern of drop, resend, drop, resend, will go on and on, a clearly undesirable state to be in. Senders need to rethink how they can *avoid* congestion in the first place, and they need to *back off* when it actually happens.

So, why have the ACK system, if it just adds to the congestion problem? In addition to promoting reliability, it turns out that we can use it much to our advantage. It can provide *negative feedback* to the senders as to how the network conditions are, so that they can lower or raise their transmission rates accordingly (similar to how cellular transmitters adjust as we saw in Chapter 1). Having this feedback is essential to preventing congestion collapse.

This is what Van Jacobson proposed in the first congestion control mechanism. It was named **TCP Tahoe** and became part of TCP in 1988. It is thought to have saved the Internet from collapsing under congestion in the late 80s and early 90s.

TCP Tahoe has been studied extensively since the 80s and improved significantly several times. But most of the essential ideas in congestion control for the Internet were in it already. Let's take a look at some of those now.

End-hosts

We could imagine a congestion control mechanism where, hop by hop, routers decide what rates the transmitters should send. This may sound intuitive, since they are the ones actually *in* the network and can probably infer it well.

But this is *not* the scheme adopted by TCP: it would require routers to keep tabs on the end-to-end connections. Remember, the modern Internet has largely been designed under the end-to-end principle: have *intelligent* end hosts and a *dumb* core network. This means that the

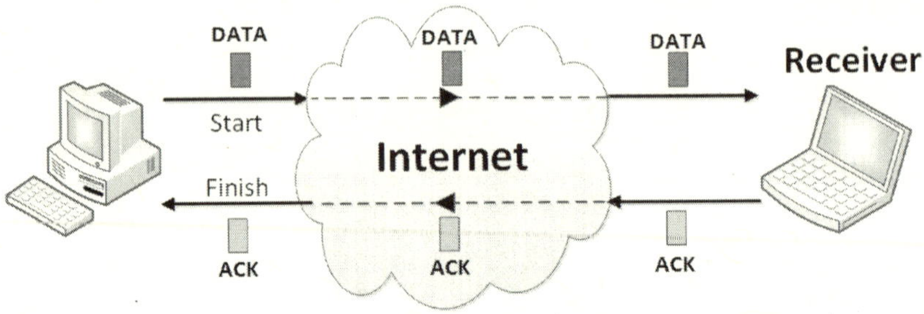

Figure 7.9 When a sender transmits a packet, the receiver sends an acknowledgement (ACK) upon proper receipt. The sender can use these acknowledgement packets to infer the state of congestion end-to-end. Specifically, the presence and timing of the acknowledgement packet is important, as we will see shortly. In this diagram, the arrows indicate the path over time: sender to receiver, then receiver to sender.

end-to-end connections and congestion control are managed by the senders, not by the network itself.

So end-hosts themselves have to decide their own rates. But how do they infer congestion? Well, the network provides "hints," feedback signals, to the senders through the acknowledgement packets that are sent back whenever a packet is properly received at the other end. This scheme is illustrated in Figure 7.9.

Specifically, the *presence* and *timing* of these ACK packets is important to tell what the network conditions likely are. If everything's going smoothly, keep sending, and perhaps at an even faster rate. But once these acknowledgements get jammed up or lost, it's time to hold back. We will elaborate more on this shortly.

Sliding-window

As we indicated, acknowledgement packets are what gives the sender the "OK" to send more. So perhaps we could require them to wait for the ACK of a sent packet before allowing them to send another. But this would be pretty slow, since it would mean you could only have one packet in the network at a time.

TCP pipelines this "one-for-one" behavior by providing a bigger allowance: rather than one packet at a time, we allow for a larger number.

Each sender maintains what is known as a sliding window, or a **congestion window**, which places a limit on the number of outstanding (*i.e.* unacknowledged) packets that it can have. The higher the size of the window, the more outstanding we can have. The window size

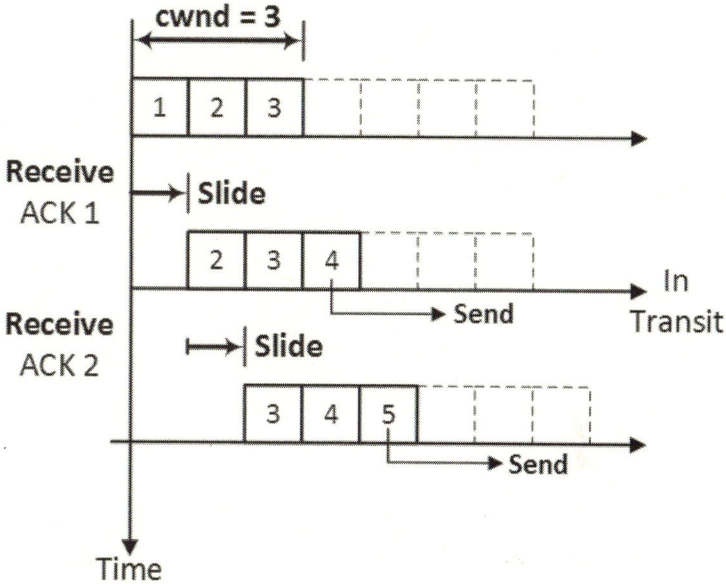

Figure 7.10 Illustration of the sliding window concept. Here, we fix the window size to three, which means that the sender can have up to three outstanding (*i.e.* unacknowledged) packets. Time is shown on the vertical axis and packets in transit on the horizontal. Initially, the sender transmits packets 1 - 3, though he has more waiting. Once he receives the acknowledgement for packet 1, the window slides, and he can send packet 4. The window slides again when the acknowledgement is received for packet 2, and he sends packet 5.

is commonly denoted by `cwnd`. If `cwnd` is 3, that means up to three packets can be sent before the sender has to pause and wait for acknowledgement packets to come back to the receiver. For each new acknowledgement packet received by the sender, the window slides one packet to the right, allowing the sending of a new packet, hence the name "sliding window." The key here is that we are limiting the maximum packets that can be sent without ensuring that everything went OK with the transmission. A diagram illustrating the sliding-window operation is shown in Figure 7.10.

Going back to Chapter 3, TCP's sliding window is similar to Netflix's DVD rental policy. They allow you to have a fixed number of movies out at a given time, depending on how much you pay per month. The more you pay, the more outstanding discs you can have, and hence your

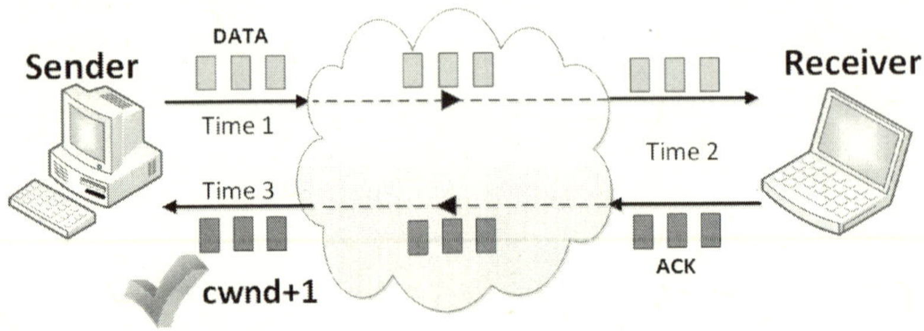

Figure 7.11 Here, we show the process of increasing the congestion window size additively. The arrows indicate packet transmission over time. At Time 1, `cwnd` is three, so the sender is allowed to transmit three packets over the network, and he does. At Time 2, the receiver gets each of the three packets one by one, and sends out acknowledgements one by one. At Time 3, the sender receives three acknowledgements, and as such, `cwnd` is increased by one, to four.

"window" is larger. If you're at your maximum allowance, you have to return one before taking another out.

Cautious growth of window size

The congestion window size doesn't stay constant. Each sender needs to change this value depending on whether they sense congestion or not.

If there's no congestion, `cwnd` should be allowed to grow, in order to efficiently utilize the network's available supply. Otherwise, the network is *underutilized.* But *increasing* the `cwnd` value is different than *sliding* the window: we are saying that it gets larger *in addition to* moving forward.

TCP increases `cwnd` linearly. Specifically, the goal is to have it grow by 1 if all the currently outstanding packets are properly received. So if the window size is three, once the sender receives acknowledgements for three packets, the window will increase to four. Then, once four more packets are acknowledged, it will increase to five, and so on. This process is shown in Figure 7.11, where we illustrate the path of three packets, and then their acknowledgements, over time.

The way we accomplish this linear increase is by having `cwnd` increase by 1/`cwnd` for each acknowledgement that comes in. To see that this is equivalent, note that if `cwnd` is 3, then we have three outstanding packets. For each one that comes back, it will increase by 1/3: from 9/3 to 10/3, 11/3, and finally 12/3, or 4. We can't send a fraction of

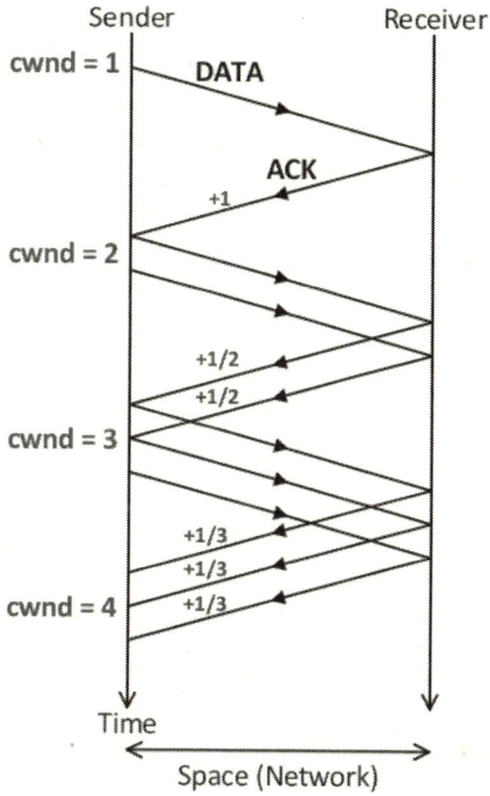

Figure 7.12 A space-time diagram of TCP packets being sent and acknowledged. Space is the network in between the two end-hosts. The window starts at 1, and when this packet is acknowledged, the sender increases it to 2. When these two packets are acknowledged, it goes up to 3, and so on. This illustrates the additive-increase feature of TCP windowing when there's no congestion. We also show how much each acknowledgement adds to `cwnd`, *i.e.*, 1/`cwnd`.

a packet, so we don't see the effect of the increase until it reaches 4. Then, we are allowed 4 outstanding packets, and for each that comes back, `cwnd` will increase by 1/4: from 16/4 to 17/4, 18/4, 19/4, and finally 20/4, or 5. We illustrate this process in the space-time diagram in Figure 7.12, where space is the network between the two end-hosts.

On the other hand, if there is congestion, we have to reduce `cwnd` to alleviate it. Otherwise, the network is *overutilized.* But rather than decreasing it linearly, TCP says it should decrease *multiplicatively.* Specifically, it should be cut in half for the next sequence of packets. So if the window is 8 right now, and we detect congestion, it should be 4 next. And then, if we still sense it, we reduce it to 2, and so on.

Increasing `cwnd` additively and decreasing it multiplicatively means that the control of packet injection into the network is conservative. It would have been much more aggressive if it were the other way around: multiplicative increase and additive decrease.

We can draw another analogy here to DVD rentals (not Netflix in this case). Suppose you are a new customer, and the company you are renting from doesn't really know you. The first time you shop with them, they allow you only to take one movie out. When you return this one, they increase your allowance to two. When you bring these two back, they allow you to take three, and so on. They are gradually rewarding you for being a reliable customer. This is the idea of increasing the window size in TCP. Now you're up to four movies, and say you fail to return them. As a means of caution (and also to punish you), they cut your allowance in half, back to two again. Hence, they have decreased your "window" size.

It's worth mentioning that when a connection is first established, TCP doesn't follow the additive increase behavior, because it wants to get `cwnd` up to a reasonable value first. This is called the **slow start** phase, which probably should be called the "fast-start" phase since we ramp up `cwnd` quickly. What we have described here is how TCP behaves in the **congestion avoidance** phase, which comes after slow-start.

Infering congestion

As we said, the TCP window is changed as the congestion conditions in the network change. But how do we know whether there is congestion or not?

If you're running a TCP connection on your iPhone, you really have no idea what the network looks like. You have no idea what path your packets are taking, which other end-hosts share links with you, or which links along the path are congested. We illustrate this problem in Figure 7.13. You only know what's going on around you, and yet you have to make an educated guess: is your connection experiencing congestion somewhere in the network or not?

The early versions of TCP made an important assumption: if there's packet loss, then there's congestion. This sounds pretty reasonable, but how do we determine whether there's a packet loss? Well, as we said, the network provides us feedback in terms of acknowledgement packets. Recall that a successful ACK means the packet was delivered. TCP

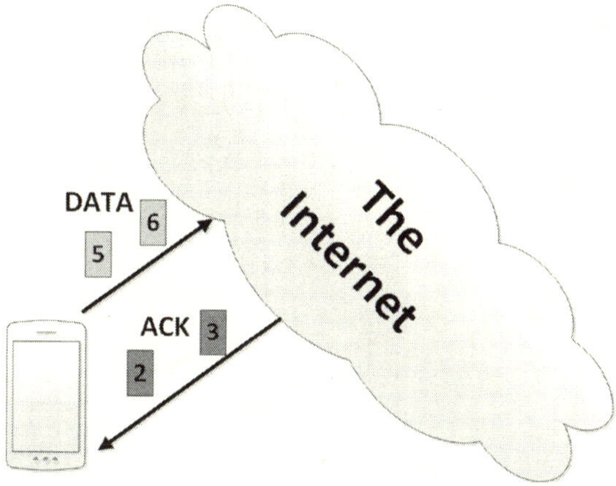

Figure 7.13 As an end-host, you have very little information about what is going on inside the network. You have no idea what paths your packets take to get to the destination, what the congestion conditions on these individual paths look like, or anything else. All you know is what packets you sent, and which acknowledgements came back. How, then, do we infer congestion?

uses two common sense approximations to make a reasonable guess as to whether the packet has been lost or not.

1. If the sender waits a long time and the acknowledgement for a packet doesn't come back, the packet was probably lost.

So how long is a "long time?" TCP keeps what is known as a timeout counter, which is based on the normal **round trip time** (RTT) between sender and receiver. The RTT is the time it takes for the packet to reach the receiver, plus the time it takes for the acknowledgement to get back to the sender. The timeout counter could be, say, three times the normal RTT.

What is the "normal" RTT? It's the RTT for a packet that does not experience congestion. And how is it determined? Each packet transmitted by the sender is timestamped, and the receiver indicates this value in the acknowledgement packet that gets sent back. Using this, the sender can figure out the RTT of each packet that he receives an acknowledgement for. Roughly speaking, the *minimum* of the recent RTT values can be taken as the no-congestion, "normal" RTT.

When a packet has been outstanding for longer than the timeout counter, it's probably reasonable for TCP to assume that a loss has occured. We have to wait a sufficient amount of time just in case the

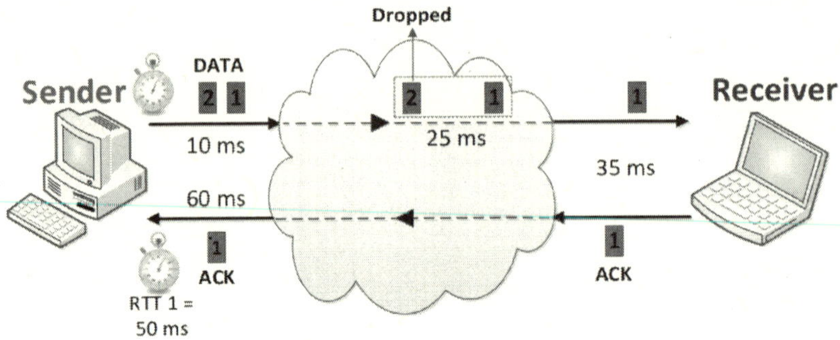

Figure 7.14 Here, we show how RTT values are found and how they are used to infer packet loss. The sender and receiver establish a TCP connection, and after 10 ms (arbitrary) the sender transmits two packets. Along the way, at 25 ms, packet 2 gets delayed, and ultimately dropped. Packet 1, on the other hand, makes it to the destination, and the sender gets an acknowledgement at 60 ms. The RTT for packet 1 is the difference between the transmit and acknowledgement time: $60 - 10 = 50$ms. The sender uses the average of this RTT and some others recently observed as the "normal" RTT. After maybe three times this, packet 2 will time out, and the sender will assume it was lost. We are conservative in defining this timeout counter just in case packets are delayed, rather than lost.

packet was delayed along the way. We illustrate how the sender times this to make a reasonable assumption in Figure 7.14.

Now, onto the second assumption.

2. If we have received acknowledgements for several packets that were sent *after* the one we are waiting for, the packet was probably lost.

TCP keeps track of this by assigning each packet it sends a sequence number. This way, it knows packet 1 was sent first, 2 was second, and so on.

Suppose the sender is waiting for an acknowledgement for packet 9. Since the window size was large enough, it also sent out a number of subsequent packets. It hears from the receiver that several, say three, of these later packets (numbered 10, 11, and 12) have been received. But what about packet 9? Still waiting, no acknowledgement. This means that it has probably been lost. We illustrate this in Figure 7.15.

Of course, packet 9 may have traversed a different path with a longer RTT (we discussed so-called multipath routing briefly in the last section on the network layer). But if as many as three later packets have already arrived, chances are that it is not just late, but lost.

Starting with TCP Tahoe in 1988 and its slightly modified variant

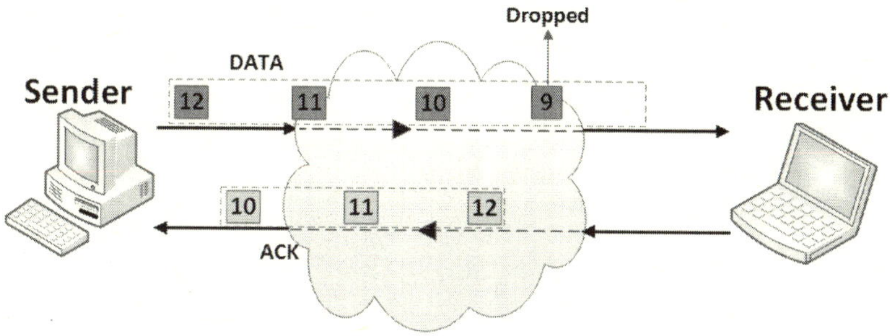

Figure 7.15 In this example, the sender has four outstanding packets: 9, 10, 11, and 12, which he sent out in sequence. Each is labeled with its corresponding sequence number. Somewhere along the way, packet 9 gets dropped, but the other three safely make it to the receiver and get acknowledged. The sender gets ACK messages for 10, 11, and then 12, but still nothing for 9. It's highly unlikely that packet 9 was delayed this long, and as a result, it's safe to assume that it was dropped.

TCP Reno in 1990, such congestion control mechanisms based on *lost* packets was standard for almost a decade. And loss-based congestion control certainly sounds reasonable enough. But there are a few problems with this.

- Sometimes, packet loss is caused by *poor channel quality*, rather than congestion.
- It's often too late to react to congestion by the time packets are being dropped.

The first problem is especially true when you're using wireless, like cellular or WiFi. As we discussed in Chapter 1, we see interference due to indirect channel gains and receiver noise in CDMA, and due to timing collisions in WiFi. Both of these can cause significant packet losses. To deal with this, there have been many proposals for "TCP Wireless."

The second problem is largely solved by switching from packet *loss* to packet *delay* as the congestion feedback signal. The first method that used this strategy was **TCP Vegas**, which came out in 1995. The benefit of using delay as the feedback signal is that it's not "all-or-nothing." With loss-based feedback, there are two possibilities: either packets are being dropped (meaning there is congestion), or they are not being dropped (meaning there is not congestion). If we look at packet delay instead, we can ask ourselves *how much* delay there is. Maybe the delay is a little, so we drop the window size a little. Maybe

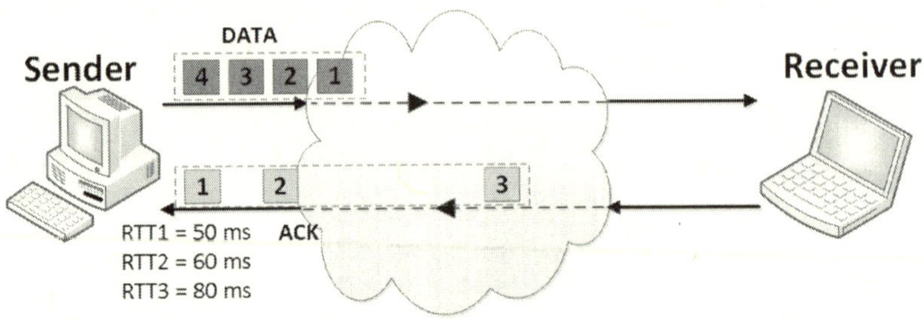

Figure 7.16 Illustration of the benefit of using delay rather than loss as the signal for congestion. Here, the sender has transmitted four packets to the receiver, close in duration to one another. The acknowledgements, however, show increasing delay in when these packets are received: first 50, then 60, and then 80 ms RTT values. This is likely due to the fact that congestion is increasing. Furthermore, the fourth packet has been lost altogether. With delay-based signaling, we would respond to this as we saw the RTTs increasing, with more severe decreases in `cwnd` as the RTTs increased further. Loss-based, on the other hand, would wait until the fourth packet is lost, which may already be too late to prevent further congestion.

it's a lot, so we drop it a lot. Or maybe there's *less delay* than we expected, so we bump the window size up. We can quantify delay by comparing the RTT value of the incoming packet to that of the "normal RTT," as we just looked at.

With delay-based feedback, we can respond to differing *degrees* of congestion, rather than waiting until it's so much that packets are being lost. We illustrate this in Figure 7.16.

Beyond TCP Vegas, there have been many other variants both proposed and implemented over the past two decades. For instance, **FAST TCP** in 2002 helped to stabilize congestion control. And **CUBIC** in 2005 combined both loss- and delay-based congestion signals, and is now the default TCP mechanism for Linux.

As mentioned in the last chapter, TCP/IP is the "thin waist" of the Internet layered protocol stack. It glues the functional modules below it, like the physical and link layers, to those above it, like the application layer. As part of that thin waist, the above elements of congestion control design in TCP led to a great success. The wonderful fact that the Internet has not collapsed, despite the incredible and unstoppable surge of demand, is partially attributable to its congestion control capability.

Now, let's look at loss- and congestion-based feedback in a little more detail.

7.2.2 Loss-based congestion inference

If you think about it, for end-to-end congestion control without any message passing from the network, an end-host really has very little to work with. Estimates of packet loss and calculations of packet delay are pretty much the only two pieces of information it can obtain through time stamping and numbering the packets.

TCP Reno, one of the main TCP variants, uses loss-based congestion control. In fact, it is widely in use today for the Windows operating system. Let's take a look at it.

The main operation of TCP Reno is quite simple. For each round trip time, we ask one question: have all packets been received properly?

- If they have been, then increase `cwnd` by 1.
- If they haven't been, then cut `cwnd` in half, *e.g.* from `cwnd` to `cwnd`/2.

We define being received "properly" according to the two congestion inference mechanisms we discussed in the last section: the packet is received (1) within a reasonable amount of time, and (2) reasonably in sequence. Hence, the sender can estimate that a packet *wasn't* received properly if either (1) a certain number of RTTs have passed (*i.e.* two or three) or (2) a certain number of future packets have been successfuly acknowledged (*i.e.* three or four) without this packet having yet been acknowledged. There are other, more subtle features of TCP as well that we will not have time to get into.

Let's look at an example. For simplicity, we assume that the RTT

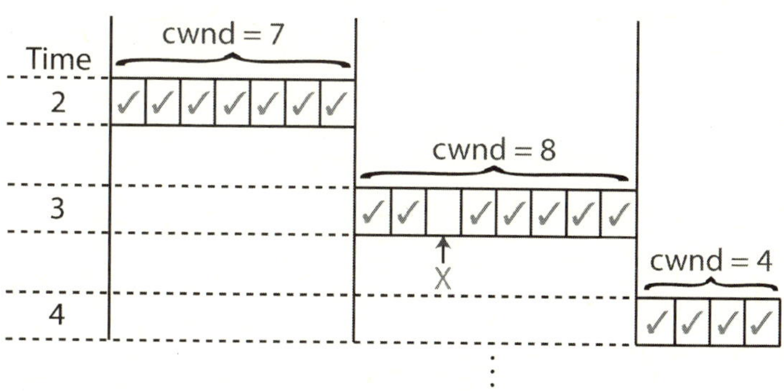

Figure 7.17 In our example, at Time 3 `cwnd` is increased to eight and the sender transmits eight packets. But one of these is detected as lost since too many future packets are acknowledged. So at Time 4, `cwnd` is reduced by a factor of two, to four.

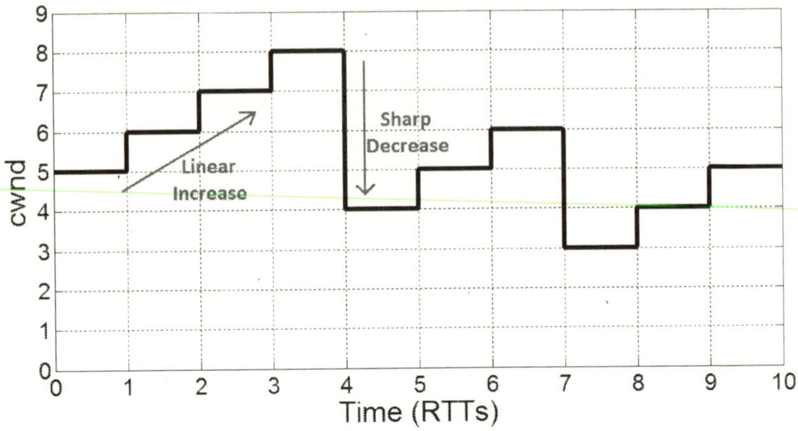

Figure 7.18 An example of `cwnd` evolution over time for TCP Reno. Each RTT, the window either increases by 1 or gets cut in half.

is constant, and define one unit of time as one RTT. In reality, the RTT is about 50 ms across the US and will vary as the congestion conditions change. When we say Time 1, it means that exactly one RTT has passed, Time 2 means exactly two have passed, and so on.

Assume `cwnd` is initially 5. All packets will be properly received during each RTT, except for numbers 4 and 7.

- Initially (at Time 0), `cwnd` = 5, so the sender sends five packets and pauses.
- At Time 1 (after 1 RTT), the sender has received five ACKs, so it does two things: it (1) slides the congestion window by five packets, and (2) increases `cwnd` by 1. Then, it sends six packets.
- At Time 2, the sender has received six ACKs. Again, it slides and increments `cwnd`, and then sends seven packets.
- At Time 3, the sender has received seven ACKs,. so it sends eight packets.
- At Time 4, the sender detects a lost packet. Too many additional packets were acknowledged and this one still hasn't come in. It thus halves `cwnd` to four, and sends four packets. We illustrate this in Figure 7.17.
- At Time 5, the sender has received ACKs for these four packets, with no losses. So it slides and increments `cwnd`, and then sends five packets.
- At Time 6, the sender has received five ACKs, so it sends six packets.

- At Time 7, the sender detects a lost packet. It halves `cwnd` to three and sends three packets.
- At Time 8, the sender has received three ACKs, so it sends four packets.
- At Time 9, the sender has received four ACKs, so it sends five packets.

Figure 7.18 shows these values of `cwnd` over time. When there is no packet loss, it grows linearly, increasing by one each RTT. When packet loss occurs (Time 4 and 7), it decreases sharply.

7.2.3 Delay-based congestion inference

Now we turn to delay-based congestion control, like TCP Vegas.

As we said, we measure delay using the round trip time. We can see two main components of the RTT by thinking about what would cause a packet to be delayed.

1. Even if the network was perfectly ideal, we are still limited by the nature of physics. Our packet has to get from point A to point B, and in doing so, it has to traverse some physical distance. And it's impossible to travel faster than the speed of light. What we are describing here is the **propagation delay**, the time it takes to just go through the links.
2. When a packet reaches a router, it spends time in the router's buffer before the output link becomes available. The more congestion there is, the more packets are waiting for the same output link, and the more time each one has to spend in the buffer (think of the traffic jam analogy again here). This is called the **queueing delay**: the time a packet spends waiting in the queue (i.e., in line) due to congestion.

We illustrate each of these delays in Figure 7.19. The queueing delay is really what we're interested in here, since that's the one that depends on congestion. The heavier the congestion, the longer the wait.

So how do we quantify queueing delay? For each packet, we need a way to measure how big (or small) its RTT is. After all, the higher it is, the higher the queueing delay.

Recall our discussion of the "no-congestion" RTT, which is the minimum of the recent RTT values that the sender has observed. This tells the sender what the delay value should be if there is (almost) no congestion. By comparing the current RTT, say `curRTT`, to the minimum RTT, say `minRTT`, he can get a sense for how much congestion there

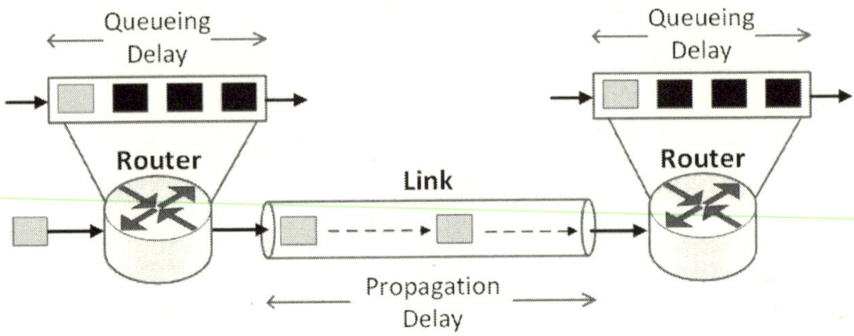

Figure 7.19 The two main delays we experience are propagation delay and queueing delay. Propagation delay is due to the fact that packets have to traverse physical distance over links to get from point to point. Queueing delay is a measure of how long packets have to wait "in line" at the routers, and is highly dependent upon congestion conditions. The higher the congestion, the more packets queued, and the longer the wait.

currently is in the network: the higher `curRTT` − `minRTT` is, the worse it is, and the lower, the better.

Rather than comparing *delays*, TCP Vegas compares *transmission rates*. That is, it looks at how many packets per second we are sending under current congestion conditions, and compares that to the case of no congestion.

The current congestion window size, `cwnd`, tells us how many packets we sent this RTT (*i.e.* packets/RTT). And the RTT itself tells us how many seconds it took to make one round trip (*i.e.* seconds/RTT). Therefore, dividing `cwnd` by RTT gives us (packets/RTT)/(seconds/RTT), or packets/second. So the current transmission rate is:

$$\frac{\texttt{cwnd}}{\texttt{curRTT}},$$

and the transmission rate under (almost) no congestion is:

$$\frac{\texttt{cwnd}}{\texttt{minRTT}}.$$

The main operation of TCP Vegas is as follows. Upon receiving *each acknowledgement*, the sender subtracts `cwnd/curRTT` from `cwnd/minRTT`. This is how much the current transmission rate (in packets per second) is lagging behind what it would be without congestion. If this difference is smaller than some threshold, say 3, then congestion isn't hurting us much. If it's larger, then our speed is lower than it should be. So, it

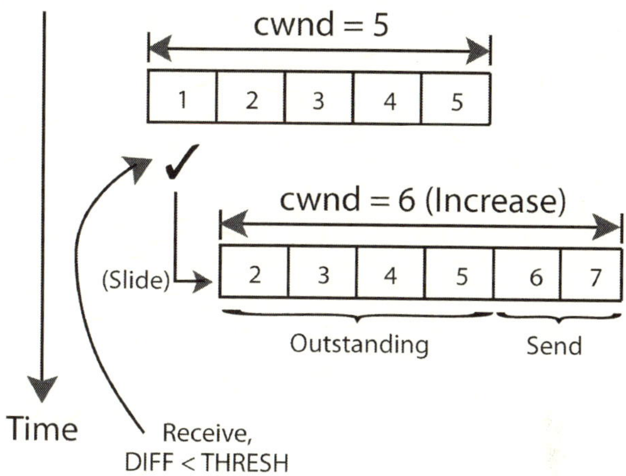

Figure 7.20 In our example, initially the congestion window is set to five, so the transmitter sends five packets. The first one arrives at 10 ms with a delay *less than* the threshold, so `cwnd` is increased by one to six. As a result, the sender transmits *two* more packets, to have a total of six outstanding.

asks the following question: how does the *difference* compare with the *threshold*?

- If it's smaller, `cwnd` is increased by 1.
- If it's larger, `cwnd` is decreased by 1.
- If it's the same, `cwnd` stays the same.

Let's illustrate this with an example. In doing so, we will assume that the minimum RTT never changes, and is fixed at 50 ms. In reality, this is not the case, and the sender will dynamically adjust this minimum value based on previous measurements. Initialize `cwnd` to 5, as before.

- Initially, the transmitter has sent out five packets, and is awaiting acknowledgements.
- At 10 ms, the first packet comes back, and the RTT is found to be 51 ms. Therefore,

$$\frac{\texttt{cwnd}}{\texttt{curRTT}} = \frac{5 \text{ packets}}{51 \times 10^{-3} \text{ sec}} = 98.03 \text{ packets/sec},$$

and

$$\frac{\texttt{cwnd}}{\texttt{minRTT}} = \frac{5 \text{ packets}}{50 \times 10^{-3} \text{ sec}} = 100 \text{ packets/sec}.$$

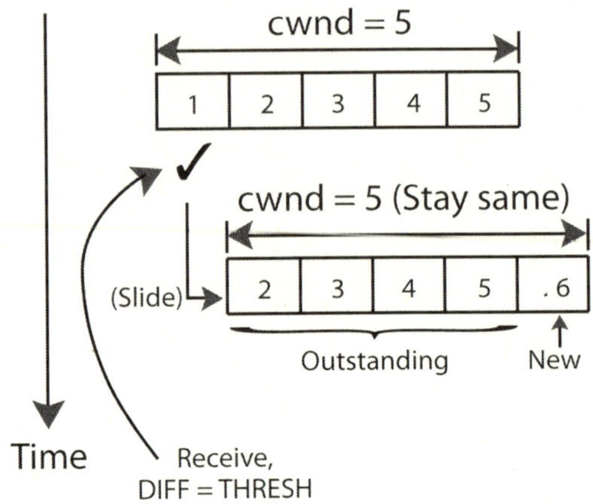

Figure 7.21 Suppose the congestion window is set at five, as in Figure 7.20. If a packet is received with a delay *equal* to the threshold, then `cwnd` stays the same. Hence, only one more packet will be sent to keep the same number of outstanding packets.

The difference is $100 - 98.03 = 1.97$ packets/sec, which is less than 3. So, we increase `cwnd` by one, to 6. Since we now have *four* outstanding packets, the sender sends *two* more, for a total of six. We illustrate this in Figure 7.20.

- At 20 ms, the second packet comes back, and the RTT is found to be 50.5 ms. So,

$$\frac{\texttt{cwnd}}{\texttt{curRTT}} = \frac{6 \text{ packets}}{50.5 \times 10^{-3} \text{ sec}} = 118.81 \text{ packets/sec},$$

and

$$\frac{\texttt{cwnd}}{\texttt{minRTT}} = \frac{6 \text{ packets}}{50 \times 10^{-3} \text{ sec}} = 120 \text{ packets/sec}.$$

The difference is $120 - 118.81 = 1.19$ packets/sec, which is less than 3. So, we increase `cwnd` by one, to 7. Since we now have *five* outstanding packets, we again send two more, for a total of seven.

- At 30 and 40 ms, we receive the third and fourth packets, who have RTTs that again tell us to increase `cwnd`. At 40 ms, it is up to 9.
- At 50 ms, we get an acknowledgement for the fifth packet, with an RTT of 50.8 ms. This makes

$$\frac{\texttt{cwnd}}{\texttt{curRTT}} = \frac{9}{50.8 \times 10^{-3}} \approx 177 \text{ packets/sec},$$

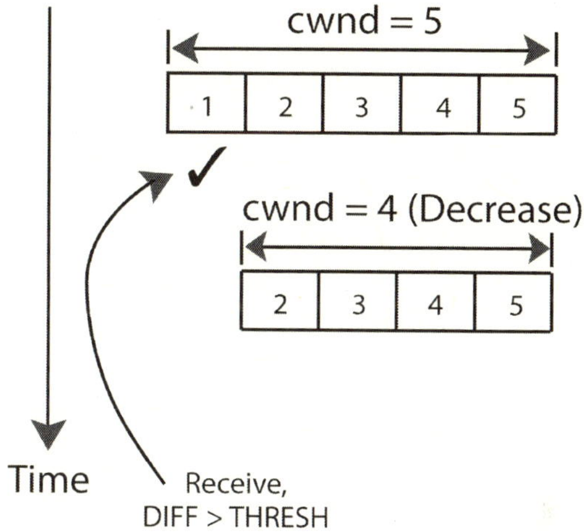

Figure 7.22 Suppose the congestion window is set at five, as in Figure 7.20. If a packet is received with a delay equal *greater than* the threshold, then `cwnd` is decreased by one. Hence, no additional packets are sent.

and meanwhile,

$$\frac{\texttt{cwnd}}{\texttt{minRTT}} = \frac{9}{50 \times 10^{-3}} = 180 \text{ packets/sec.}$$

This difference is roughly three, and as such, we *do not change* `cwnd`. It stays at 9. We send out *one* packet, for a total of 9 outstanding. In Figure 7.21, we show what happens visually when the difference is the same as the threshold (note the `cwnd` value doesn't correspond to this case).

- At 60 ms, the sixth packet comes back, with an RTT of 51.8 ms. Thus,

$$\frac{\texttt{cwnd}}{\texttt{curRTT}} = \frac{9}{51.2 \times 10^{-3}} = 173.7 \text{ packets/sec,}$$

while from last time, the speed with no congestion is 180 packets/sec. The difference is 180 − 173.7 = 6.3 packets/sec, which is *greater* than 3. As such, we *decrease* `cwnd` to 8. Since we have 8 outstanding packets, we don't send any more at this time.

- At 70 and 80 ms, we get the seventh and eighth packets back. Witnessing that we are again above the threshold, we continue to de-

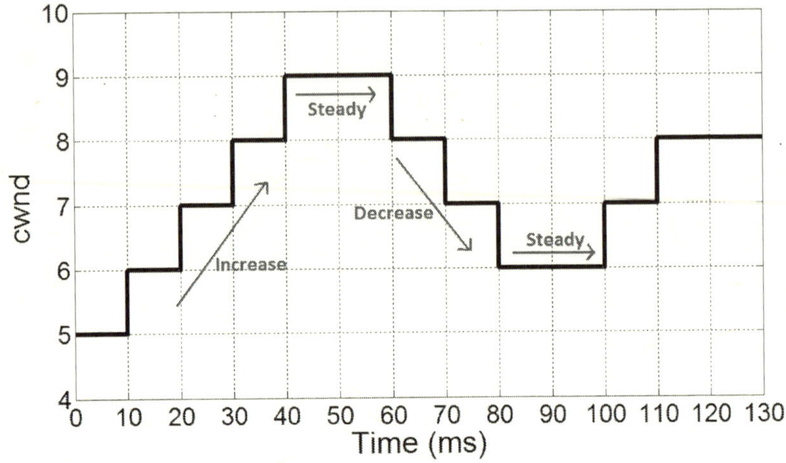

Figure 7.23 A view of the congestion window changing over time for TCP Vegas.

crease `cwnd`, first to 7, and then to 6. We send no new packets during this time.

- At 90 ms, we get an acknowledgement for the ninth packet, with an RTT of 51.3 ms. So,

$$\frac{\texttt{cwnd}}{\texttt{curRTT}} = \frac{6}{51.3 \times 10^{-3}} \approx 117 \text{ packets/sec},$$

and meanwhile,

$$\frac{\texttt{cwnd}}{\texttt{minRTT}} = \frac{6}{50 \times 10^{-3}} = 120 \text{ packets/sec}.$$

Once again, the difference is exactly the threshold. So `cwnd` stays constant at 6, and we send one more packet for a total of six outstanding.

- At 100 and 110 ms, the 10th and 11th packets come back, and the difference is lower than the threshold. `cwnd` is increased to 7 and 8, respectively.
- At 120 and 130 ms, the difference is exactly the threshold, so we keep `cwnd` constant at 8.

You can probably see a pattern here, of rise, even, fall, even, rise, and so on. We show this evolution of our example in Figure 7.23.

Indeed, this is the behavior we would expect for delay-based congestion control. If we're underutilizing the network, the RTT is too

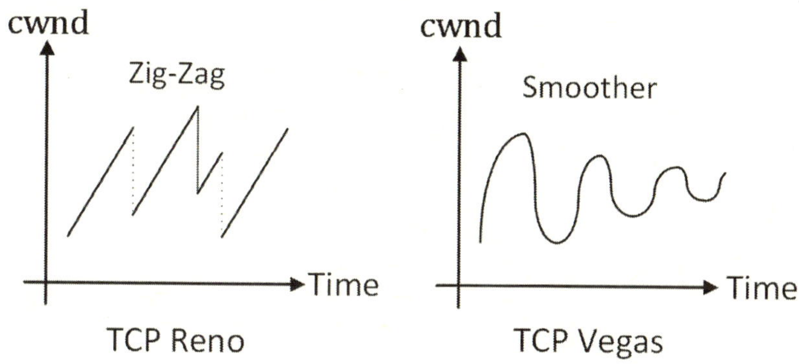

Figure 7.24 Typical evolutions of `cwnd` values in TCP Reno on the left and in TCP Vegas on the right. TCP Reno uses loss as the congestion signal whereas TCP Vegas uses delay as the congestion signal. The zig-zags between overshooting and under-utilizing capacity tend to be smaller in Vegas (if the algorithm is fine-tuned properly).

fast, indicating we could be sending more. At the same time, if we're overutilitzing the network, the RTT is way too slow, indicating that we should send less. And if the difference is the same as the threshold, we won't change `cwnd` at all.

Why don't we increase or decrease `cwnd` when we're at the threshold? Because the speed is at a *reasonable* level. We don't want to increase `cwnd` for fear that we will cause more delays to occur, and we don't want to reduce it because we wouldn't be sending as much as we could be each RTT. We're at a good place. Remember, the goal is *not* to get the current rate to be equal to the rate under no congestion. That would mean we had absolutely no queueing delay. There are other people sharing the network with us, and we have to "play nice" with them. Each of us has to accept a *reasonable* amount below the no-congestion case.

We can make a comparison between this and power control that we looked at all the way back in Chapter 1: if we can find a point at which everyone is achieving this acceptable transmission speed, then we will have reached an *equilibrium*. Just as we showed that distributed power control could find equilibrium power levels for everyone to be satisfied with while sharing the air, distributed congestion control can find equilibrium transmission levels to do the same. At the equilibrium, everyone stops changing their window size simultaneously. Unfortunately, we will not have time here to get into the mathematics behind achieving sending-rate equilibria over multiple users.

Zooming out, Figure 7.24 shows a typical evolutions of TCP Reno and Vegas' congestion window sizes over time. You can see that TCP Reno, which uses loss-based signaling, is constantly ramping up to a rate that is too aggressive, leading to congestion, and then cutting back multiplicatively to a much lower level. This "zig-zag" between a rate that is too aggressive and one that is overly conservative is generally reduced when we use TCP Vegas, a delay-based TCP variant. While loss is a simple "yes or no" signal, delay generally allows us to react to congestion quicker and smoother, as long as the algorithm is tuned properly.

Recap .

Over the past few sections, we've walked through some remarkable intellectual landscape behind the design of the Internet. We started briefly with a historical perspective, which introduced us to the fundamental concepts behind it: packet switching, among others. We also described in detail two layers of the protocol stack: the network layer, a *connectionless* module which handles hop-by-hop routing decisions, and the transport layer, a *connection-oriented* module which deals with end-to-end functions like congestion control. Our goal was not to have you master the engineering detail or mathematical models: that would likely require multiple college-level courses, going into high detail on each of the layers shown in Figure 7.1. Rather, we hope you understand the underlying principles behind routing, congestion control, and how it's possible to support the billions and billions of devices that are connected to this ever-increasing increasing network of networks.

7.3 Multicasting peers

We just discussed the notion of end-to-end connectivity at the transport layer and how that relies on hop-by-hop routing from the layer below. These two ideas represent the basic "building blocks" of the Internet and how we communicate with one another over it.

The idea of a "building block" is important here. Over the years, people have decided to take advantage of the TCP/IP foundation to create diverse services *on top* of the Internet. These are the **applications** that we actually see as users sitting on the endpoints of the Internet. Emailing, web browsing, and more recently, file sharing, are

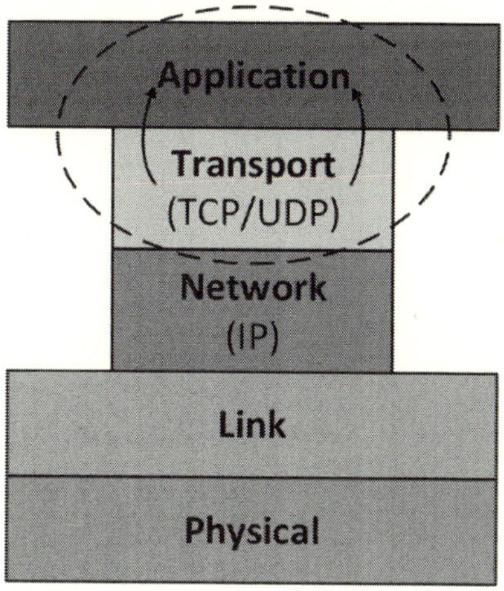

Figure 7.25 The last section focused on the transport layer. In this one, we will move up to the application layer and discuss P2P.

all examples of applications, and all have their own methods which they use to run. They lie at the fifth layer of the protocol stack.

In this section, we are going to "move up" the stack once more, as shown in Figure 7.25. Specifically, we will focus on peer-to-peer (**P2P**) applications that illustrate some key principles of networks scaling up. Two main examples of this are file sharing (like BitTorrent) and video chatting (like how Skype used to work). In exploring P2P, we will see how important the idea of "network on network" really is.

7.3.1 P2P basics

First, we will illustrate the differences between peer-to-peer and how the Internet was originally planned. In doing so, a simple analogy will aid our understanding.

Client-server

Suppose you're working for a company, doing some engineering work. As part of your job, you are required to submit reports for each of the projects you are working on. You have three different supervisors, Allie, Ben, and Charlie, each of whom is in charge of one of the projects that has been assigned to you.

One day, you get an email from each of your advisers asking you for a project update. Your task is to write three different reports, one for

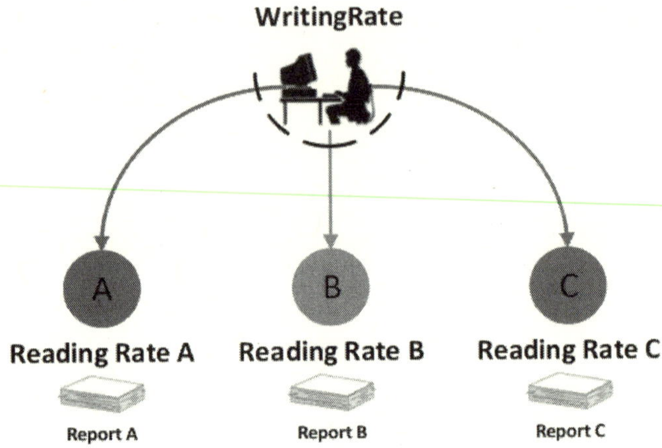

Figure 7.26 In this scenario, you have three supervisors who have all requested reports from you. They have each told you to send portions of the reports as you finish them, so they can proofread as you go along.

each adviser. Each has to proofread the corresponding report before sending it up the corporate ladder as required.

Since the deadline is approaching very rapidly, Allie, Ben, and Charlie want to make the process go as quickly as possible. Each time you finish a section, they ask you to send it off, so they can proofread as you're writing the next section. So you write a section, send it to Charlie, write another, send it to Ben, write another, send it to Allie, and this way, they can proofread while you write the next batch. Hence, the writing and proofreading processes are going on simultaneously.

We show this scenario in Figure 7.26. Clearly, how long this process will take depends on a few things:

1. How long each report is. The shorter the report, the faster you can write it (within reason), and the faster they can proofread it.
2. How fast you can write a section of a given length.
3. How fast each of them can proofread a section of a given length.

Let's see if we can figure out the total time it will take to get the report out. To do so, we need to figure out how the above three are related.

- *Writing the reports*: You have to write three reports. Let's assume that each has the same number of words, and that you write the same number of words per minute throughout. With the report length in

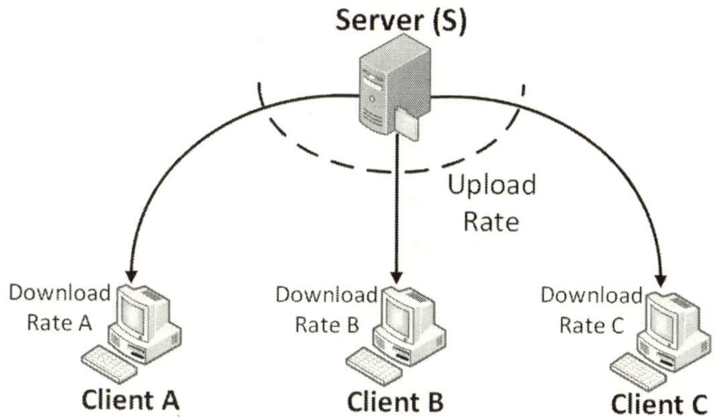

Figure 7.27 Illustration of the client-server Internet architecture. The server has a given upload bandwidth that it splits equally among the clients that are requesting from it. At the same time, the clients have download rates, which is a measure of how quickly they can receive the content. Here, we show three clients (A - C) making a request from a server for some file.

words, and the writing rate in words/second, we can divide the two to give us the time it takes to write them (in seconds):

$$\text{Time to write} = \frac{3 \times \text{ReportLength}}{\text{WritingRate}}.$$

- *Proofreading the reports*: Once a supervisor gets a section, he/she has to proofread it. But there's one person for each report, so we only have to look at the *slowest* person, meaning the one that takes the longest to proofread. Maybe this person is Ben. If Ben proofreads some number of words/second, then the time it takes for the proofreading to finish is given by:

$$\text{Time to proofread} = \frac{\text{ReportLength}}{\text{Ben'sReadingRate}}.$$

- *Total time*: If we first wrote all the reports and then sent them out to be proofread, the total time would be the sum of the writing and proofreading times. But as we said, we are going section by section, so writing and proofreading are happening *simultaneously*. Therefore, the total time is the *larger* of the two, or whichever one will finish last.

This situation is analogous to the **client-server** model for transferring content through the Internet. In this model, we have a bunch of

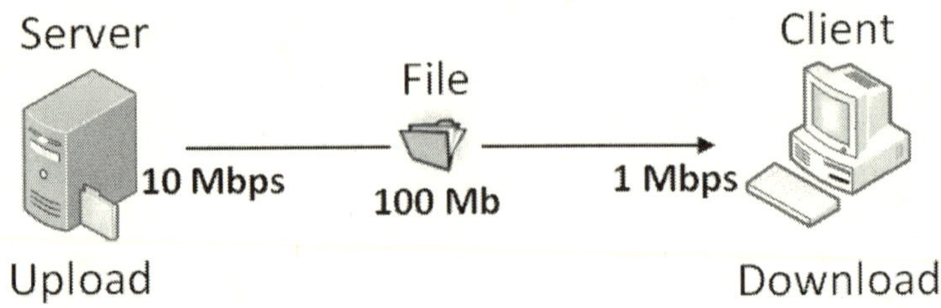

Figure 7.28 Suppose you are requesting a file from a server. The server is dedicating an upload rate of 10 Mbps to you, and your download rate is 1 Mbps. Since the download is slower than the upload, we are limited to 1 Mbps for the session. Hence, to transfer a 100 Mb file, it will take 100Mb/1 Mbps = 100 s.

client machines, like your desktop, laptop, smartphone, etc., which request some piece of content from a central server, as shown in Figure 7.27. Where this server is depends on what website you are downloading from (*e.g.* Google, YouTube), as different companies own different servers. The server has to *upload* the file to each of the clients, which happens at a certain speed called the **upload rate**. At the same time, the clients have to *download* the file as it comes in, at a certain speed called the **download rate**. The upload and download rates are both given in bps. So our "unit" of transfer here is a bit, rather than a word.

In our analogy, the supervisors take place of the clients (requesting content), and you take place of the server (distributing content). The various reports take place of the file that is being requested by each client, which has to be replicated at each for **multicasting**. Finally, the upload rate becomes the the writing speed (per report), and the download rates becomes the proofreading speeds.

We illustrate the concept of uploading and downloading in Figure 7.28, for a server with a single client. Suppose you are downloading a file of size 100 Mb (millions of bits). Maybe the server is dedicating 10 Mbps to your session, and maybe your computer can download up to 1 Mbps. In this case, the download speed is *less* than the upload speed, so that's what limits the time it takes: your computer can't handle all that the server's offering. Hence, it would take 100 Mb/1 Mbps = 100 s to complete the transfer. On the other hand, if the download speed was larger than the upload speed, then we would be limited by what the server was pushing out, and it would take 100 Mb/10 Mbps = 10 s.

Now let's find the total time in the context of the client-server model. We have a certain number of clients (NumClients) with different download rates and a server with a given upload rate. We want to transfer a file of a certain size (in bits). The total time is given by the larger of the the upload and download times. The longest download time is limited by the client with the *slowest* download rate. It follows that the total time is the larger of these two numbers:

- *The upload time*, or

$$\frac{\text{NumClients} \times \text{FileSize}}{\text{UploadRate}}, \text{ and}$$

- *The longest download time*, or

$$\frac{\text{FileSize}}{\text{MinDownloadRate}}.$$

The client-server model is the way that the Internet was thought of traditionally. Whenever you request a file, it would come from a server somewhere, and the number of servers is *way less* than the number of clients. And indeed, that's still how a lot of Internet applications work, whether you are receiving an email (which comes from the email server) or going to a random webpage (which downloads from its web server).

But there's an intrinsic problem with the client-server model, which was not anticipated originally: What happens when the number of clients becomes *very large*?

We illustrate this in Figure 7.29. When the number of clients increases, we become limited by the upload rate of the server very quickly, as opposed to what we see in Figure 7.28 for only one client. Indeed, the server has to send out more and more copies of the file, which takes longer and longer: the server's capacity is split equally among more and more peers.

What can we do to fix this? Well, one possibility is to increase the upload rate of the server. More bps means less time to send the file. But this can become very expensive as well as difficult in terms of technology, especially if we have a very large number of clients. And naturally, the number of clients is going up as more and more people use the Internet!

You can imagine the increased strain on the server in terms of our analogy, if your company kept giving you more supervisors. If three were asking you for reports in one week, maybe you could manage. If they gave you another, perhaps you could increase your "typing speed" to send out quicker. But put this to five, six, seven reports in a week,

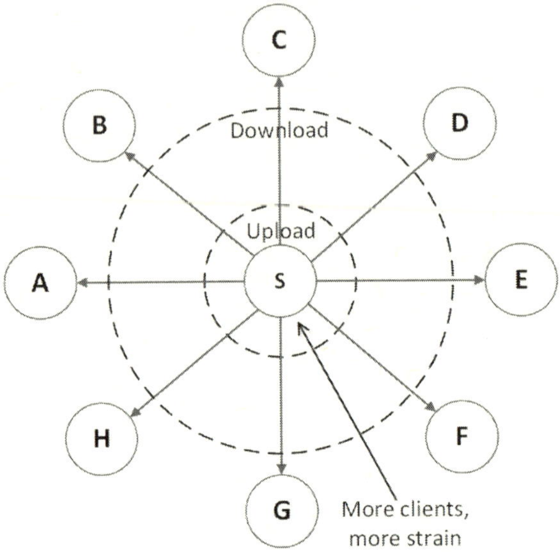

Figure 7.29 As the number of clients goes up, the server will have that much more data to send out. As a result, the upload rate that each client gets will go down. The only way we can fix this with this model is by improving the upload rate of the server, which can become very expensive, especially as the number of clients grows large.

and you would get to a point where you couldn't possibly get all of them out in time. You'd just have to tell them you needed longer.

So what's the other alternative? Can we possibly add *more servers* to fix the strain? We illustrate this in Figure 7.30. Each server could take some of the client demand (they would probably pool their resources together, so that one with more upload capacity would take more clients). But there are problems with this as well. First, it's not cheap to keep building servers (as well as all the infrastructure that comes along with it). Second, it would be hard to tell exactly when it was time to build a new server. So then, what should we do?

P2P

The client-server model works well for many applications. But as the number of Internet users goes up, there are going to be more and more clients requesting files from a server at any given time. This is especially true for downloading significantly large files.

Back to our analogy again. As a *team*, if you really wanted to get the reports done as quickly as possible, the supervisors wouldn't just

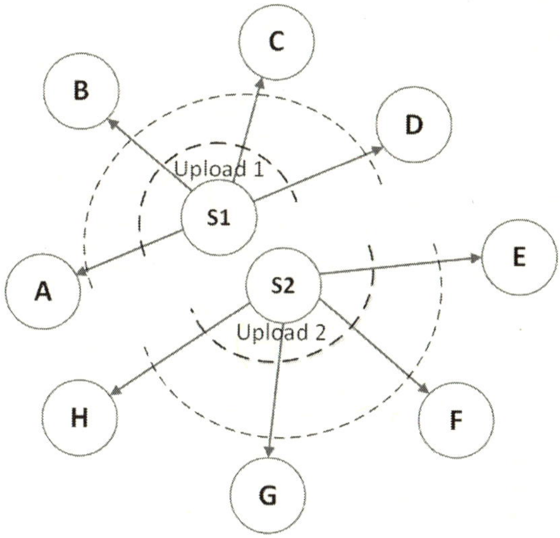

Figure 7.30 Another potential way to fix the upload capacity problem is by building more servers. Then each one can take some of the clients (or they can pool their resources together). But there are *scalability* issues here as well.

proofread the reports; they would *write* portions of them as well. You'd each be writing different sections of the reports, filling everyone's down time. And the supervisors wouldn't just focus on their reports: everyone would be helping everyone out, just in case some people work slower than others. We show this process in Figure 7.31: now, everyone is writing and reading, and in the end, three reports are generated.

So now, we have more than just *your* writing speed: we have Allie, Ben, and Charlie's as well. Four people working together to write and proofread three reports. The only problem is we have to coordinate it such that no two people write the same section; this would just be wasted writing time. But if we evenly divide the number of sections with no overlap, we'd be augmenting your writing speed with all the others. So we'd be writing three reports at the *combined* rate of You + Allie + Ben + Charlie:

$$\text{WritingTime} = \frac{3 \times \text{ReportLength}}{\text{YourWritingSpeed} + \text{Allie'sWritingSpeed} + \cdots}$$

Suddenly, this doesn't sound so bad anymore. Whenever we *add* a supervisor, naturally we have to write one more report. But the addition also brings something to the table: *additional writing power*. This notion

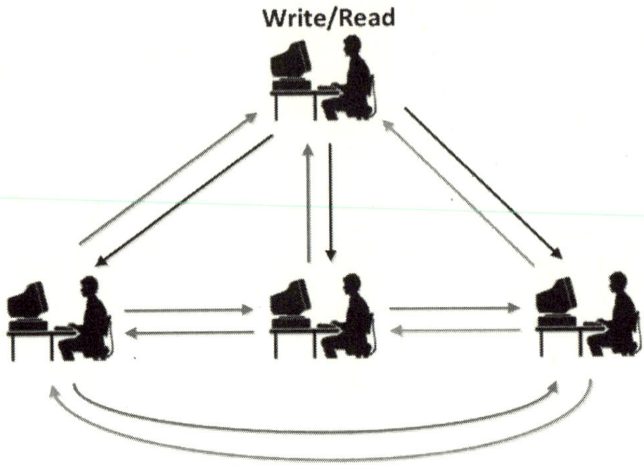

Figure 7.31 If we kept increasing the number of supervisors you had, eventually you would not be able to write fast enough to get the work in on time. The fastest way to get this done without adding any co-workers is to have the supervisors themselves write sections of the report. You can all work together to get the job done.

of adding nodes and having them help is an example of the **positive network effect** (as opposed to the negative network effect we saw, for example, with the tragedy of the commons in Chapter 5). But we could still have a problem: some people may bring very little writing speed to the table. Maybe they are busy, or maybe they are unknowledgeable. For whatever reason, this is another report your team would have to write without any extra speed.

The additional contribution we see here is analogous to the way the client-server problem is fixed in Internet applications. Rather than having our devices *only download* a file, why not have them *upload* what they have obtained to others as well? Computers are perfectly capable of doing both at the same time. When they are, we call them *peers* rather than clients.

This is the very idea behind peer-to-peer (**P2P**) networks, which are becoming more and more common, especially for file sharing. For example, consider the setup in Figure 7.32, where we have a group of peers who all want to get some file. The file originates from a server, whose job is to "seed" the file within the peer network, so they can all share among themselves. The best way for the server to do this is to give each peer a *different* portion of the file (just as we gave each person different sections to write). This way, they are all sharing unique pieces

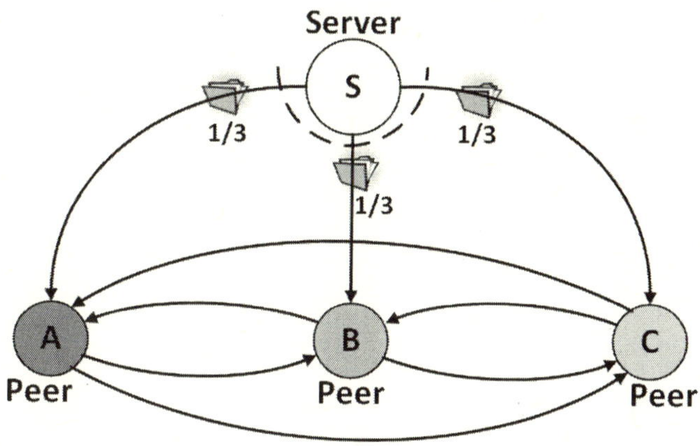

Figure 7.32 In the P2P model, we turn the *clients* into *peers* and have them both download *and* upload the content. We use the available upload capacity of these devices to help distribute the file throughout the system quicker. Here, all the peers are uploading and downloading to/from all others, though that doesn't have to be the case. Assuming all the peers have the same rates, the server will split the file evenly among them (into thirds).

with each other, and there's no overlap in what they are uploading. Of course, how much the server gives to each peer will depend on how much each can handle: those with higher upload rates will get more, and those with lower will get less (you'd want the supervisors who wrote quicker to do more sections). But if all the peers are equal (as shown here), then we split the upload evenly.

7.3.2 Examples

Let's consider two examples to see how we can quantify the benefit of P2P over client-server.

First, suppose we have a server which has a total upload capacity of 12 Mbps. The task is to distribute a file, 12 Mb in size, to each of three computers. We will assume each of the computers have the same download rates, of 24 Mbps, and the same upload rates, of 8 Mbps.

Client-server

What happens under the client-server model? The server sends the file to each computer, as shown in Figure 7.33. Since there are three of them, each gets 12 Mbps / 3 = 4 Mbps of upload capacity from the server. The total time to upload a file of 12 Mb is

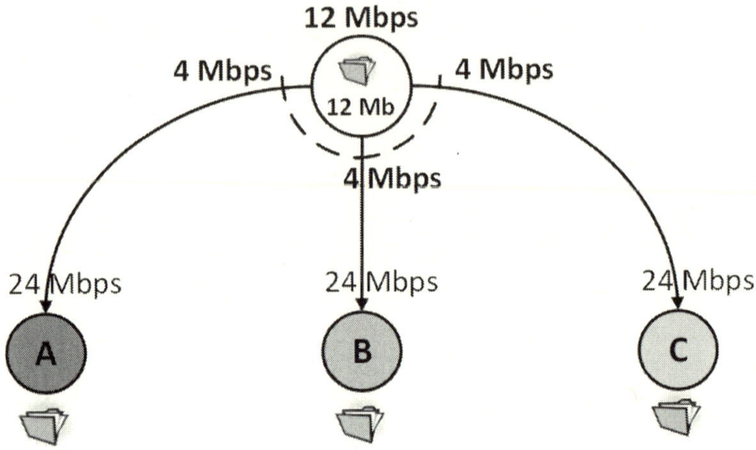

Figure 7.33 Illustration of the client-server model for our example. Since the server has an upload rate of 12 Mbps, each computer (A - C) sees 12/3 = 4 Mbps. To transfer a 12 Mb file to each of them, it takes 12 Mb x 3 / 12 Mbps = 3 s.

$$\text{Upload time} = \frac{3 \times 12 \text{ Mb}}{12 \text{ Mbps}} = 3 \text{ s},$$

and the time it takes for each computer to download a file of 12 Mb is

$$\text{Download time} = \frac{12 \text{ Mb}}{24 \text{ Mbps}} = 0.5 \text{ s}.$$

Since 3 is greater than 0.5, we are limited by the upload time. Hence, it takes 3 seconds to transfer the file using the client-server method.

P2P

Now, what about when we add peering into the mix? In this case, we don't have the server doing all the upload work. As a result, we'd like to divide the labor of transmitting the file between each of the four nodes.

The best way to think about this process is that the server has to decide how much of the file it is going to assign each peer to upload. Ignoring the download rate for a moment, this would be determined by the upload rate of each peer. The higher a peer's rate, the more the peer would be allocated to upload. But in our example, *each peer has the same capacity*, of 8 Mbps, and hence, we divide the file evenly into

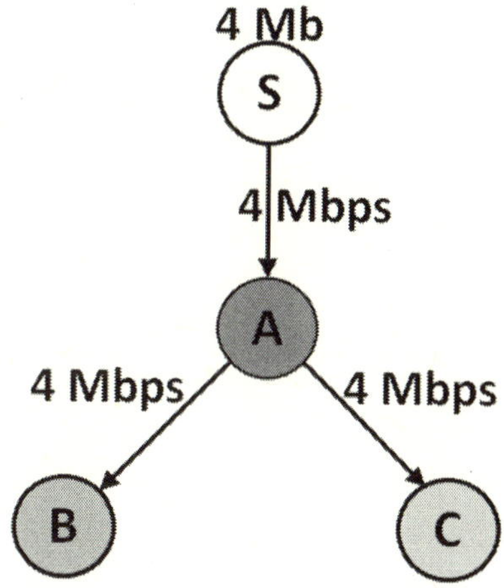

Figure 7.34 This is one of the trees we use to transfer the 12 Mb file to all the peers. This one corresponds to the first third of the file. The server dedicates one third of its capacity, 4 Mbps, to peer A. At the same time, A divides its 8 Mbps upload capacity evenly between B and C, giving 4 Mbps each.

thirds. So Peer A gets the first third, Peer B the middle, and Peer C the last.

At this point, we ask each peer to distribute their portion to the other two peers. But why wait until they have finished downloading the whole third? This should happen *simultaneously*: as soon as they start downloading it, they should be uploading it to the others, too, to make the process go as quickly as possible.

Basically, we're defining a number of **trees** that each portion of the file is traversing. We show the tree for the first third of the file in Figure 7.34. What exactly is a tree? Formally, it's a special type of undirected graph for which there is exactly one path connecting any two nodes. It gets its name because visually, it appears to be an upside-down tree, where the root is at the top and the leaves are at the bottom. In light of this, every tree has a root node (top-most) and leaf nodes (bottom-most).

In Figure 7.34, the server transmits it directly to peer A, dedicating one-third of its upload capacity (4 Mbps) to this tree. In turn, the 8 Mbps upload rate of A is split evenly between B and C, giving them 4 Mbps each. How long will this take to happen? For A, it's easy: we send a 4 Mb chunk at a rate of 4 Mbps, which will take 1 second. But

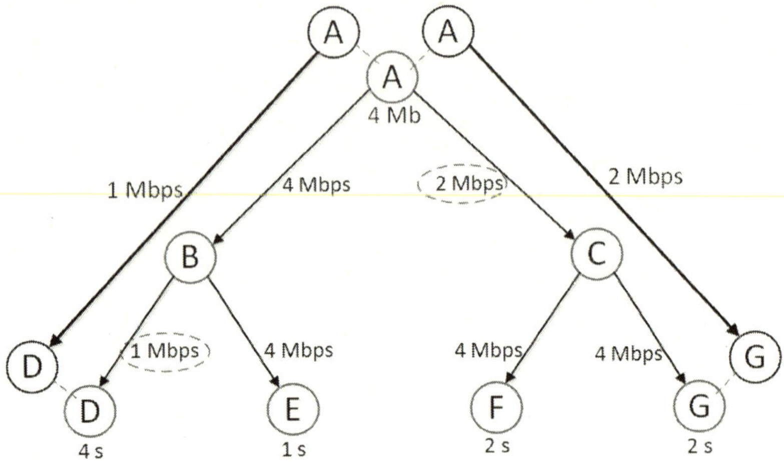

Figure 7.35 When we transmit down a tree, we start at the root node and end at the leaf nodes. How long it takes for each leaf node to get the whole file depends solely on the *weakest link* to the leaf. For D, we have links (A, B) and (B, D), where (B, D) is only 1 Mbps. This 1 Mbps limits the transmit time of a 4 Mb file to 4 seconds. For E, on the other hand, both links are 4 Mbps, so it only takes 1 s. On the other side of the tree, both nodes are limited by link (A, C), only 2 Mbps, so each takes 2 s to finish.

what about B and C? Each of them has *two* links of 4 Mbps from the source. Do we add these capacities? Take the largest/smallest?

The answer is quite intuitive: we're always limited by the *weakest link* whenever sending between two points. In Figure 7.34, all the links are the same, so 4 Mbps is the weakest. We show a more general case in Figure 7.35. In this tree, A is the root, and D, E, F, and G are the leaves. How long it will take to send to a given leaf depends solely on the *minimum* capacity assigned to any of the links on the path to the leaf. For D, we have (A, B) with 4 Mbps followed by (B, D) with 1 Mbps. So we are *bottlenecked* by 1 Mbps, and the time will be 4 Mb/1 Mbps = 4 s. We can go through a similar process for the rest.

This "weakest link" concept is very similar to what you experience when you use a funnel, as shown in Figure 7.36. When you pour liquid into the top, it doesn't matter how fast you do so, or how large of an area there is at the top: the rate at which the liquid exits will be determined *completely* by the bottom, where the liquid is pouring out from. The larger the bottom, the faster the liquid will exit (as long as it's smaller than the top). This is where the term bottleneck comes from: you're always limited by the speed of your *weakest link*.

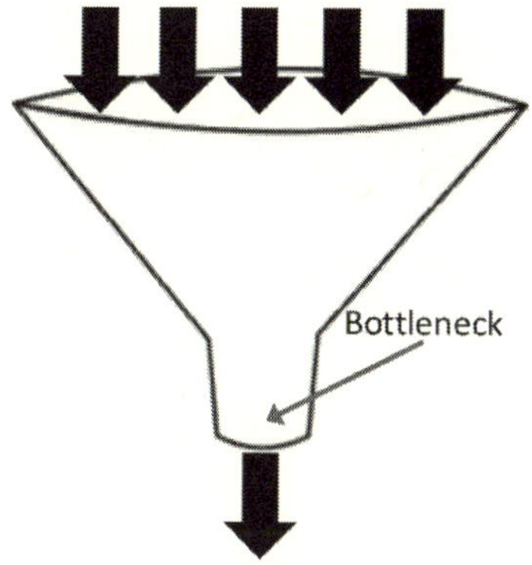

Figure 7.36 The idea of being limited by the weakest link is quite similar to what you experience with a funnel. When you pour liquid into the top, it doesn't matter how much you put in or how fast you do so: the speed at which the liquid exits the funnel will *always* be determined by how large the bottom is.

So then, when transmitting from one node to another, all we really care about is the weakest link along the path. The capacity of the weakest links are what we use when constructing an **overlay network** graph, in which we only label the start and finish points of each path. We show two such overlays in Figure 7.35, between nodes A and D as well as A and G. We just draw a link directly between them, and label the bottleneck capacity, which tells us how long it takes us to go from point to point. This overlay is purely a logical concept, and exists on top of the physical **underlay** graph (the entire tree). The *links* in the overlay (here, (A, D) and (A, G)) are *entire paths* in the underlay ((A, B, D) and (A, C, G)).

In P2P networks, it's common to look at overlay graphs because they're usually more convenient. And you can probably relate well to this in real life. Suppose you're travelling from your house in Princeton, NJ to somewhere in Stanford, CA. You probably think to yourself "I'm going to take a plane from here to Stanford." Now this is absolutely true, because you do get on a plane, and it represents the (hopefully) longest portion of the trip. But in reality, you're doing a lot more than that: you get into your car, drive to the train station, take the train to the airport, take the plane to another airport, ... and eventually arrive at the destination. The "trip" makes a logical connection between your house and the destination, and we tend to think of it that way. This is the overlay. But the underlay actually consists of a lot more: all the

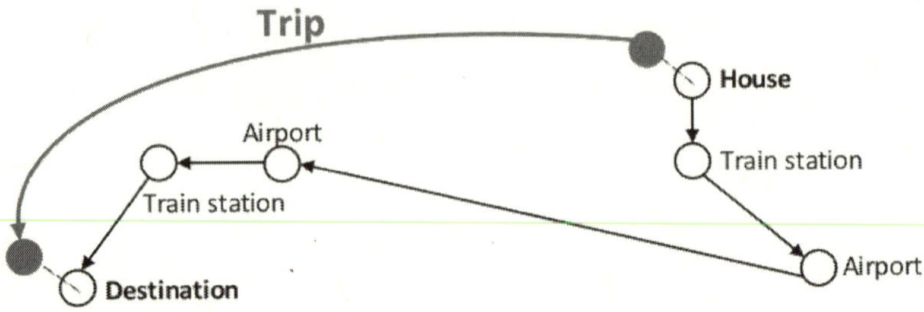

Figure 7.37 Traveling offers an interesting analogy to overlay networks. You would think of a "trip" logically as connecting your house and wherever you're going. But really, there are a lot of intermediate steps. In this case, the overlay is the trip, while the underlay actually contains the physical paths you take, like all the roads, tracks, and airspace.

roads, tracks, and paths you traverse to get there. The "trip" contains all these "roads," just as a "link" in the overlay contains "paths" in the underlay. We illustrate this in Figure 7.37.

With the bottleneck and overlay concepts in our back pocket, we can return to our example. What we are doing is finding three different trees to send each third of the file down. And why do we need three different trees? There are two main reasons:

- We don't want to overload any of the nodes with too much to upload, to avoid any bottlenecks.
- We want to make sure we are using as much upload speed as possible, without wasting any.

If we give a node too much, then it will hold up the time and become the bottleneck in the system. We would rather *distribute* the upload as much as possible. As a result, we define a **multi-tree** structure, like we discussed before, where the server sends one-third of the file to each of A, B, and C. We illustrate this in Figure 7.38. Our three trees are as follows:

1. *First third*: The first chunk of the file is uploaded from the server directly to A. In turn, A splits it and uploads to B and C.
2. *Middle third*: The second chunk goes directly to B. In turn, B uploads to A and C.
3. *Final third*: The final chunk goes to C, which uploads to A and B.

We make an interesting observation here. The server is using all of its upload capacity: 4 Mbps to each peer. Each peer, in turn, is using

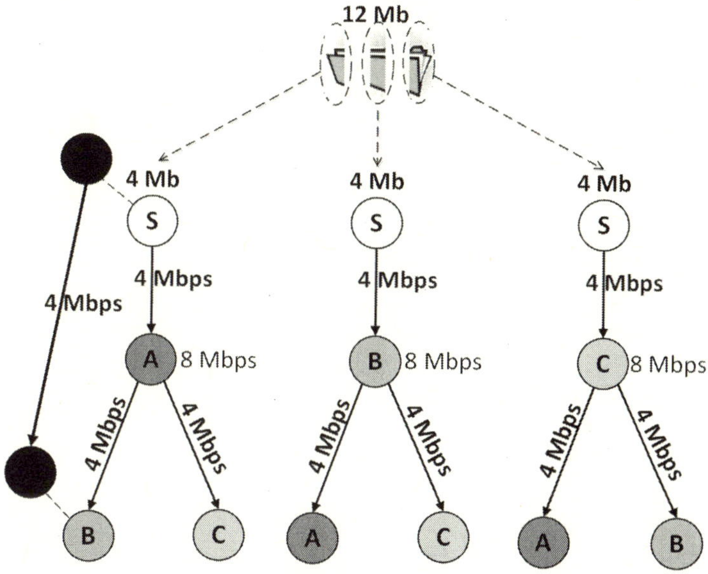

Figure 7.38 We can view the peering process as if we divide the file into segments, or chunks. With three identical peers, the server divides the file three ways, and sends the first 4 Mb to A (left), the second to B (middle), and the third to C (right). As each peer downloads its piece, it simultaneously uploads it to its neighbors.

all of its upload speed: 4 Mbps to each other peer. As a result, *we are doing the best we possibly can.* In general, we know that we are getting the best performance possible if we are using all of the upload rates to their full capabilities. This is unless, of course, we are limited by someone's download rate, so let's take a look at that for a moment. In this diagram, each node is required to download 4 Mbps from the server and 4 Mbps from each other peer, for a total of $4 + 4 + 4 = 12$ Mbps. And since each peer can download up to 24 Mbps, we know this can be handled.

So how long does the process take? This is a pretty simple example, because all the trees are symmetric: each link is 4 Mbps, and hence, 4 Mbps is the bottleneck link rate. As a result, each tree finishes in 4 Mb/4 Mbps = 1 s. And they're all going on at the same time, so the total time to transfer the entire file is 1 second. Compare that to the client-server case, and you will see that we've reduced the transfer time by *a factor of three*, which is the number of clients! Hence, the network *scales itself* with more peers by leveraging the presence of nodes to provide additional upload capacity.

7.3.3 The general case

We went through a very simple example to illustrate how much of a benefit we can get from P2P networks. But in general, when we have peers with all different upload/download rates, finding the rates to assign to each can become quite a challenging problem.

But one thing we can do is figure out the best that we can possibly do. As we've seen, there are three things that can limit the total transmit time:

1. *The server's upload time*: Since the server is the only place where the file is at first, it has to transmit the entire file one time. Who gets what depends on how we assign the upload capacities, but either way, the server has to send everything exactly one time. The time it takes it to do this depends on its upload speed and the size of the file:

$$\text{Server upload time} = \frac{\text{FileSize}}{\text{ServerUploadRate}}.$$

   We saw in our example that with a file size of 12 Mb and a server upload rate of 12 Mbps, it took $12/12 = 1$ second to transfer the whole file one time.

2. *The system's upload time*: While the server has to send out the file once, at the end of the process each peer needs to have a copy of the file. This means we have to transfer a total of NumPeers $\times$ FileSize bits. And how much upload capacity do we have available to do this? Not only the server's, but the peers' as well: ServerUploadRate + PeerAUploadRate + PeerBUploadRate + $\cdots$. Therefore, the time it takes to transfer each copy of the file is as follows:

$$\text{System upload time} = \frac{\text{NumPeers} \times \text{FileSize}}{\text{ServerUploadRate} + \text{PeerAUploadRate} + \cdots}.$$

   This highlights the *scalability* of P2P networks: as we get more peers, we get more upload capacity, too, unlike with the client-server model. In our example, we had three peers, each with upload rate of 8 Mbps. Therefore, the system upload time was: $(3 \times 12 \text{ Mb})/(12 \text{ Mbps} + 8 \text{ Mbps} + 8 \text{ Mbps} + 8 \text{ Mbps}) = 36 \text{ Mb}/36 \text{ Mbps} = 1$ s. So, the server

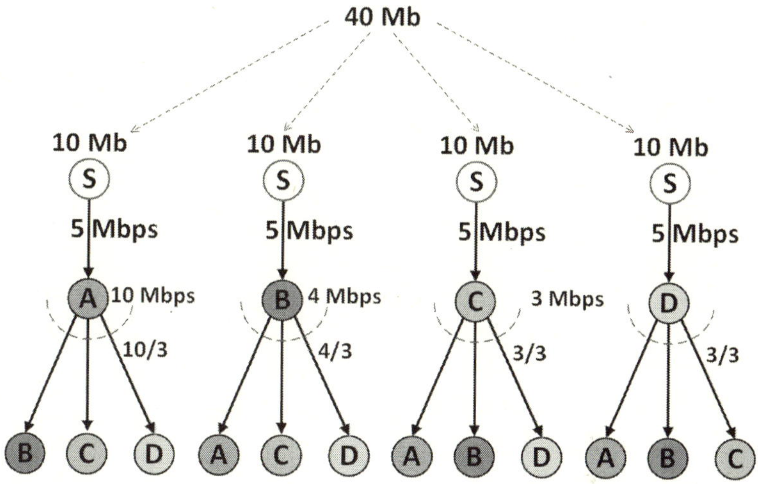

Figure 7.39 Applying the multi-tree idea to the second example. We have four trees, where the server uploads to one of the peers (A - D) in each. We divide the file into quarters (10 Mb each) and send it down each tree. The slowest trees to finish will be the third and fourth, since C and D only have upload rates of 3 Mbps.

upload time was the same as the system upload time.

3. *The slowest download time*: And of course, each peer has to download one copy of the file. We are concerned with the one that has the *slowest* download rate, which will take the longest:

$$\text{Slowest download time} = \frac{\text{FileSize}}{\text{MinDownloadRate}}.$$

In our example, each peer had 24 Mbps download rate, so each of them was the lowest. Therefore, we have: 12Mb/24 Mbps = 0.5 s.

The total transfer time will be the *largest* of the above three numbers, which gives the limiting case. And the largest here is 1 second, as we thought.

These three numbers give us the *best* we can possibly do in a P2P network. But as we said, sometimes it's tough, if not impossible, to achieve this. Why? There are two main reasons.

1. *Peers aren't always the same*: Different peers have different upload and download rates, so each of them would have to bear a different

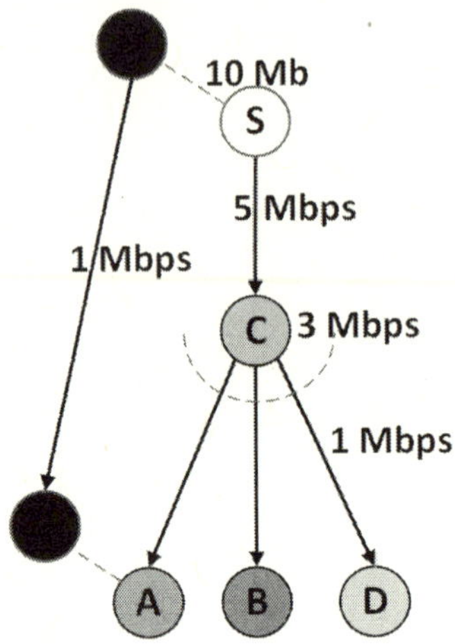

Figure 7.40 The third (and fourth) trees in Figure 7.39 will finish the slowest, since they contain the slowest links. Each of (C, A), (C, B), and (C, D) are only given 1 Mbps apiece, and are therefore the bottlenecks.

brunt of the upload capacity. Our example assumed they were all the same.

2. *Not all peers are connected*: Sometimes it's not possible to establish a connection between two peers, even if we wanted to. Our example assumed we could connect them all.

We will consider another example quickly to illustrate the first point. Suppose we have a server and a set of four peers (A – D). The server has an upload rate of 20 Mbps. The upload rates of the peers are 10, 4, 3, and 3 Mbps, respectively. Assume all the download rates are high enough that they don't matter, and that we are trying to transfer a 40 Mb file.

Let's try applying our multi-tree approach here. Since we have four peers, we will have four trees, where the server uploads directly to one of them in each case. We illustrate this in Figure 7.39. As before, we split the file size equally: 10 Mb for each tree. Since the server can support 20 Mbps, it gives 5 Mbps to each tree. A – D each split their upload rates equally in their respective trees, for 10/3, 4/3, 4/3, and 3/3 Mbps, respectively.

To find the time it will take to finish transferring the file, we have to figure out which tree will take the longest. In each tree, we are limited by the bottom links, because they each have capacity lower than the 5 Mbps that the server is pushing ($10/3 < 5$, $4/3 < 5$, $3/3 < 5$). Of these,

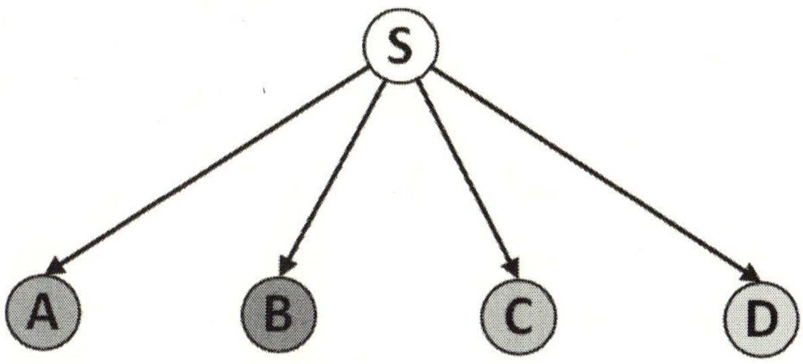

Figure 7.41 We can squeeze more capacity out of our multi-tree by adding another one to transfer portions of the file to each peer at once. This way, we can use up any excess server capacity (assuming there is enough download capacity left).

the 1 Mbps in the third and fourth trees are the smallest, and hence are the bottlenecks for the entire system. In the overlay graph for the third (or fourth) tree, as shown in Figure 7.40, we draw a 1 Mbps link from the root to the leaf nodes. To send 10 Mb down this tree, it would take 10 Mb/1 Mbps = 10 seconds, and hence, that's how long it takes to complete the entire transfer across the graph.

Let's compare that with the best case scenario. As we said, the server has to send out the file once. With a size of 40 Mb and an upload rate of 20 Mbps, the ideal server upload time is 40 Mb/20 Mbps = 2 s.

Further, we need to get a copy of the file to each peer. With 4 peers who have upload rates of 10, 4, 3, and 3 Mbps, the ideal system upload time is

$$\frac{4 \times 40 \text{ Mb}}{20 \text{ Mbps} + 10 \text{ Mbps} + 4 \text{ Mbps} + 3 \text{ Mbps} + 3 \text{ Mbps}} = \frac{160 \text{ Mb}}{40 \text{ Mbps}} = 4 \text{ s.}$$

Taking the larger of these two gives us a best-case of 4 seconds (remember, we assumed the download time was large enough). But our multi-tree in Figure 7.39 will take *10 seconds*! That's a pretty big difference (2.5 times). Obviously, our approach is not optimal here.

On the bright side, if the right peering relationships are established (and the download speeds are large enough), then an approach *similar* to the one shown here will always achieve the best case. Without going into detail, that would require two additions to our method. First, we would adjust the amount the server gives to each tree based on the capacity of the peers. Second, in the case the server has leftover upload capacity not used in the trees, we could add another tree where the server transmits directly to each peer, as shown in Figure 7.41.

Enough math. Let's take a look at an application that uses techniques very similar to what we discussed here: BitTorrent.

7.3.4 BitTorrent

P2P started becoming popular around 1999, with the advent of file-sharing applications. People became enthralled with the idea that they could download files from each other, *for free.* Figure 7.42 shows just how widespread P2P was by 2006 and how quickly its traffic spiked up. For example, Kazaa and Gnutella were widely used P2P file- and music- sharing systems back in the early 2000s.

What happened to Kazaa? Why don't we hear about it anymore? Well, a lot of people using it would *just download*, without contributing anything. This was either because (1) they just didn't have a very good Internet connection to be able to upload to people, or (2) they were selfish, not wanting to contribute anything to their peers. Indeed, why would I let people upload from me, when I can just download at will, and nobody is *incentivizing* me to upload?

These people, who only download, are known as *free riders.* It's like that person in your class group who doesn't do any work, but expects it to be done for him. And perhaps that's a good philosophy for the free-rider, at first. But what happens when *nobody* does the work? Well,

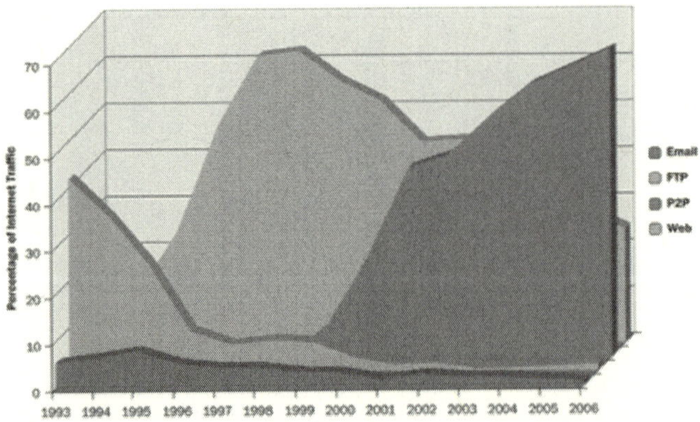

Figure 7.42 After 1999, P2P traffic began to skyrocket. Indeed, by 2006, CacheLogic Research showed P2P as well over 50% of Internet traffic.

Rank	Upstream		Downstream		Aggregate	
	Application	Share (%)	Application	Share (%)	Application	Share (%)
1	**BitTorrent**	**36.8**	Netflix	33.0	Netflix	28.8
2	HTTP	9.83	YouTube	14.8	YouTube	13.1
3	Skype	4.76	HTTP	12.0	HTTP	11.7
4	Netflix	4.51	**BitTorrent**	**5.89**	**BitTorrent**	**10.3**
5	SSL	1.65	iTunes	3.92	iTunes	3.43
Total	**Top 5**	**57.55**	**Top 5**	**69.61**	**Top 5**	**67.33**

Figure 7.43 A 2012 report by Sandvine showed BitTorrent consuming over 35% of upstream bandwidth in the US. Its aggregate usage was over 10% of all traffic.

nothing gets done, and there's nobody to take it from anymore. This was the problem with Kazaa: there were a lot of free riders who did not contribute nearly as much as they consumed.

Along came BitTorrent in 2001, used heavily for file sharing (including a lot of movies). To say the least, it has been quite successful. At one point in the mid 2000s, P2P took up *more than half* of the Internet traffic, and BitTorrent alone was responsible for 30%. Can you believe that? Almost one in every three packets sent through the Internet was from BitTorrent alone!

P2P sharing of multimedia content is still very popular today, with over 250 million users on BitTorrent. A recent report still put it as the most used application in the US in terms of *upstream* bandwidth, as shown in Figure 7.43.

Tit-for-tat, this for that

So, what made BitTorrent so successful? Naturally, people use it because it's free, but let's hold off on that for a minute. There's a lot going on that's very similar to what we discussed in this section.

BitTorrent uses P2P for resource sharing. It shares the upload capacities as well as the content stored by each peer. Of course, this content probably originated from a server in the first place, but we can ignore that: if you want a file, there's probably someone using BitTorrent who already has it.

In this way, we expect that content sharing will *scale itself*: every time a new peer comes into the picture, she should bring some upload power to the table, as well as the files she has. She has to give in order to receive, and this is her way of internalizing the negative externality she imposes on the network by downloading in the first place.

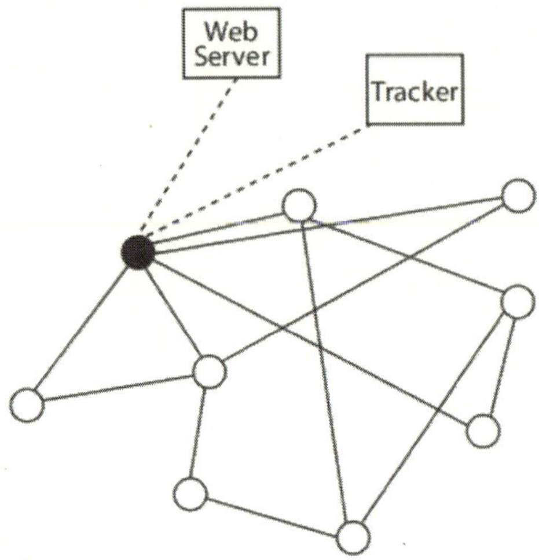

Figure 7.44 A typical overlay graph of what your BitTorrent network looks like. You (black circle) talk to a tracker which gives you a list of 50 of your peers that have chunks of the file you want. Then, from these 50 peers, you select 5 to be exchanging with at any given time.

BitTorrent has a very clever way of aligning these incentives. Let's take a look at its operaton now.

- *Web server*
  When you want a file, you first talk to one of BitTorrent's web servers. This server will give you some information about the file: its name, size, and the number of **chunks** that it has. Note that BitTorrent doesn't transfer based on *whole* files, but based on chunks. As we saw, each one can traverse different trees, which allows us to speed up the process.
- *Tracker*
  The server also gives you the address of a *tracker* that you should talk to. As the name suggests, this keeps track of peers on BitTorrent who have chunks of the file you need. When you talk to it, you receive a set of 50 (or so) of these peers that you can exchange chunks with. At first, you pick five of these to be in your peering group. These are the people you will download from, and upload to in return, as illustrated in Figure 7.44.
- *Exchanging chunks*
  Once you select these five, you exchange information with them, so you can find out what chunks each of you have, and which ones you

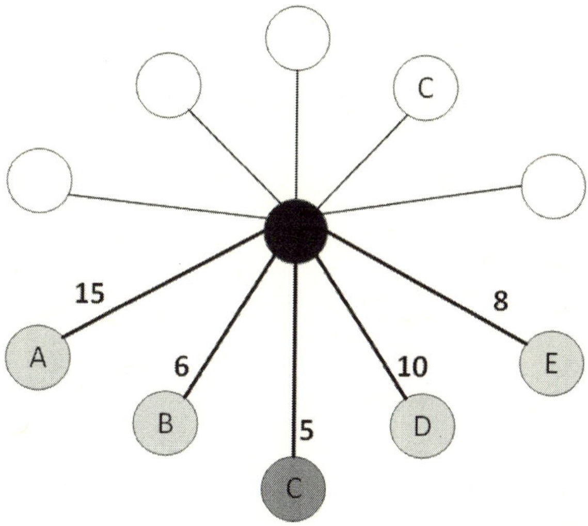

Figure 7.45 Here, we show a peer executing the tit-for-tat strategy. Since the last time he updated his list, we indicate the number of chunks he has received from each of his peers (A – E): 15, 6, 5, 10, and 8. Since 5 is the least, C would be ousted from the list, and the other four would be kept. Another peer is also taken at random from the set of 50.

need. BitTorrent uses the policy of *rarest chunk first*: when a peer looks at the list and determines which chunks to download, it starts with the ones that very few peers have. By equalizing the availability of chunks, this mitigates a potential problem we could run into where most of the peers have most of the chunks, but very few have the rarest, and each peer is waiting to download these.

And here's where the beauty of BitTorrent comes in: every so often, you *update* your list of five peers, according to two criteria.

- Four of them are the top four contributors to you since the last time you updated your list. These are the people that you received the most content from, the ones that gave you the most upload. This is called **tit-for-tat**: I'll only give to you if you've been generous enough to me.
- The remaining peer is selected at random from the set of 50.

The first feature is precisely how the free-rider problem is mitigated in BitTorrent. If a peer is not contributing to you, you can kick him out of your group when you update your list. Nodes who decide not to contribute will be punished, because nobody will want to include them. We illustrate the tit-for-tat strategy in Figure 7.45.

The second feature avoids unfairness to nodes that have little upload capacity to contribute. These peers will get "passed around" a lot, rather than ignored completely: they will be in one group, get released, go into another, get released, and so on. Also, randomization is generally a good idea to avoid getting "trapped" with your current set of peers. For instance, there might be someone out there not in your current group who is better than anyone who is, and who will contribute more to you. This gives you a chance of finding that person.

And the other thing that's attractive about BitTorrent is that it's *free*. There are two main reasons why it can stay this way:

- It's *self-contained*. It leverages the peers uplink capacities, so they can send chunks of files to *each other*.
- It's *scalable*. As more peers enter BitTorrent, there's more upload capacity to be used. Additionally, they bring their own files into the mix. And we don't really have to worry about free-riders here.

For both of these reasons, there's no need to deploy large media servers to transfer content to peers, and hence, no real expenses to be had (in terms of hardware, at least).

P2P showcases a major success of the evolution of the Internet: make the basic design simple and allow overlay constructions. The architecture of the Internet focuses on providing simple, ubiquitous, stable, and economical connectivities, leaving the rest of the innovations to *overlays* to be constructed in the future. Note that beyond file sharing, different types of applications have been built using P2P overlays, like video streaming (e.g., Skype), television broadcasting, and on-demand media distribution.

Summary

This concludes our two-chapter discussion of the fundamentals of the Internet and different levels of the protocol stack. In Chapter 6, we began our study with a discussion of packet switching and distributed hierarchy, and moved on to talk about the connectionless, hop-by-hop network layer. In this chapter, our focus was on modularization in the Internet, and we went into detail on important tasks that are assigned to layers above the IP layer. At the transport layer, we discussed congestion control and the management of end-to-end connectivity. Then we moved on to multicasting at the application layer, with a specific focus on peer-to-peer, highlighting its scalability as compared to the traditional client-server model.

It is worth mentioning that the "division of labor" presented here is not the only one that has been tried. For instance, load-sensitive routing at the network layer was once considered, which would assign congestion control to layer 3 rather than 4. Also, multicasting at the IP layer, as opposed to the application layer, was extensively studied. But today, most congestion control is done in the transport, and most multicasting in the application layer, as portrayed here.

8 Bigger And Bigger

The networks that we've been talking about are *huge*. How can they remain efficient, let alone operate, at such large scales?

In this final chapter, we will investigate some popular questions related to the ever-increasing size of networks. First, is it still a "small world" in a growing network? Then, we will look at the implications of the recent tendency to use cloud storage: how do data centers scale up? And finally, we will ask a question that was the subject of debate for many years: does the Internet have a single router that becomes an "Achilles' heel," without which it would cease to function?

8.1 Small world

How is it that we are connected in six steps or less? As we will see, this depends on the ways in which social networks are structured, and on the ways in which you search for such short distances.

8.1.1 It's a small world, after all

Since Stanley Milgram's 1967 experiment, the **small world** phenomenon, or the **six degrees of separation**, has become one of the most widely

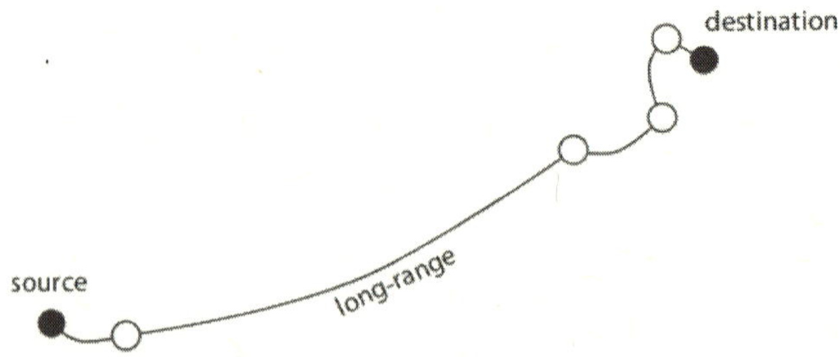

Figure 8.1 A picture illustrating the Milgram experiment in 1967. A key phenomenon is that there can be one or two long-range links in these short paths between Omaha and Boston. It turns out that out of all the letters that arrived, the median number of hops was six, leading to the widely publicized six degrees of separation.

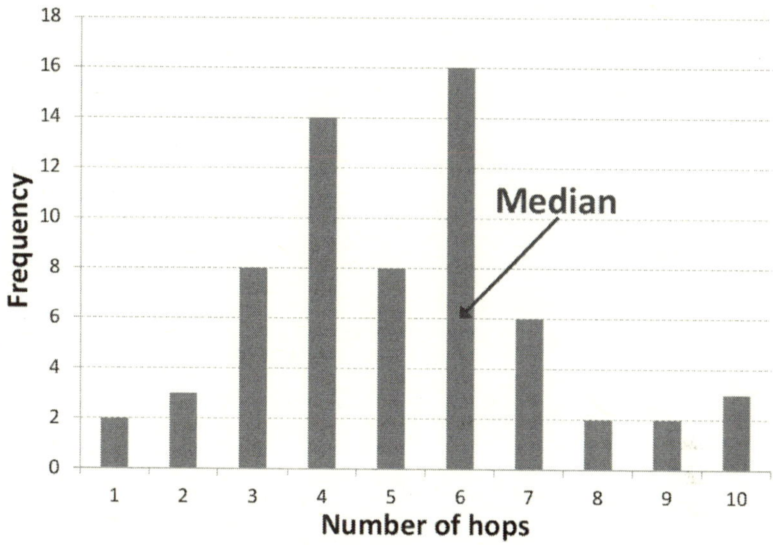

Figure 8.2 Bar graph of the number of hops from different sources in Omaha to the common destination in Boston in Milgram's experiment. The median value is six, meaning that the search path included five people other than the source and destination (in the median case). This led to the famous six degrees of separation.

told stories in popular science books. Milgram, an American social psychologist, asked 296 people living in Omaha, Nebraska to participate. He gave each of them a passport-looking letter, which was destined to a suburb of Boston, Massachusetts. The recipient's name, address, and occupation (stockbroker) was shown.

The goal was to send this letter to one of your friends, which he defined as someone you knew on a first name basis. If you did not know the recipient by first name, you had to send the letter via others, starting with sending it to a friend (which constitutes one hop), who then sent it to one of his friends (another hop), until the letter finally arrived at someone who knew the recipient by first name and sent it to the recipient. Obviously, knowing the recipient's name and address was important, but as it turned out, knowing his occupation played an important role too. This way, a person could say, "Well, I don't know the recipient personally, but I know someone who lives in Boston, so I'll forward it to him," or "I know someone who is a stockbroker, so he has a better chance of knowing this circle than I do." You can guess that the letter will go through maybe one or two long range links (e.g.,

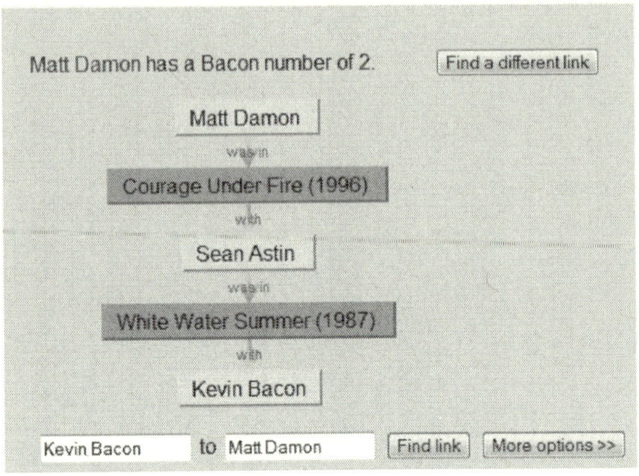

Figure 8.3 Illustration of the website you can visit to find an actor/actress' degrees of separation from Kevin Bacon, in terms of co-star relationships. Matt Damon has a "Bacon number" of two.

to get it to the general location of Boston), and also some short ones. This is illustrated in Figure 8.1.

Of these letters, 217 were actually sent out and 64 arrived at the destination, a seemingly small arrival rate of 29.5% but actually quite impressive, considering that a later replica of the experiment via email had only a 1.5% rate of arrival with roughly the same average number of hops (at a first glance, you would probably think the one via email would have a much *higher* arrival rate). The other letters might have been lost along the way, and needed to be treated carefully in the analysis of this experiment's data. But, out of those 64 that arrived, the average number of hops was 5.2, and the median was 6, as shown in Figure 8.2.

Researchers have long suspected that the **social distance**, or the average number of hops it takes to reach anyone in a population, grows very slowly as the population size grows. And Milgram's celebrated experiment suggests just that. From the 1970s to the online social media era, much empirical evidence has suggested the same: from the study of Erdos number, which looks at the number of hops between research collaborators as co-authors on papers, to looking at co-star relationships on IMDb.

In addition to popular science textbooks, the "small world" concept has permeated into culture, appearing in movies, TV shows, and even songs. One recent example of this was in 2007, when Martina McBride hosted the TV show "Six Degrees of Martina McBride." In it, six aspir-

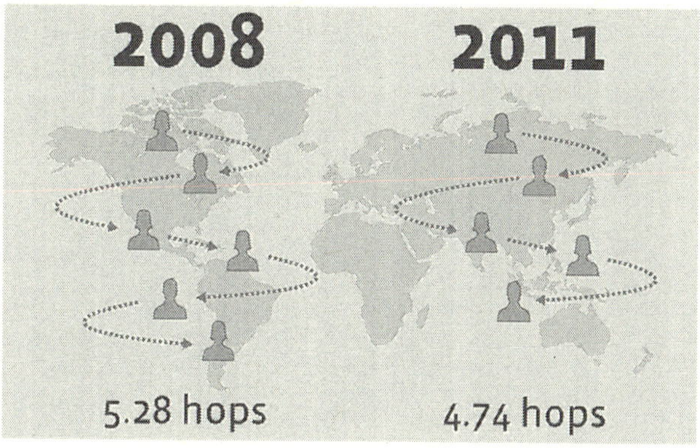

Figure 8.4 In 2008, the degrees of separation on Facebook was computed to be 5.28. That number dropped to 4.74 by 2011 (image from the Daily Telegraph, 2011).

ing country singers competed in an attempt to connect themselves to her using less than six hops. The winners were awarded a studio session with McBride and a shot at a record deal.

Another example is known as the "Six Degrees of Kevin Bacon," based on the idea that a lot of people in Hollywood are close to the famous actor Kevin Bacon through co-starring relationships. Bacon himself has a so-called "Bacon number" of 0, someone who has acted with him has a 1, someone who has acted with someone who acted with him gets a 2, and so on. You can visit a website called "The Oracle of Bacon" (`http://oracleofbacon.org/movielinks.php`) and enter any actor/actress' name to get the degrees of separation he/she is from Bacon. In Figure 8.3, we see that popular actor Matt Damon has a Bacon number of two.

Should we be surprised by this seemingly universal "small world" observation of social networks? There are two points we have to take up here.

- One is *structural*: Social networks are formed in such a way that there exist short paths.
- The other is *algorithmic*: With very limited local information, a node can navigate through a social network and find one of these short paths to a destination.

The second kind of "small worlds" is more surprising than the first and requires more careful modeling of **social search**. For example, a report in November 2011 computed the degrees of separation on Face-

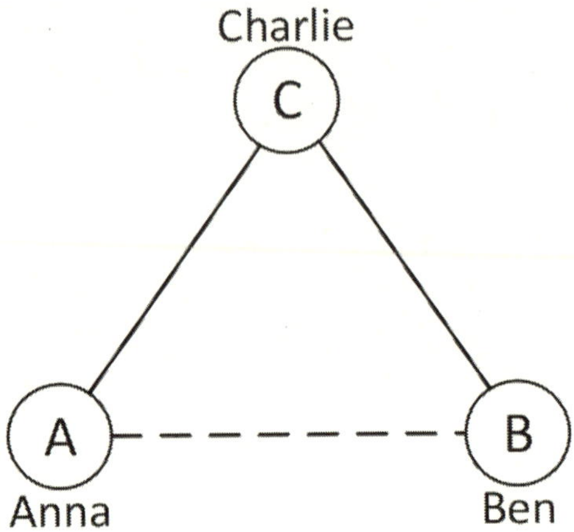

Figure 8.5 An illustration of triad closure in social networks. If Anna knows Charles and Ben knows Charles (the solid lines), it is likely that Anna and Ben also know each other (the dotted line). If so, the connected triple forms a triangle. As we will see, the clustering coefficient can quantify the relative number of triangles in any graph.

book to be 4.74, down from 5.28 in 2008 (see Figure 8.4). But that report was concerned only with the *existence* of short paths, not the more relevant and more surprising *discoverability* of short paths from local information. As we will see, it is also more difficult to create a general explanation for an algorithmic small world than it is to create a purely structural one.

But first, we focus on the existence of short paths. On the surface, it seems fascinating that you can likely reach anyone in six steps or fewer. Then, on second thought, you may reason that, if I have twenty friends, and each of them has twenty friends, then in six steps, I can reach $20 \times 20 \times 20 \times 20 \times 20 \times 20$ people. That is already 64 million people. So of course six steps often suffice.

But then, giving it further thought, you realize that social networks are filled with"triangles," or **triad closures**, of social relationships. This is illustrated in Figure 8.5: if Anna and Ben both know Charlie, Anna and Ben likely know each other directly too. This is called **transitivity** in a graph (not to be confused with transitivity in voting). In other words, the catch of the "64 million" argument above is that you need your friend's friends to not overlap with your own friends. Otherwise, the argument fails. But of course, many of your friend's friends are your own friends too: there's *a lot* of overlap. The phenomenon of people who are alike tending to form social links among each other is called **homophily**. As we will discuss later, it can be quantified by the clustering coefficient. Now, six degrees of separation is truly surprising, bringing us to the key question we will explore in this section:

- Can you really reach anyone in six steps?

But Milgram-type experiments suggest something even stronger: not only are there short paths, but they can also be discovered by each individual node using very limited information about the destination and only its own, local view of the network. Compared with sending packets through the Internet as we discussed in Chapter 6, social search is even harder since nodes do not pass messages around to help each other construct a global view. Some help is embedded in the address and occupation of the recipient, and possibly in the name that can reveal something about the destination's sex and ethnicity. From this, people in Milgram's experiment must have constructed some sort of *distance metric*. For example, New York is closer to Boston than Chicago is, on the geographic proximity scale measured in miles. Or, a financial advisor is perhaps closer to a stockbroker than a nurse is, on some occupation proximity scale, which is more vague but nonetheless can be somewhat quantified. Suppose each person uses a simple, "greedy" algorithm (based on a local view) to forward the letter to his friend who is closest to the destination, where "closeness" is defined by some combination of these "distances." Is it a coincidence that this social search strategy discovers a short path?

In the rest of this chapter, we will walk you through some models that have been constructed over the years to explain the small world phenomenon.

8.1.2 Six steps are plausible

Let's now see if we can come up with a type of graph that has the same properties as the apparent "small world" phenomenon. Such a model will help us explain what the networks look like that give rise to it in the first place. Indeed then we can say that six degrees of separation makes sense.

Quite intuitively, we need our graph to be "small." But what exactly do we mean by this?

Average shortest path

There are actually several ways to measure how "big" a graph is, and we touched upon them in the last chapter when we discused centrality

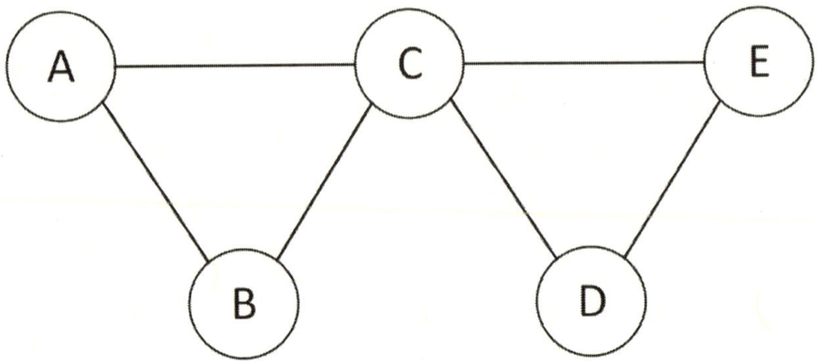

Figure 8.6 A small example that we will use to illustrate computations of two metrics: clustering coefficient, and (average) shortest path distance.

measures. One way is using the **diameter**, which is the length of the longest shortest path between any pair of nodes. "Longest shortest" probably sounds strange. To clarify, we first find the shortest path lengths between all pairs of nodes, and then we take the largest of each of these. Then we hope that the diameter is quite small with respect to the number of nodes in the network.

But the diameter is an extreme measure. For instance, it could represent the shortest path between one pair of nodes that happen to be quite far apart. This doesn't encapsulate the picture for *all* pairs, it just gives us the worst case. When we think of a network as a small world, we are more concerned with the *median* or *average* of the shortest paths between all node pairs. This is found by applying the following equation (which is the reciprocal of closeness centrality, discussed in Chapter 4):

$$\text{Avg. Shortest Dist.} = \frac{\text{Sum of shortest path lengths}}{\text{Number of node pairs}}. \tag{8.1}$$

Let's apply Equation (8.1) to the example network in Figure 8.6. To do this, we need to find all the shortest path lengths, just as we did when computing closeness centrality in the last chapter. Starting at Node A, we have the following:

- A to B: There is a direct link, making the shortest path of length 1.
- A to C: Again, the shortest path is a direct one, which has length 1.
- A to D: Now, we have to go through an intermediate node, C, to get from A to D using the least amount of links. Path `(A,C,D)` has length 2.
- A to E: Similarly, path `(A,C,E)` has length 2.

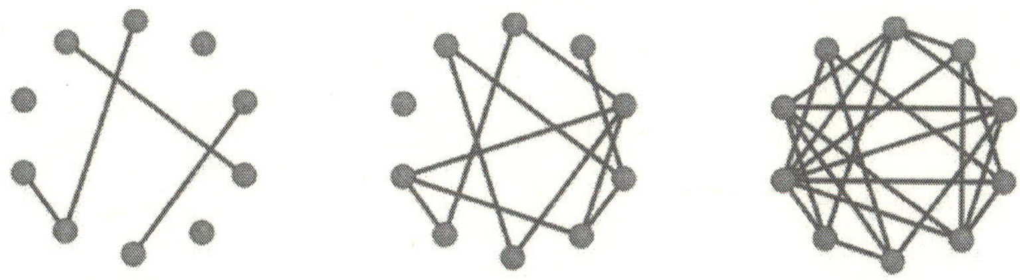

Figure 8.7 Process of random graph formation, illustrated for 10 nodes. As the probability of connecting any two links increases, there are more in the end, as we would expect. Such graphs have not only a small (average) shortest path distance, but also a small clustering coefficient.

We can follow this logic to find the shortest path lengths from B to C (1), B to D (2), and B to E (2); from C to D (1) and C to E (1); and from D to E (1). Note that we don't have to consider reverse paths, like from E to A, because that will be the same as from A to E since the links are undirected. There are a total of $4 + 3 + 2 + 1 = 10$ node pairs (e.g., A and B, A and C, and so on). Then we find the average as indicated in Equation (8.1):

$$\frac{1+1+2+2+1+2+2+1+1+1}{10} = \frac{14}{10} = 1.4.$$

So what does this tell us? With a judicious choice of path, a person can get to any other by taking 1.4 steps on average. This is pretty intuitive in this example, as there are only five nodes and the shortest paths have lengths of either 1 or 2. But as the number of nodes grows, knowing what this length is can be important, and also surprising.

Random graphs

Let's now continue to search for an explanatory model. Consider the following situation. Suppose we are given a set of say, 10 nodes. Then, we iterate through each pair, and we randomly pick whether or not to establish a link between the two nodes. A link is established with some probability, and the higher this probability is, the more links we will see in the resulting graph. This is the basic idea behind a **random graph**, which we illustrate in Figure 8.7. On the left, the probability of link establishment is only 10%, so we see a small number of links. As this number increases to 50% on the right, we see many more links: roughly half of the pairs are directly connected in this case. In fact, the number of links is *expected* to be proportional to whatever this probability is (we say *expected* here because link establishment is a random process).

Of course, this process of network formation does not sound like most networks. It would require something along the lines of taking 10 random people who don't know each other, putting them in a room, and telling them to decide randomly whether or not to become friends. Probably not going to happen.

Clustering coefficient

It turns out that random graphs also don't have the exact properties we need to explain small world. On the one hand, they will in general have small (average) shortest distances: long range links that connect nodes far away are possible, and with just a few of them, we can lower the shortest path distances. But the *homophily* of the graph is still off: there's no tendency for friends of friends to be your friends as well. This is important in real networks.

How do we quantify this? We use something called the **clustering coefficient**, which should not be confused with the cluster density that we discussed in Chapter 4. It is a measure of how many triad closures we have in the graph, relative to the number of **connected triples**. We define a set of three nodes X, Y, and Z as a connected triple if there is a path connecting them:

- If there are links `(X,Y)`, and `(Y,Z)`, then `(X,Y,Z)` is a connected triple, regardless of whether there is a direct link between X and Z or not. For instance, in Figure 8.8, we say that `(A,B,C)` and `(B,C,D)` are both connected triples, even though the latter has no direct link from B to D.
- If there are links `(X,Z)` and `(Z,Y)`, then we have a connected triple `(X,Z,Y)`. From Figure 8.8, we see that `(A,C,B)` is a connected triple, but `(B,D,C)` is not since there is no link `(B,D)`.
- If there are links `(Y,X)` and `(X,Z)`, then `(Y,X,Z)` is a connected triple.

But if we have links connecting each pair in a "circle," e.g., `(X,Y)`, `(Y,Z)`, and `(Z,X)`, then we have not only these three connected triples, but also a triangle. We call this triangle `XYZ` a triad closure. In Figure 8.8, `ABC` is a triad closure, but `BCD` is not.

The clustering coefficient is the ratio of the number of triangles to the number of connected triples. It is computed as follows:

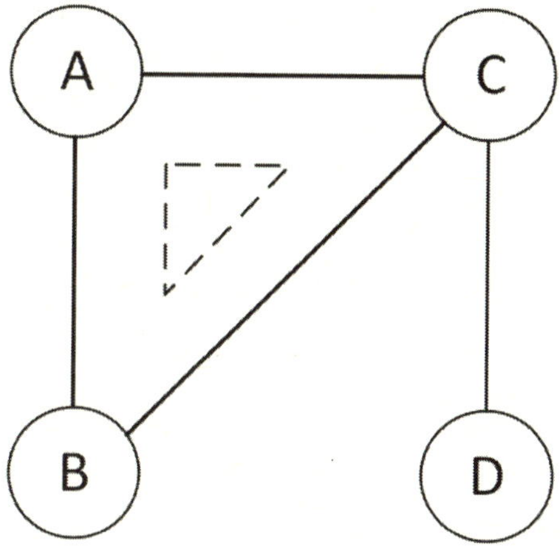

Figure 8.8 Small graph to illustrate the notion of a triad closure. Here, `ABC` is a triad closure that consists of three connected triples `(A,B,C)`, `(A,C,B)`, and `(B,A,C)`. On the other hand, B, C, and D do not form a triad closure, because they only have one connected triple `(B,C,D)`.

$$\text{Clustering Coeff.} = \frac{\text{Number of triad closures}}{\text{Number of connected triples}/3}. \tag{8.2}$$

So, the more triad closures we have, the higher the clustering coefficient: this indicates we have more overlap in who knows whom. The division by 3 *normalizes* the metric, meaning that it will be between 0 and 1. A graph of a single triangle, for instance, has 3 connected triples and 1 triad closure, making the clustering coefficient $\frac{1}{3/3} = 1$.

In Figure 8.8, we have one triangle `ABC`, and five connected triples: `(A,B,C)`, `(B,A,C)`, and `(A,C,B)` from the triad closure, as well as `(A,C,D)` and `(B,C,D)`. Therefore, the clustering coefficient is

$$\frac{1}{5/3} = \frac{3}{5}.$$

It might be hard to see why there are exactly three connected triples in a triad closure. For instance, why not count `(B,C,A)` and `(C,A,B)` as well? We elaborate on this reasoning in Figure 8.9. Basically, we can obtain a *unique* connected triple by removing any of the three links `(A,C)`, `(B,C)`, and `(A,B)` one at a time. Removing the first leaves A connected to B, and B to C. Hence, we obtain `(A,B,C)`. Removing the second and third we obtain `(B,A,C)` and `(A,C,B)`, respectively. These are the only unique triples in the triad: `(B,C,A)` uses the same links

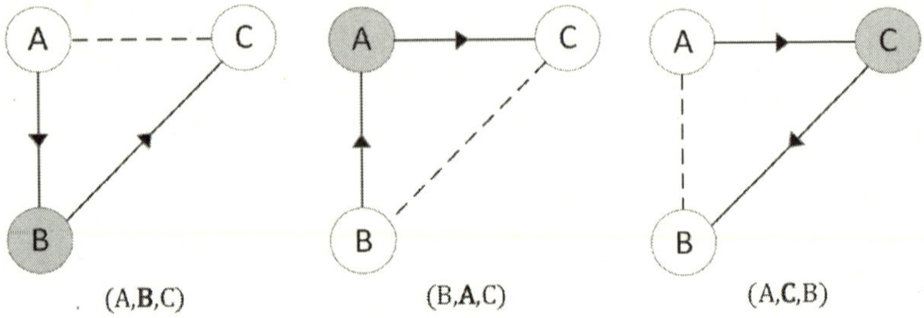

Figure 8.9 Explanation of why there are three connected triples in each triad closure. We can remove any one of the three segments, and obtain a unique connection among the three nodes A, B, and C. On the left, we remove `(A,C)` and keep `(A,B)` and `(B,C)`, making B the midpoint of the connected triple `(A,B,C)`. Similarly, we have `(B,A,C)` in the center and `(A,C,B)` on the right. We have kept the center point colored/shaded in each case.

as `(A,C,B)`, and `(C,A,B)` uses the same as `(B,C,A)`. We only want to count each of them once.

Let's return to our first example in Figure 8.6 to calculate the clustering coefficient. There are two triad closures: `ABC` and `CDE`. Each of these triangles has three connected triples: `ABC` has `(A,B,C)`, `(B,A,C)`, and `(A,C,B)`; `CDE` has `(C,D,E)`, `(D,C,E)`, and `(C,E,D)`.

But there are more connected triples we have to account for: `(A,C,D)`, `(A,C,E)`, and `(B,C,E)`, which are not in any of the triangles. These add to the denominator in Equation (8.2), but not to the numerator: they represent social connections that are not part of triad closures (e.g., my friends are not necessarily your friends). We have superimposed these three on the graph in Figure 8.10. To summarize, we have two triad closures, and a total of 9 connected triples. Hence, by Equation (8.2) the clustering coefficient is

$$\frac{2}{9/3} = \frac{2}{3}.$$

Equipped now with an explanation of what the clustering coefficient implies, we can continue to show why a random graph is not a good explanation of the small world phenomenon. In a random graph, the coefficient, roughly speaking, is given as follows:

$$\frac{\text{Average degree of a node}}{\text{Number of nodes} - 1}.$$

Remember that the degree of a node is simply the number of links

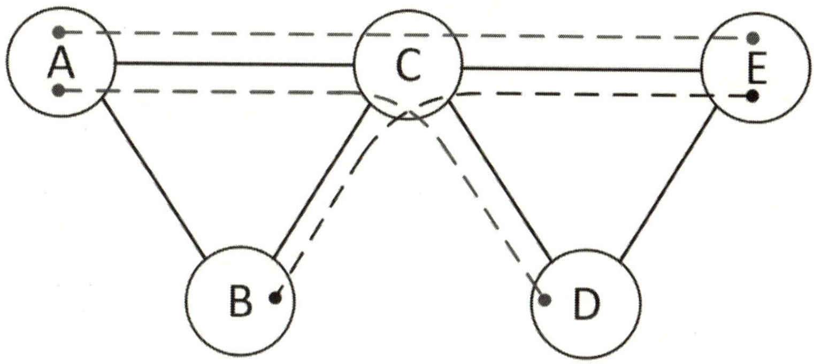

Figure 8.10 In this graph, there are two triad closures: `ABC` and `CDE`, each of which have three connected triples. But outside of this, we have `(A,C,D)`, `(A,C,E)`, `(B,C,D)` also as connected triples, for a total of 9. This leads to a clustering coefficient of 2/3.

it has (i.e., the number of "friends"). This equation should make sense intuitively, because as the degree of each node goes up, this should increase the amount of triangles: there's more of a chance that if A is connected to B and C, that B and C will also be connected. In the extreme case where everyone is connected to everyone, the expected degree becomes one less than the number of nodes (since a node won't connect to itself), and hence the clustering coefficient is 1.

If we consider numbers here in terms of Facebook, there are roughly 1 billion users (nodes), and each person has, on average, say 300 friends. Modeled as a random graph, this would give a clustering coefficient of

$$\frac{300}{10^9 - 1} = 3 \times 10^{-8}.$$

This is just way too small for realistic social networks.

Regular graphs

It is important to keep in mind what we are looking for in coming up with a general graph model that can explain small world: one that has

- a small average shortest path length, and
- a large clustering coefficient.

Let's turn our attention to a different type of graph then. As opposed to a random model, we can consider a very regular structure, as illustrated in Figure 8.11. This is a **regular graph**, in which we have

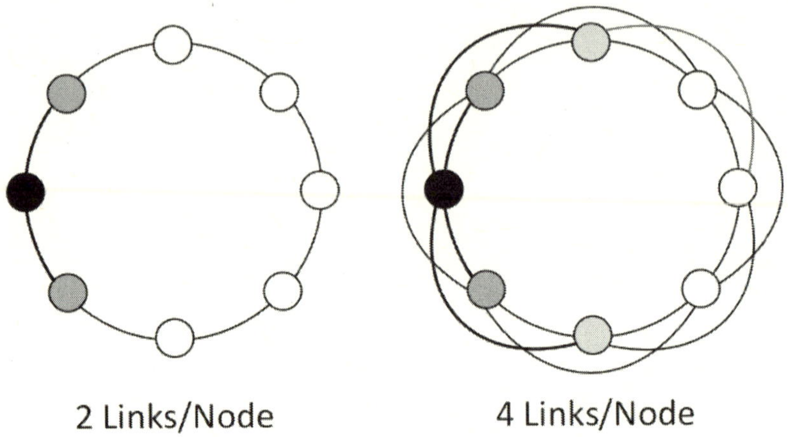

Figure 8.11 Examples of a regular graph, with 8 nodes, each having a certain number of links to other nodes (links/node). The number of nodes and links/node need to be even, due to symmetry of the graph with respect to each node: all nodes have half of their links going to the left, and the other half to the right. With 2 links/node, we see that each node is connected only to the ones directly to the left and right, and the graph reduces to a simple ring. As this number increases, the structure becomes more complicated: when there are 4 links/node, each node has 2 links to the left, and 2 to the right. The clustering coefficient increases rapidly as this number increases.

a certain number of nodes living on a ring. The structure of the graph can be determined by two numbers:

- the total number of nodes living on the ring, and
- the total number of links that each node has (denoted "links/node").

Both must be even numbers. The number of links is *symmetric* about each node: half of the links go to the left, and the other half the right. When there are 2 links/node, the regular graph reduces to a simple ring, in which each node is only connected to those right next to it. As this number increases, the structure becomes more complicated: nodes connect to more and more of their (closest) neighbors. When there are 4 links/node, each connects to its two immediate neighbors to the left, and the two to the right. For 6 links/node, we have three neighbors on each side, and so on.

The best way to envision this graph is just to think of a group of people standing in a circle. Each person is told to become "friends"

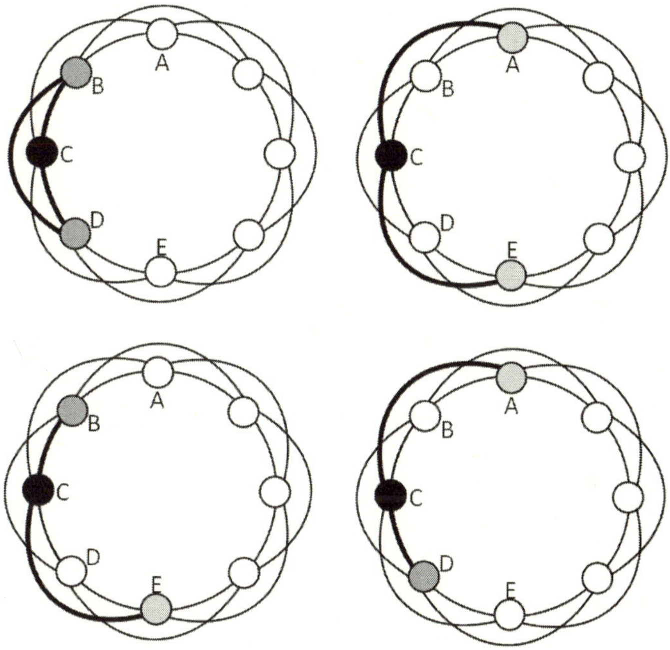

Figure 8.12 Here we highlight different sets of nodes that are necessary to compute the clustering coefficient. All these diagrams are centered around node C, and are the same for each other node as well, since the graph is perfectly symmetric. At the top left, we see that `BCD` forms a triad closure centered around C. The rest of the figures show connected triples `(A,C,E)`, `(B,C,E)`, and `(A,C,D)`, which are centered around node C but do not form any triad closures. We see then for 4 links/node, the clustering coefficient is 1/2.

with a certain number of others, such that half of them are from their closest neighbors standing to the left, and half from the right.

So how can we evaluate the clustering coefficient for such a graph? The case for two links per node is pretty simple: there are no triangles. Hence, Equation (8.2) says that the clustering coefficient is 0. This is sort of disappointing, since we need it to be large.

But let's look at what happens when the number of links/node increases. Consider 4: now, some triangles start to appear. Specifically, we have one triad closure centered at *each* node. This is shown at the top left of Figure 8.12 for C's case: `BCD` forms a triad closure, giving us one triangle and three connected triples. But there's more: each node has three additional connected triples centered on it that are *not* in

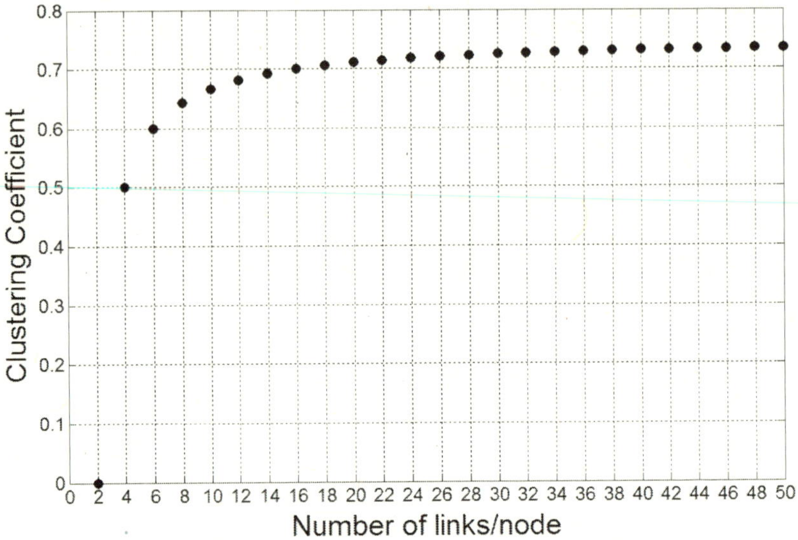

Figure 8.13 Plot of the clustering coefficient for a regular graph as a function of the number of neighbors each node has. The clustering coefficient increases from 0 to above 1/2 very rapidly, and for very large links/node, it approaches 0.75.

any triangles. We show segments (A,C,E), (B,C,E), and (A,C,D) in the rest of Figure 8.12. This gives a total of six connected triples so far.

Is there a need to look at the rest of the nodes too? No. The nice property about regular graphs is that all the nodes have exactly the same number of links, and are structured the same. Hence, each node has exactly the same number of triangles and connected triples centered at it. What this implies is that the clustering coefficient does *not* depend on the total number of nodes: for every node we add, we will add the same number of triangles and connected triples. The clustering coefficient is independent of the number of nodes we have.

So, for 4 links/node, we have 1 triangle and 6 connected triples per node. Hence, the clustering coefficient is

$$\frac{1}{6/3} = \frac{1}{2}.$$

This is quite a dramatic increase from zero for a pure ring.

We can generalize the above argument to an arbitrary number of links/node. Without going into mathematical detail, in general, the clustering coefficient can be found as

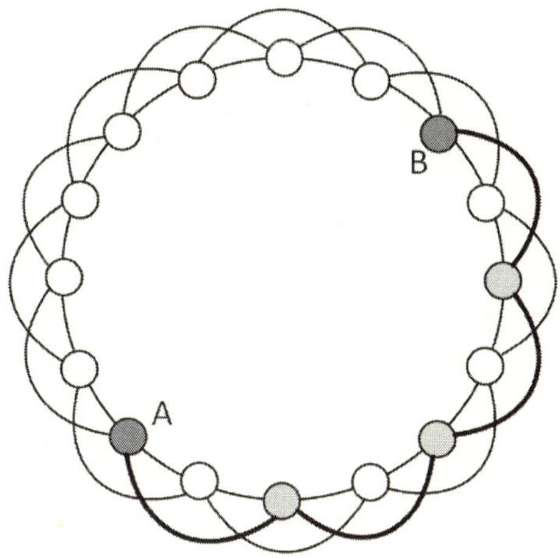

Figure 8.14 Even though regular graphs have large clustering coefficients, their average shortest path lengths are too large for small world. For instance, in this graph with 16 nodes and 4 links/node, it takes four links to get between nodes A and B. We have to traverse a lot of short-range links to make it to the opposite ends of the graph.

$$\frac{3 \times (\text{links/node} - 2)}{4 \times (\text{links/node} - 1)}.$$

We can try two test cases to verify this: with 2 links/node, we get zero. This is the smallest case possible where as we said, we only have a circle of links, so indeed the clustering coefficient should be 0. With 4 links/node, it jumps up to $(3 \times 2)/(4 \times 3) = 1/2$, as expected.

We plot the clustering coefficient against the links/node in Figure 8.13. As the variable becomes very large, the clustering coefficient approaches the largest it can be for a regular ring topology: 3/4. This is *many* orders of magnitude larger than that of a random graph.

The regular (ring) graph model stands in sharp contrast to the random graph model: it has a high clustering coefficient, which is realistic for social networks. But unfortunately, it also has a large average shortest path length: each node is only connected to its closest neighbors, and hence there are only *short-range* connections, as can be seen in Figure 8.14. It requires traversing many intermediate nodes to get to those that are "far away."

You may think that if we make the number of links/node large relative to the total number of nodes, then aren't we establishing more di-

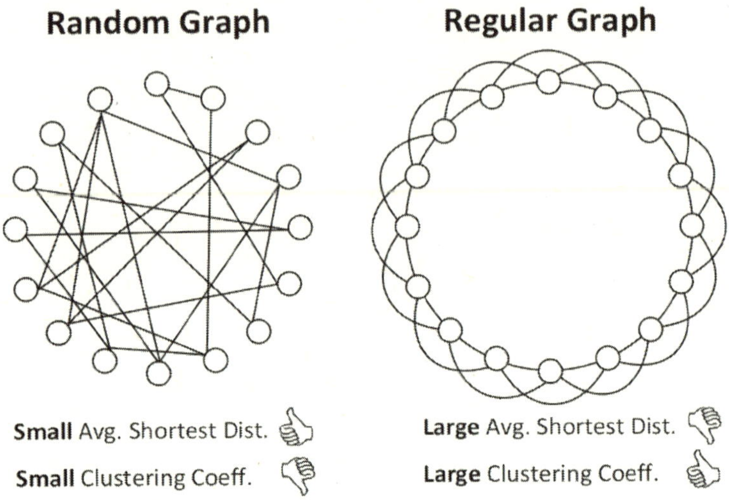

Figure 8.15 Summary of the key differences between the random (left) and regular (right) graph models. While the former tends to have small average shortest distance and small clustering coefficient, the latter is exactly the opposite.

rect connections between them, and lowering the shortest path lengths? This is technically true, but you have to think about the average number of shortest distances in such a model. If this number is large relative to the total number of nodes, then this means people are directly connected to the majority of others in the network. This is obviously not true in real life: on Facebook, you are only "friends" with a very minute fraction of the total population. So such a model cannot explain small world.

On the other hand, random graph networks are exactly the opposite: they have a small average shortest path distance but also a small clustering coefficient. We illustrate the key differences between the two in Figure 8.15. Is it possible to have a *hybrid* graph that combines properties of random and regular graphs to get the best of both worlds: large clustering coefficient and small average shortest distance?

Watts-Strogatz Model

We just discussed why we cannot rely on increasing the number of links/node in the regular graph to reduce the average shortest path length. So let's think practically here. We need a large clustering coefficient, and the regular graph provides this quite nicely. At the same

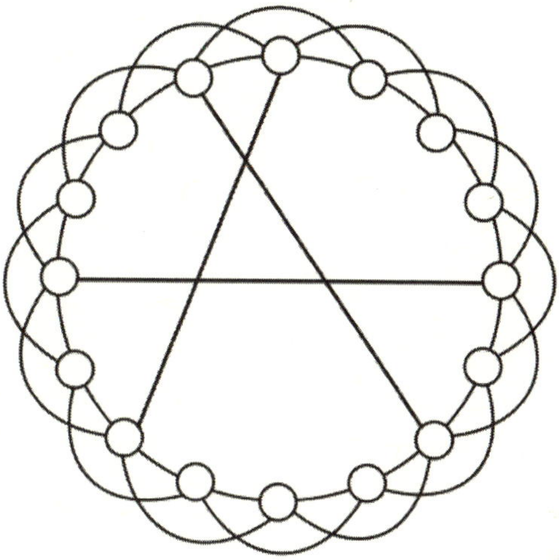

Figure 8.16 An example of the Watts-Strogatz model. It adds random links, possibly long-range ones, to a regular ring graph. Whenever there is a short-range link, there is now a probability of establishing a long-range link randomly connecting two nodes. The resulting graph has both a large clustering coefficient and a small (expected) average shortest path length, which provides an explanatory model for small-world.

time, we need some long-range links to reduce the distance. So what if we add a *few* links between nodes on opposite ends of the graph?

This is the basic idea behind the **Watts-Strogatz Model**, which was first published in 1998 by Duncan Watts and Steven Strogatz in *Nature*. As shown in Figure 8.16, we start with a regular graph with a reasonable number of links/node, and *randomly* add some long range links between nodes.

This model became the most fundamental explanation of small world networks with large clustering coefficients. It has two parameters (besides the total number of nodes):

- links/node, which is the number of nearest neighbors each node has, as in the standard, regular graph.
- `prob`, which is the probability that we connect some random pair of nodes, for each link that is in the regular graph.

To see how long-range links are added, suppose there are 32 links in a regular ring graph. Then, for each of these 32 links, we may (or may not) establish a long-range connection between two random nodes. The

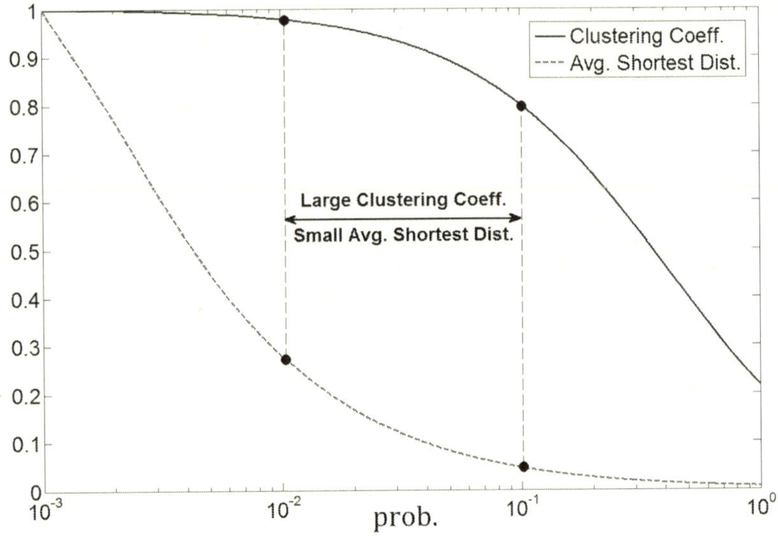

Figure 8.17 A numerical example of the impact of `prob` on the clustering coefficient and the average shortest distance in the Watts-Strogatz model. We divide the traces by their maximum values to limit the plots to between 0 and 1, for a fair comparison. When `prob` is small, e.g., between 0.01 and 0.1, Clustering Coeff. barely decreases at all, but Avg. Shortest Dist. is substantially smaller. This is how it obtains small-world with a high clustering coefficient.

higher `prob` is, the more likely it is that each of these 32 "trials" will result in "success," therefore establishing more long-range links.

The key point is that with additional links, we get to preserve the large clustering coefficient of a regular graph while achieving the small world effect. With a little randomization, the average shortest path's distance can be reduced substantially.

Naturally, adding random long-range links will reduce the clustering coefficient, because it will increase the number of connected triples and most likely will not create more triangles. Remember, we are connecting nodes potentially many links away from each other, so it's likely that they do not "know" the same people.

So, how much randomization do we need for distance reduction, and how much can we *tolerate* for preserving the clustering coefficient? Without going into mathematical detail, it turns out that as long as `prob` is small (e.g., around 0.1), the impact on the clustering coefficient will be almost negligible. Yet the average shortest-path distance will decrease dramatically. We illustrate this in Figure 8.17, where we simulate the Watts-Strogatz model with 600 total nodes and 6 standard links/node, while varying the probability of establishing random links. We have plotted the clustering coefficient and the average shortest path lengths as functions of this probability (divided by their maximum values, to limit the vertical axis to between 0 and 1 for a fair comparison). When `prob` is in the range of 0.01 to 0.1, the former is large and the latter is small, as we wanted.

You are probably wondering why it is that `prob` reduces the average shortest distance so dramatically, but does not impact the clustering coefficient much. Fundamentally, it has to do with the very definition of our metrics:

- The shortest path is an *extremal* quantity. We only care about the *shortest* path, and as such, there is no need to reduce *all* the paths' lengths. All you need to do is add a few long-range links, and even add them randomly, and then the shortest paths will be much shorter.
- In contrast, the clustering coefficient is an *average* quantity: it is defined as the number of triangles divided by the number of connected triples, and as we saw, we can do this as an average at each node. So adding a small proportion of non-triangular, connected triples does not hurt the clustering coefficient that much.

Therein lies the magic of small world with a large clustering coefficient: we have triad closure relationships with most of our friends, but a very small fraction of our friends are outside our normal social circle. All Milgram needed to observe six degrees of separation was this very small fraction of long-range links.

8.1.3 Six steps are locally discoverable

Why is it that we defined the shortest path as an extremal quantity, but the clustering coefficient as an average quantity? Rather than finding the shortest paths between all nodes, shouldn't we have to find the average path lengths, to be "fair?" (Be sure not to be confused by the fact that in either case, we average all the lengths in the end). Well, we don't think that's necessary. For any two nodes in the graph, just the fact that there *exists* a short path should suffice, since they can just use this to communicate. There's no reason to guarantee that *all* paths are short.

But there's an issue with this. If we are relying on the fact that there *exists* a short path, we have to make sure that people can *find* these

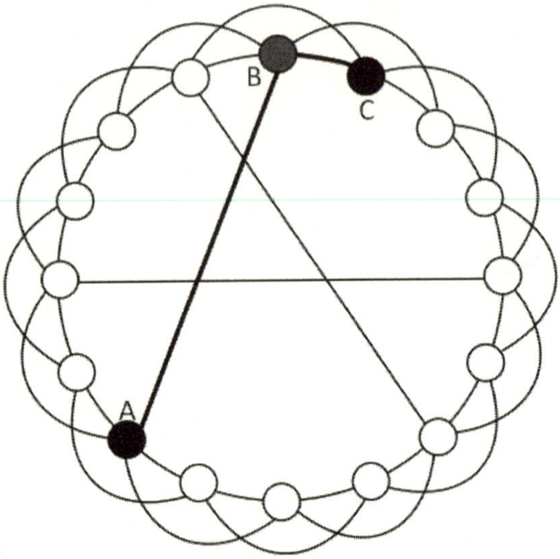

Figure 8.18 In this Watts-Strogatz graph, we have a few long-range links. We see that there exists a short path from A to C: `(A,B,C)`. But how does A *know* what this short path is? It's not clear that he would know B and C are neighbors, and hence *finding* the best next step (here, B) can still be difficult.

paths. This process of social search is not always easy: people don't *know* the network structure, and hence if they aren't directly linked to their destination, it can be difficult to determine where to go next. For instance, in Figure 8.18, for A and C to communicate, how do they know that they are both connected directly to B?

As we saw with the Internet in Chapter 6, devices called routers explicitly pass messages to each other so they all know where to send information. Unfortunately, we don't have this convenience in social networks.

Greedy social search

If you were one of those people participating in Milgram's experiment, or one of those along the paths initiated by these people, how would you decide on the best next hop to use by just looking at the destination's name, address, and occupation? You would probably take into account some combination of geographic proximity (relatively easy to see) and occupational proximity (relatively hard to determine) to come up with some "social distance." Then, you would look at all your friends whom

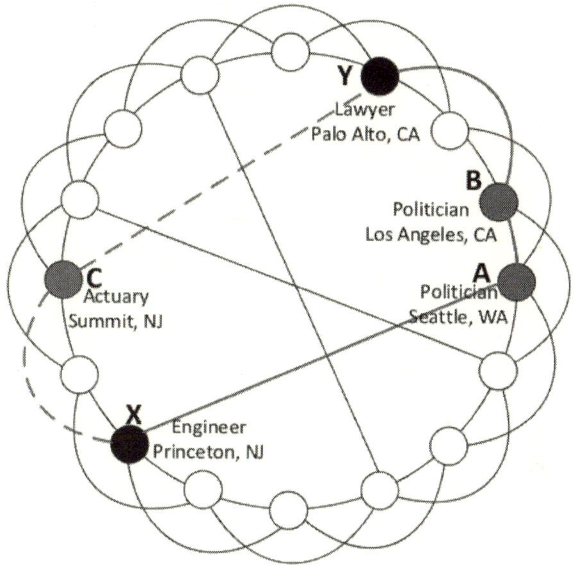

Figure 8.19 Illustration of the process of greedy social search. X, an engineer from NJ, needs to send a message to Y, a lawyer in CA. He looks among his neighbors to see who is the "closest:" a politician (A) in WA is the best in terms of geographical and occupational "distance" that he has. Then the politician runs greedy search and sends to B, and B knows the destination personally, relaying the message to Y. What the engineer does not know is that one of his neighbors (C), an actuary in NJ, happens to be linked directly to the destination. While the shortest path is 2, the discovered path by greedy search is 3. Our hope is that paths discovered will be similar in length to the shortest path, as it is here (2 is not much smaller than 3).

you know by first name, and pick the one closest in characteristic to the destination based on this distance.

This process is called **greedy social search**, in which you only take into account what you know *local to you*, and make the best possible decision from what you know. So, if Node X wants to get to Node Y, she would look at all her neighbors, and determine who is the best to send to. Then this Node, say B, would look at the characteristics of Y, and determine who to send to. Then this Node, say C, would do the same, and so on, until finally the message arrived at Y. Naturally, this is not awlays going to result in the optimal route, since nobody has a global view of the network, but it will probably suffice. We illustrate this in Figure 8.19, in which X takes path `(X,A,B,Y)` to get to the destination, but the shortest is `(X,C,Y)` in reality. Node A runs greedy search to

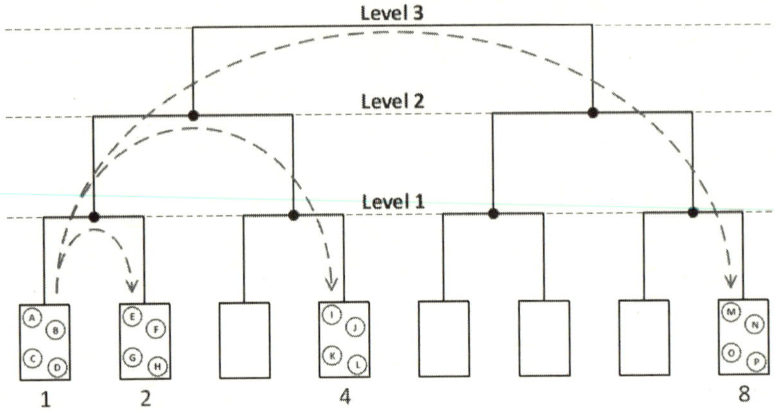

Figure 8.20 An example of the Watts-Dodds-Newman model. At the bottom of the tree, we have eight leaves. Each leaf has complete, direct connections between all people inside, e.g., A, B, C, and D are all connected in Leaf 1. Each person also has a certain chance of having a link to those outside of its direct leaf, but the probability of this being the case decreases as the first level of common ancestry between the nodes gets higher. For instance, A and E's first level of common ancestry is Level 1, A and J's is 2, and A and M's is 3. There's a greater chance of A and E being connected than A and J, and this is in turn more likely than A and M.

determine that a Politician in Seattle, WA is "closer" to a Lawyer in Palo Alto, CA than any of his other neighbors, but in reality, it just so happens that an Actuary in Summit, NJ has a direct relationship with the destination.

We wonder whether the average length discovered by such a greedy search can be close to the average shortest path length: if so, then in addition to being *plausible*, we can say that short paths are *discoverable*.

Watts-Dodds-Newman Model

There have been several models for social search in the past decade beyond the original Watts-Strogatz model. We will examine one here in particular: the **Watts–Dodds–Newman Model**. Here, people live in different "leaf nodes" of a **binary tree**, as depicted in Figure 8.20. In this tree, there are eight leaf nodes (labeled 1 – 8) at the bottom, and within each, all people are connected directly to one another. Moving up the tree, we have increasing levels of *ancestry*, which include more and more of these leaf nodes. There are four ancestry nodes at level 1, each of which contains two groups of leaf nodes (1 & 2, 3 & 4, and so on). There are two at level 2, both of which contain four groups of leaf nodes (1 – 4 & 5 – 8). Finally, the third level encapsulates all the leaves. Adding more and more leafs, which must be done in a multiple of two to preserve symmetry, just adds ancestry levels. So 16 nodes requires 4 levels, 32 requires 5, and so on.

All the people within the same leaf are directly connected, as we said. Thinking in terms of Milgram's experiment, we expect them to all be close in terms of both geographical and occupational proximity, and clustered accordingly. To analyze how far two leaves are from each other, we have to look at the level of *first common* ancestry. Person A (in Leaf 1) and Person E (in Leaf 2) are first connected at Level 1. Person A and Person I (in Leaf 4) are first connected at Level 2, and so on. As this level increases, the probability that the two nodes know each other gets smaller. So A and B, being in the same leaf, are guaranteed to know each other. A and E are not, but are more likely than A and I, which is in turn more likely than A and M (in Leaf 8). The higher this level becomes, the *further* the people are from one another, in terms of geography and occupation. But it's still possible that they know one another: this encompasses the random-adding of links in Watts-Strogatz.

This model sounds quite intuitive, but we still have to show that the path discovered by greedy social search exhibits small world. And what exactly does this mean? Well, without going into mathematical detail, it suffices to say that the discoverable path length must grow *very slowly* as the number of nodes in the graph increases. So as we increase the number of leaf nodes from 8 to 16, 32,...,1024, 2048 (powers of two) to millions and millions, this path length must remain relatively small.

As it turns out, if we specify how the probability of people knowing each other will decay as the level of common ancestry goes up, we can achieve this. In particular, the probability must decrease by *exactly one half* between one level and the next. For example, if there are 3 levels, the probabilities at 1 - 3 must be roughly 57.1%, 28.5%, and 14.3%, since $28.5 = 0.5 \times 57.1$, and $14.3 = 0.5 \times 28.5$ (note they must add up to 100%). From our three-level graph in Figure 8.20, this means that the probability A and E are linked must be 57.1%, that of A and I 28.5%, and that of A and M 14.3%. If we changed the probability of decrease from 50% to, say, 60%, these three would change to 64.1%, 25.7%, and 10.3%, again since $25.7 = (1 - 0.6) \times 64.1$, and $10.3 = (1 - 0.6) \times 25.7$.

In Figure 8.21, we show just how important this one-half decrease is. In this example, we assume that there are 50 people in each leaf-node, and that each person is, on average, connected directly to 100 people

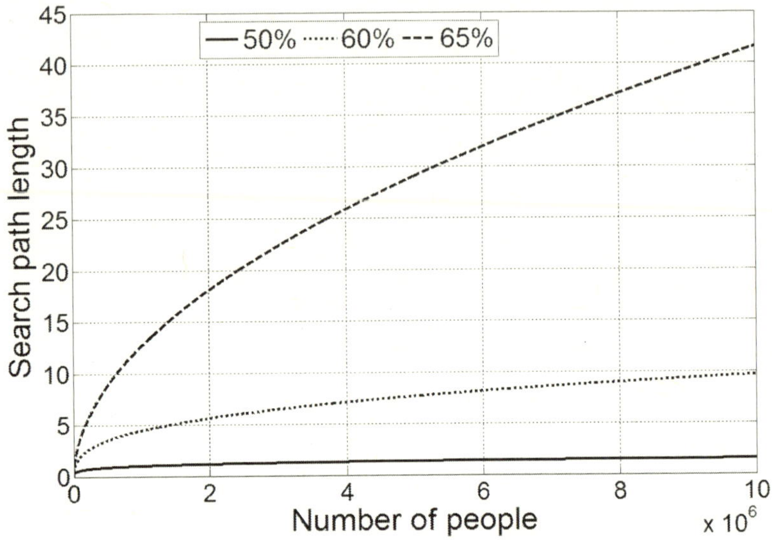

Figure 8.21 Impact of changing how the probability of people knowing each other goes down with increasing ancestry, in the Watts-Dodds-Newman model. For 50%, we see the search path length barely increases as the number of people increases to 10 million. For 60%, the increase is perhaps still tolerable, but at 65%, it increases drastically. This highlights the importance of keeping it at 50%.

(so 50 in their leaf, and 50 outside). We plot three cases: the ones when the decreases are 50%, 60%, and 65%. In each, we have plotted the search path length as the total number of people increases to 10 million. For 50%, the search path length only increases to about 1.5 at 10 million people, which is close to nothing. For 60%, it increases to about 9.7, which a lot higher than 1.5. And for 65%, it goes all the way up to 41.6, which is just way too large.

In this numerical example, as long as we keep the decrease at 50%, we can guarantee that the greedy search length will exemplify small world for the Watts-Dodds-Newman model. We can thus conclude that Milgram's observed six degrees is indeed discoverable in realistic networks.

8.2 Big "cloud"

Many times in this book, and probably in many other places, you've seen things represented as "clouds." Usually the meaning is very simple:

Figure 8.22 We use a cloud to represent things we are trying to abstract and view at a high level in diagrams. They take a "black box" role in our minds, where we don't concern ourselves with what's inside. In this section, our focus is on determining what makes up the *literal* "cloud service," which got its name from the fact that we view it as a cloud.

everything that's out there "inside" the network that we don't need to know the details about, at least right now. You just have to know that it's there, and how it fits into the diagram that you're looking at.

In this way, we treat "clouds" as "black boxes" (see Figure 8.22). But did you ever stop to think what was inside?

Clearly, the answer will change depending on what it is we are looking at. When we put a picture of a cloud to represent the Internet, we meant it to be some large, complex interconnection made up of millions of routers, switches, and other devices. If we drew a cloud to stand for a large corporation like AT&T, looking inside would reveal about a quarter of a million employees, each of whom are completing just a small fraction of the work output that the company completes each day. And if we drew a cloud to represent a lego set, we would be abstracting all the small little pieces that go into making the whole from the sum of the parts.

There's a new and emerging innovation that we hear about and use so often, but probably just take for granted. Conveniently, we have come to call it "the cloud," a quite-fitting name since we rarely think about what's inside of it, or where it is for that matter. Where do Dropbox and Google Drive store all the data you entrust them with? How does iCloud allow you to sync all your music, videos, contacts, and more whenver you have an Internet connection, wired or wireless? As we uncover the contents of the cloud, we'll see just how fitting the "building block" concept really is for answering these questions.

The trend with cloud is that consumers can actually use *networks* as *computers*. Not only can it provide us with storage (like our hard

Figure 8.23 Today, we use cloud services for both storage and computing. Cloud gives you memory and processing capabilities outside of your computer, perhaps from across the country. Our phones, laptops, desktops, iPads, and any devices with Internet connection are increasingly taking advantage of it.

drives), but it also allows us to run computationally-intensive applications on all of our devices, as shown in Figure 8.23. Perhaps the epitome of this is Google's ChromeBook, introduced in June 2011, which is a "cloud laptop:" you really only have a web browser (Chrome) with Internet connectivity. All the processing, storage, and software lie somewhere in Google's servers, and you access them remotely. All the computation power is no longer in front of you, but somewhere you don't even know. This truly represents the ultimate version of online computing.

Beyond individual consumers, even large corporations have been shifting to cloud services. They are able to run all their applications and software by renting out portions of **data centers**, which are the places that cloud storage and applications are kept and executed. Data centers are useful because they house many servers, which are basically large computers with lots of storage and processing capability. Today, large data centers are typically over 300,000 square feet, house hundreds of thousands of servers, and cost hundreds of millions of dollars to build. Cloud providers operate these data centers, and rent them out to consumers, ranging in size from you to Target.

These servers are one of the "building blocks" of cloud storage. As

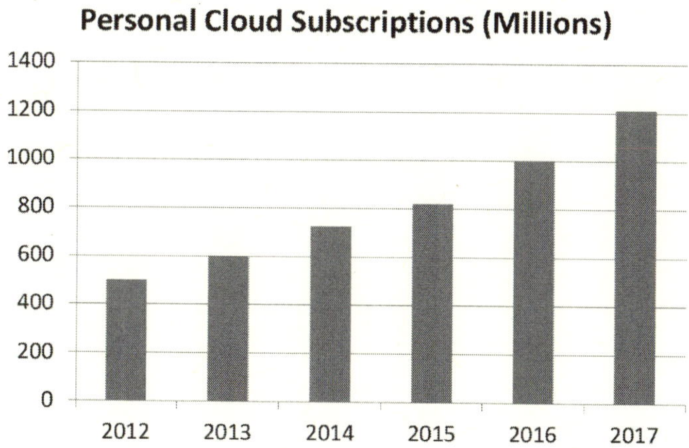

Figure 8.24 A recent study by iHS Suppli forecasts the number of personal cloud subscriptions by year. In 2012, there were about 500 million, and by 2017, there are expected to be well over a billion.

a company has many employees, a datacenter has many servers, each of which only does a small fraction of the overall work. But without them, we wouldn't have computing in data centers at all.

There are three major cloud providers as of 2012: Amazon's EC2, Microsoft's Azure, and Google's AppEngine. A pioneering player in cloud services is actually Amazon, even though to most consumers Amazon stands for an online retail store. Indeed, they own and operate their own data centers, which you can rent as a cloud consumer.

For many years, it had been a goal in the computing and networking industries to one day allow users to rent resources inside the network, in such a way that everyone would benefit economically. Thanks to advances in technology, cloud services are taking off and evolving fast. In Figure 8.24, we show results from a recent study on personal (not counting enterprise) cloud subscriptions. In 2012, there were about 500 million, and by 2017, it is anticipated to be well over a billion.

There are a lot of terms people use when describing cloud computing, that can be quite confusing. But we can get an overall sense of the industry, from supplier to user, in Figure 8.25. In general, there are three key components of the "food chain."

1. **Cloud providers** build and manage the data centers and their associated hardware, like servers. But there's more to them than just half a million servers. We also need to connect these servers together, which is done using network switches. Switches are like

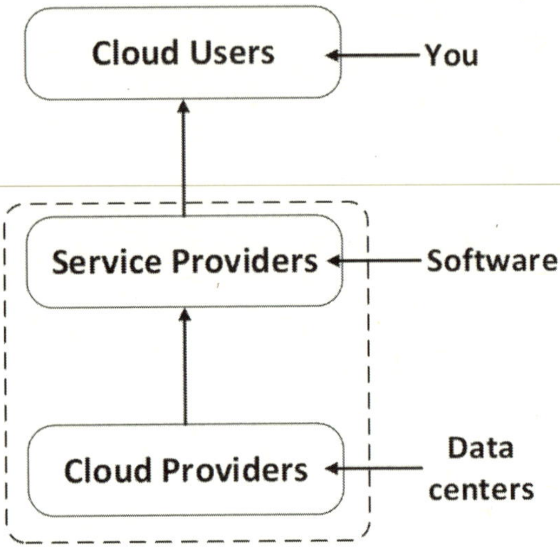

Figure 8.25 Here, we show the cloud service "food chain." The cloud providers operate the data centers, which includes hardware such as servers and switches. Service providers run software using the data centers, that can be used by the end users. Finally, cloud users include both consumer (like you) and enterprise (like Target) customers, who use the software/hardware offered by the providers.

routers, except they typically only go up to the *link layer* in the protocol stack (see Figure 7.5), and are used for *local* connections (*i.e.*, among a bunch of servers in a data center). We will focus a lot on switches in this section. Finally, data centers need storage resources, like memory devices.

In this way, we see that a data center basically has a network of its own. While we look at them as "black boxes" or "clouds" in the graph of the Internet, they themselves each consist of large networks of servers and switches. In this graph, the servers and switches become nodes, and the links the cables that connect them.

2. **Service providers** offer software and applications that run in data centers. This software is, in turn, used by consumers. For example, an iPad application developer may decide to use Amazon's cloud resources to deliver its services. And sometimes, the service provider *is* the cloud provider. For example, portions of the iCloud app run on Apple's own data centers.

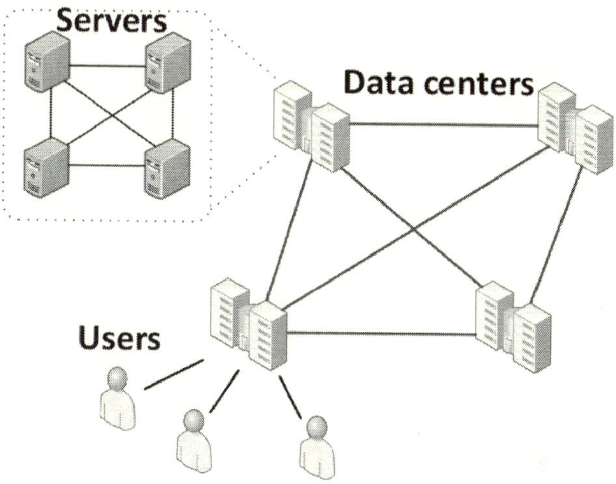

Figure 8.26 There are many different dimensions to data center networking. First, data centers have to be connected among themselves. Then, within data centers, we have to connect the servers together. Finally, all of this has to be accessible by end users. We will focus on intra-data center networking mainly.

3. **Cloud users** are consumers and enterprises that use services running in data centers. You can get the content you want (e.g., documents, books, music, video) or the software you need (e.g. Office software, an iPhone app) from the cloud. And you get it *on demand*, anytime, anywhere, and on any device, as long as you have an Internet connection. When you think about it, it's really no different from having it right on your computer, except you don't have the burden of storing and running it with your own hardware. You're just paying someone else to do that (and those are exactly the cons as well!).

We'll return to our goal of dissecting data centers in just a few minutes. But now that we understand the players involved in cloud services, let's first take a look at what exactly defines them, and some pros and cons associated with having such centralized computation and storage.

8.2.1 How do we define a "cloud service?"

To make the overall cloud food chain work, we need all of the following ingredients:

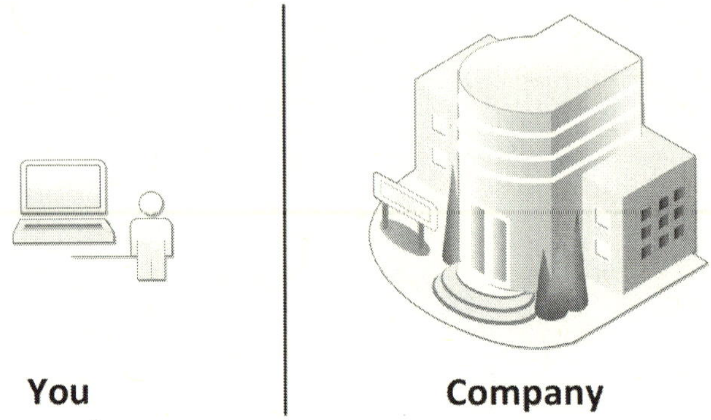

Figure 8.27 Cloud providers have to be on-demand in scale. What this means is that they have to serve customers of many different sizes, ranging from you to a large company (like Target).

1. *Large-scale computing and storage.* The computing is provided by the servers, and the storage by the memory devices. In fact, any process (i.e. application, program, or storage) you would be running would only take up a very small portion of just *one* of these. As a result, the devices are usually shared among many people, where each process gets just a slice of the resource.

2. *Networks.* Data centers are networked from within (*intra*), across themselves (*inter*), and to the end-users. We show this in Figure 8.26. The networking dimension naturally will be the focus of our discussion here, and specifically, we will look at how we build a network *within* a single data center.

3. *Software for user-interface.* The software is what you actually see and interface with as the end user, like the folder you place your document in using Dropbox. Software also handles digital rights management, security and privacy, billing and charging, etc.

If we combine these three things, i.e., our data centers have enough hardware, we have the right networking, and the software works well, we can expect a functioning cloud service. What's really key here is that as a user, you can get these services *on demand*, both in time and scale.

1. *On demand in timing*: You can change your requests for resources at short notice. When you need to store something, it's there for

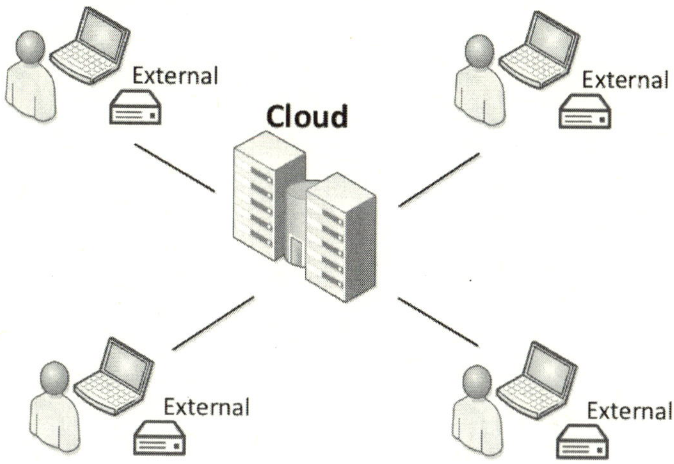

Figure 8.28 We illustrate how cloud is different from personal storage. With your own external drive, you have full control over your space, and you know exactly where it is. It's all yours. With cloud storage, you are sharing the facility with many other users, and you really don't know where in the "cloud" it's going. Plus, as more people rely on the cloud, a single outage could do more damage.

you, even if you only need it for a short time.

2. *On demand in scale*: It supports many users requiring many different levels of resources. For instance, it's dealing with customers as small as you, or as big as Target (and in all seriousness, Target runs its web and inventory control in a rented cloud, even though they are the second-largest retail store chain in the United States). We show this in Figure 8.27.

8.2.2 Can we count on the cloud?

Cloud services face many challenges even through the present day, even though the cons are increasingly outweighed by the benefits. But let's briefly discuss them before moving on.

The main negative of using cloud storage comes from the fact that it's a *shared* facility, largely outside of your control. When you use your own computer for computing and storage, you naturally have the whole thing to yourself. Nobody is renting out part of your processor or using your storage except for you. But that's not the case when you

Figure 8.29 A good analogy between personal and cloud storage is choosing between riding in a taxi or on a bus. With the latter, you pay less but don't really know what kind of seat you're going to get, or who you're sharing the bus with. This echoes the performance, security, and privacy concerns with cloud storage.

use cloud; there are many other people that you are contending with for resources. This is illustrated in Figure 8.28.

This debate of shared vs. dedicated resources runs far and deep, echoing the dichotomy we saw between packet and circuit switching in Chapter 6, and that between CDMA and TDMA in Chapter 1. Once you are in a shared facility, the performance guarantee will always be compromised. It's like choosing between riding a bus or taxi. When you ride the bus, you pay much less, but there's no *guarantee* as to where you are going to sit, or even if you're going to have a seat at all. And there's no telling how many people you are going to ride the bus with. That is the price you pay to enjoy the benefits of cloud services: it's probably cheaper and more convenient than buying and carrying around an external disc drive with you, but you have to share the cloud with people, it's not yours to keep. We describe this analogy in Figure 8.29.

Along these lines, there are also problems of privacy and security. Who can gain access to your information that you store in the cloud? What type of people are working in the data center, and how easy would it be for them to compromise the integrity of what you're storing there? The problem is that while they provide assurance, you can never be entirely sure who can grab it. While your computer is entirely under your control, the cloud is not.

This issue of control echoes the performance concern in the cloud: when can we count on it to be available? For instance, Amazon's cloud has had a few outages: one in April 2011, another in June 2012, and most recently on Christmas Eve in 2012. The two latter disrupted Netflix service for significant amounts of time, since they use Amazon cloud. On December 24, 2012, virtually all Netflix customers were left without service from about noon to late in the evening. This shows just how important the cloud is becoming to portions of our lives, and how much we rely on these outages not to occur. The best way to avoid getting caught in one of these outages is to distribute your use across multiple cloud providers, so when one is down, you can still use the others (hopefully), i.e., redundancy and diversity for robustness.

8.2.3 Why the cloud?

We just discussed some rather ominous features of the cloud: massive outages, security issues, etc. So why does it make sense? The key benefits are similar to some of the things we discussed with packet switching in Chapter 6. We will summarize them here.

First, what's in it for the cloud users? Well, remember when we talked about *resource pooling*? The idea was that we could look at the pipes in the Internet as one entity, and fill up demand on any of these pipes as needed. In this way, all the resources were being spread and shared across many users. And that's exactly what's in it for users of a cloud service: there are many of them using it, so the *costs* of building and maintaining the services is shared across many users. It's like when you ride a bus, and everyone is chipping in a portion of the costs it takes to keep that bus running. Or when you rent a car, and you pay a daily fee to the company to keep that car going. In this way, you are paying *operational* costs, not *capital* costs when using cloud services: never do you have to *buy* the rental car.

So it's cheaper for you, but why would cloud *providers* be interested? It's all in the economic advantages of running such a large storage facility, where everything is consolidated and you have complete control. As any course in economics will tell you, things get cheaper when you buy them in bulk, a principle called **economy of scale**. Such economic benefits for the cloud provider occur both on the supply side, and on the demand side.

- On the *supply side*, the cloud provider has to get all the resources to run the data center. This includes all the servers, switches, memory devices, land, employees, electricity, and so on. Because of its large scale and bargaining power, the provider can get all these at *significantly discounted prices*. Even when compared to a medium-sized data center with thousands of servers, a large scale one with hundreds of thousands can often achieve over a five-fold improvement in

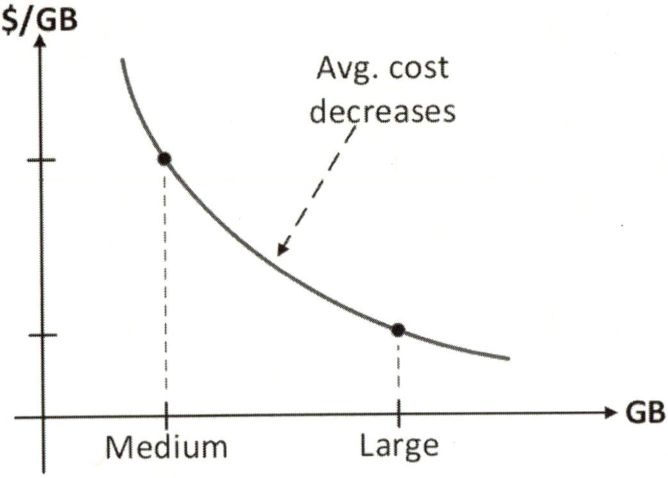

Figure 8.30 The principle of economy of scale. When a data center increases from medium to large size, in terms of the number of GB it has, the average cost it pays per GB is much lower. It's like when you have the choice to buy one thing for $20, or two of them for $30. The average cost for the latter is $15, which is lower. This is one reason why data center operators would prefer to be larger.

the cost it pays per unit. We show the principle of economy of scale in Figure 8.30.

Remember, we measure storage and data size in terms of GB. So if we add up all the costs, and divide by the GB that the data center offers, we get a cost per GB. The large scale data center pays a much lower cost per GB than the medium sized one (though it has to buy more, meaning it pays a higher *overall* price).

- On the *demand* side, scale helps again, but this time, through *statistical multiplexing*, another concept we discussed in Chapter 6. In the context of the Internet, this meant that many users would be sharing a single pipe, which meant that if a pipe had leftover capacity, someone else could use it so there would be nothing wasted. This also helped because users want things at different times: maybe User A has high demand this second, and User B has low demand, so they can share the pipe. Then the next second, User A has low demand, but User B has high demand. And if they both had low demand, we could fill up the pipe with a lot of low demand users.

  And this is exactly how it helps for data centers. We show this in

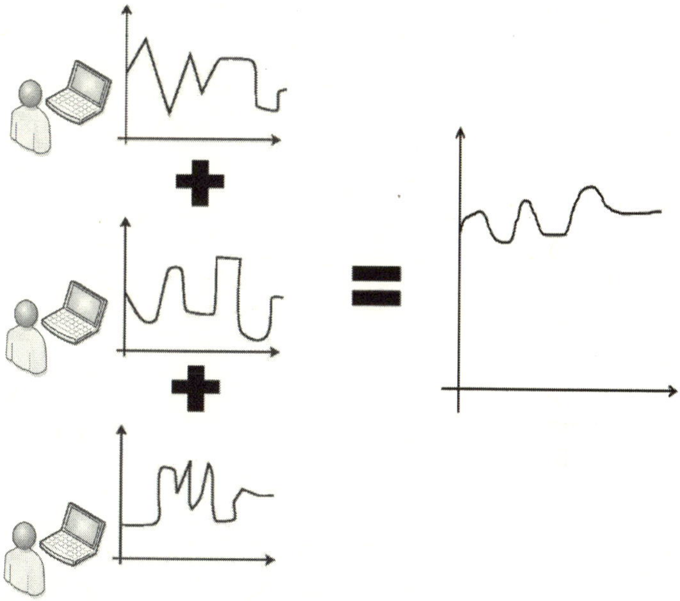

Figure 8.31 Over time, any single person's demand for cloud is likely to vary a lot. But if we look at a lot of users in aggregate at once, they will tend to peak at different times, which makes the overall demand a lot smoother. As long as users' demands do not peak together, cloud providers can leverage this smoothening effect to reduce the amount of capacity they have to provide.

Figure 8.31. People's demand fluctuates over time, but since there are a lot of them, we can always find a way to keep the data center reasonably utilized. We expect the overall demand, when we look at all the users together, to fluctuate much less than a single person's. This reduces the amount of capacity the data center operators have to provide, as long as people don't peak together. Of course, the overall pool may still exhibit time-of-day fluctuations, so this isn't a perfect solution. In fact, the average demand on a server in a data center is often below 20% of its capacity, meaning they could tolerate 80% more.

8.2.4 Building blocks

Cloud is all about *scale*. Today's large-scale data centers are indeed huge, so big that electricity and cooling costs sometimes are more than

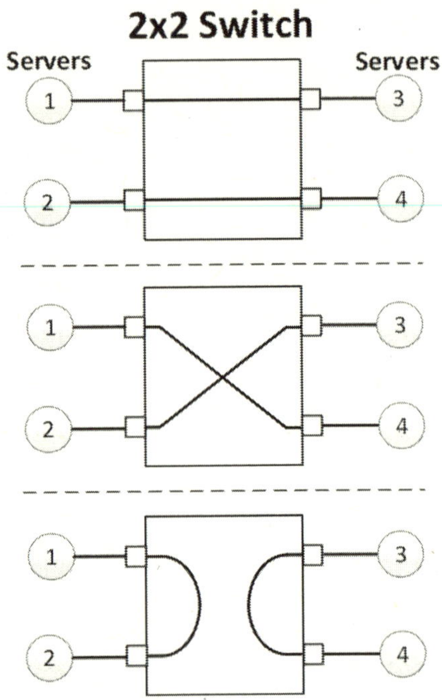

Figure 8.32 The function of a switch is to connect some of its ports together so the inputs can communicate. In this case, we show a 2x2 switch (two on the left, two on the right), with a server at each port. At the top, we have one set of connections: 1 to 3, and 2 to 4. In the middle, we have another: 1 to 4, 2 to 3. And at the bottom, we have the last case: 1 to 2, 3 to 4. The switch will change between these three as instructed.

half of the total cost. If an iPhone is one of the smallest computers we use, then a data center is one of the largest.

We have made an important assumption in the last section that it's actually *feasible* to scale up a data center. But is it possible to get the economy of scale we saw in Figure 8.30, if we need *more expensive equipment* to make a data center larger?

The answer is, probably not. So let's try to dissect how a data center is networked in the first place.

We need to provide at least one path between each server, so they can talk to each other if necesssary. To do this, we can't just connect a bunch of wires between them; we need something to actually control where the data is flowing. That's what a switch is for.

As we said in Chapter 7, a switch is a layer-2 device that deals with local networking. We show the main idea in Figure 8.32. Like a router,

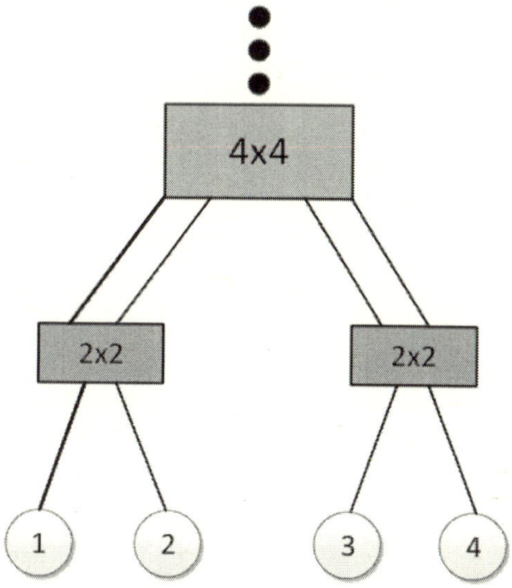

Figure 8.33 Here, we show how to start building a tree up using four servers. The servers are grouped in pairs of two at the bottom, and each pair is fed into one side of a 2x2 switch. The other side of each 2x2 switch is fed into one port of a 4x4 switch. So, the bottom side of the 4x4 handles two 2x2 switches, and the other side has four ports to connect to a higher up switch.

a switch has some ports, which it can connect together depending on what it's needed to do. Here, we show a 2x2 switch (two inputs on the left, two on the right), where we have three possibilities: connect Servers 1 and 3, and 2 and 4 (top), connect servers 1 and 4, and 2 and 3 (middle), or connect Servers 1 and 2, and 3 and 4 (bottom). The idea is that it can "switch" between these three states, using its own internal hardware, depending on what's being asked of it. A router does essentially the same thing, just based on IP addresses.

So the servers are what we want to connect, and the switches control who is connected right now. But obviously, a data center is going to have more than two sets of servers that want to be connected at any one time. Therefore, we have two options.

1. Add more ports to the switch.
2. Add more switches.

Let's think about this first option for a moment. We're talking about hundreds of thousands of servers in a data center, each of which needs to be connected. And each of these connections probably requires up

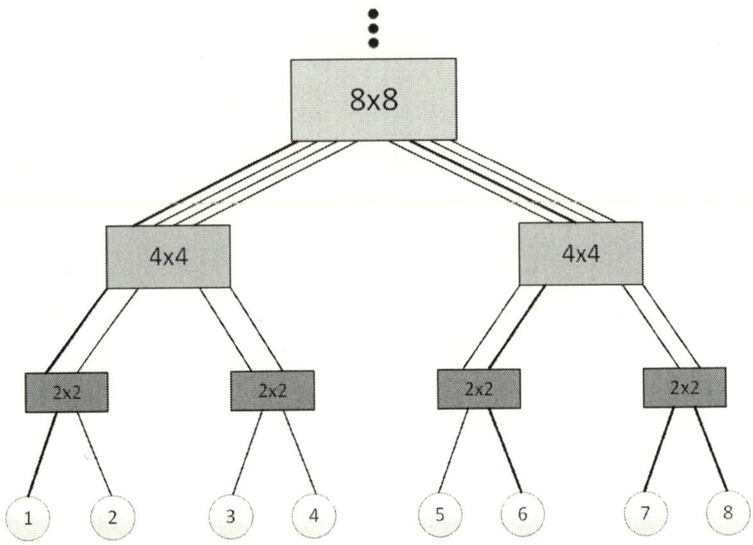

Figure 8.34 Here, we depict a tree connecting 8 servers, using four 2x2 switches, two 4x4s, and one 8x8. We would feed this 8x8 to a 16x16, and so on. The idea is that we can connect any server to any other, by finding their lowest common ancestry switch (as we discussed in Chapter 4 for social networks). So, 1 and 6 would go all the way up to the 8x8 (as highlighted), while 2 and 3 would go to the left 4x4, and 7 and 8 would just go to their immediate 2x2.

to 1 Gbps (one billion bits per second). That means that we'd have to support hundreds of thousands of giga-bits each second using one switch. This is simply infeasible.

Trees again...

So, let's try to add more switches. We can imagine a tree-like structure, in a sort of hierarchy, that we can keep adding onto as we have to add more servers. We see this in Figure 8.33, where we have four servers. At the bottom of the tree (the leaves), we have our four servers, which we divide into two groups of two (1 and 2, 3 and 4). Each pair forms the bottom inputs of a 2x2 switch, shown in the middle. The other sides feed up to the bottom of a 4x4 switch, which joins the two 2x2 switches.

One 4x4 switch isn't so bad. We could build that. But what happens as we keep going up the tree? We show the next step in Figure 8.34, for 8 servers. Now, we have two of the structures from Figure 8.33, each connected by one-half of an 8x8 switch. Once you wrap your mind

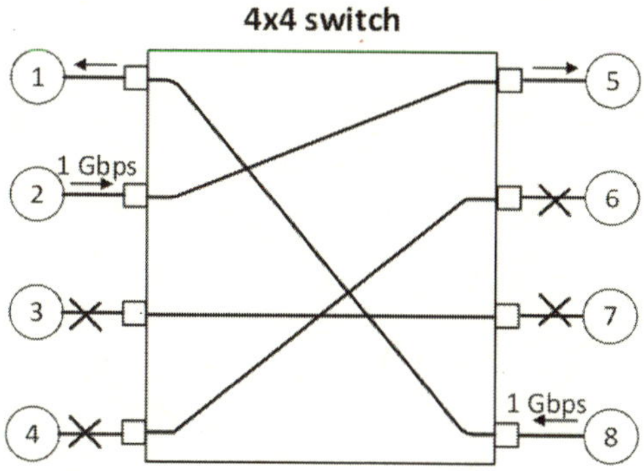

Figure 8.35 Here we show a 4x4 switch that can handle a maximum of 2 Gbps. If ports 2 and 8 are each demanding 1 Gbps (*i.e.*, demanding their capacity), then the switch can't handle anything else. Basically, you can look at it as if the other two connections are useless to the network at this time.

around this, you realize that we are doubling the number of servers each time, to maintain symmetry.

But now, what happens when we want to increase to 16 servers? Well, we'd have a 16x16 switch at the top, connecting two 8x8s. And 32? We'd have a 32x32 at the top. How about 64? 128? 262,144 (2^{18})? The problem is, frankly, we still have the same problem. We'd need an enormous switch at the root of the tree, and we'd be using at least half (the bottom side) of it to make the structure work.

As the number of leaf nodes (servers) increases to 100,000 and more (a large data center), it becomes technologically *impossible* to build the root switch. So what do we do? One possible way around this is to start *oversubscribing* as we climb up the tree. Basically, this means that the switches have more "subscribers" than they could handle, where here, the subscribers are servers. As long as not all the servers need to use their 1 Gbps at once, we'd be OK, and could handle all the connections. Sometimes, the **oversubscription ratio** runs as high as 1:200 in a large data center. This means that for every one connection we could support in the worst-case demand, we actually have 200. We illustrate an oversubscription ratio of 2 in Figure 8.35.

But, there are two problems with this. First, it's still expensive to build these big switches. Second, what happens if all the servers wanted to fully utilize their connection speeds at the *same time*? Well, we'd have 200 times more congestion than we could handle. And this would defeat the whole idea of pooling the servers' resources together, because

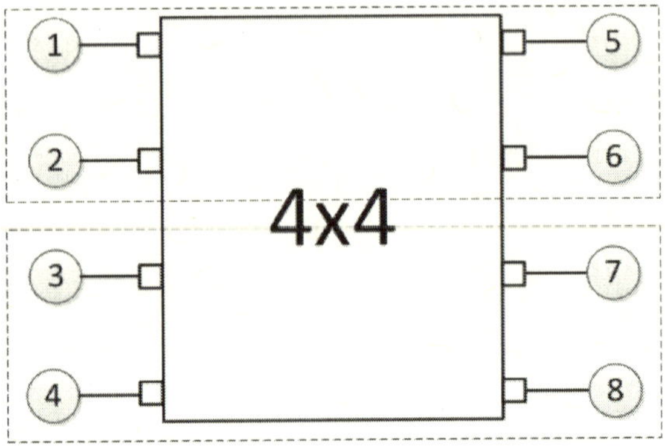

Figure 8.36 We want to be able to represent a 4x4 switch using 2x2's, because the hardware is much cheaper, especially as we scale up. The easiest way to think about this is to split the 4x4 into two 2x2's.

we'd have quite a lot of them that couldn't communicate when they needed to.

With higher costs, we'd need the customers to foot a higher bill, so they wouldn't be too happy. And the data centers wouldn't be seeing all of their benefits, either. What can we do?

The problem is with the structure we're using. No matter how you use it, a tree is just not scalable, because the root nodes will always have increasing demand as the tree gets larger and larger. You may be wondering, well, in the last chapter we used trees for P2P and came up with a scalable solution, so why can't it work here? The difference is that with P2P, our connections were *logical concepts*, rather than *physical links* as they are here. We simply can't build multiple trees, because we don't have the hardware to do so.

But there's still a solution: we can build a *large switch* out of a lot of *small ones* through rich connectivity.

This is the most crucial point we will illustrate in this section. So let's try now to see how that's done. We'll start by decomposing a 4x4 switch, and make it into a bunch of 2x2s. Once we extend this to the 8x8 case, the rest will be easy to see.

From tree to fat tree

In Figure 8.36, we show a 4x4 switch. We have eight ports, four on the left, and four on the right. How can we represent this using 2x2s?

The easiest way to do this is just to cut it into two halves, a top half

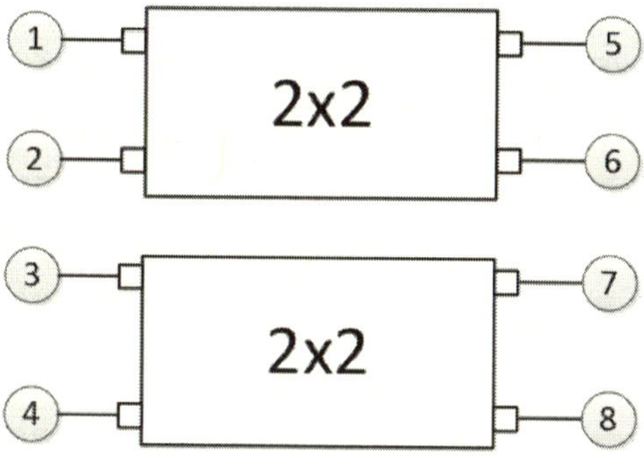

Figure 8.37 Illustration of representing a 4x4 switch using two 2x2s. We lose some of the functions, like being able to go from top left to bottom right, and bottom left to top right, but we'll return to that later.

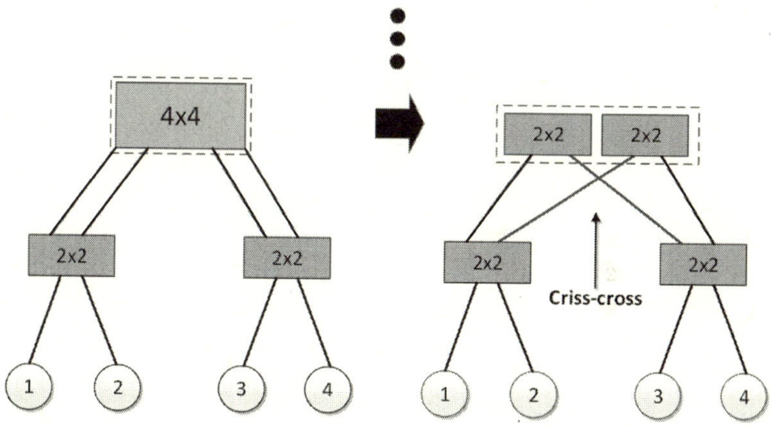

Figure 8.38 Here we show how to replace a 4x4 with two 2x2s in the tree structure. It's relatively straightforward, but we have to pay attention to the criss-cross pattern. The reason this is necessary is so Servers 1 and 2 can connect with Servers 3 and 4 without having to go higher up in the tree. We'll find this useful as we move up to the 8x8.

and a bottom half. In this way, we can have one 2x2 connecting ports 1, 2, 5, and 6, as well as another connecting 3, 4, 7, and 8.

The result is shown in Figure 8.37. Again, this is a pretty simple, but very effective idea. The question is now, how can we fit these 2x2s into the structure in Figure 8.34? This procedure is relatively straightforward, and we illustrate it in Figure 8.38. It's basically just a "cut and paste," with one minor detail: we have to make sure the two lower

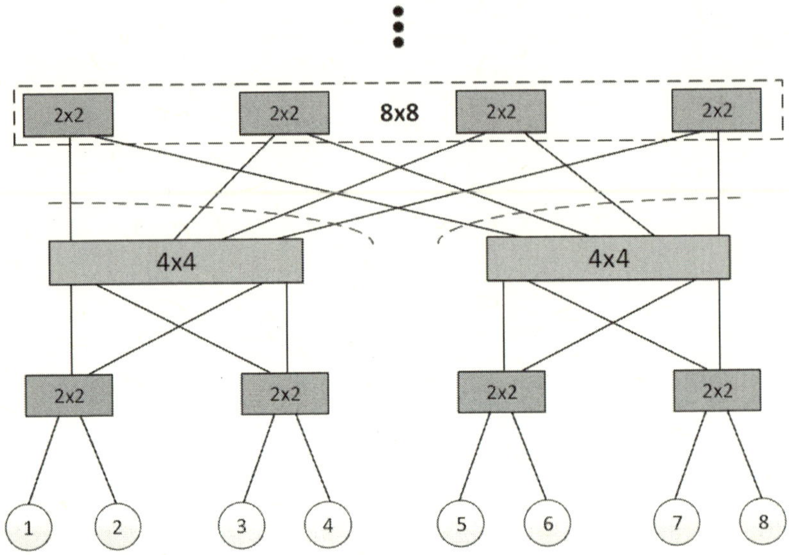

Figure 8.39 Here, we show how to replace the 8x8 with four 2x2 switches, treating the 4x4s as "blocks" again for the moment. We want to make sure that each of the 4x4s gets connected to each of the 2x2s, again to make sure we can get from any leaf to any other. To this end, each of the top 2x2s connect to *both* 4x4s.

2x2's can connect with one another directly. To do this, we enforce the "criss cross" pattern among the inner lines. This way, it's easy for, say, Server 1 to connect with Server 4: Server 1 goes up to the immediate 2x2 switch, then crosses over to the right side of the 4x4, and comes back down. If we didn't enforce this pattern, we'd have to go higher up in the tree, violating the lowest common ancestry principle. In fact, we wouldn't be preserving the tree's structure at all if we didn't.

So then, moving on up, how do we do an 8x8? Simple. Rather than replace it with *two* 2x2s, we put *four* in its place. If we again abstract the 4x4s as "blocks," the result of replacing the 8x8 is shown in Figure 8.39. In this case, we want to connect the 4x4s to the 8x8 in such a way that all the bottom servers can talk to each other. We enforced this just a minute ago with each half, by criss-crossing the connections in Figure 8.38. To keep this going as we move up a level, we make sure *each* 4x4 is connected to *every* 2x2 in the level above. So, one line from each 4x4 takes up one of the connections on each of the 2x2s.

We'd continue this as we moved up the tree, too. When we hit the first 16x16 switch, we would have two 8x8s coming into it. The 16x16

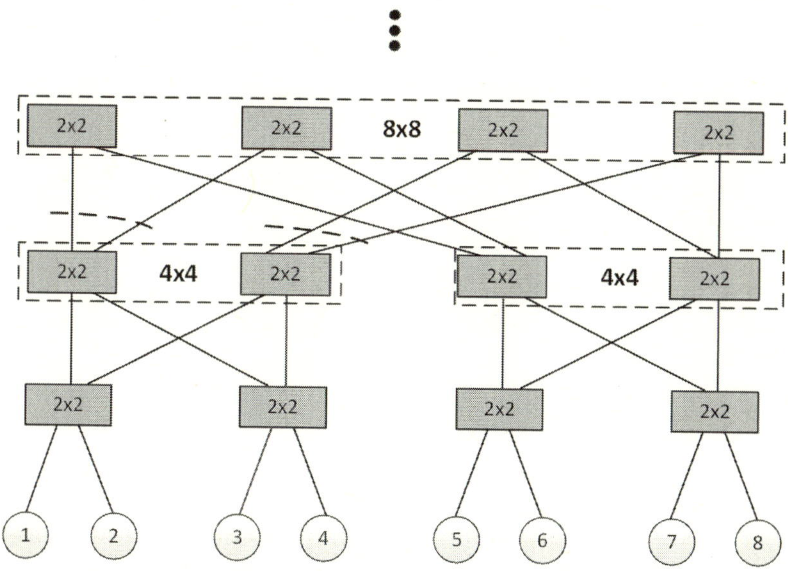

Figure 8.40 We've taken the tree structure and represented it with all 2x2 switches. A key point in connecting it is that each 2x2 spreads its connections to different switches, to make it so any server can talk to any other. In this way, we preserve the least common ancestry property from Chapter 4. This type of graph is known as a fat tree.

would be represented by $16/2 = 8$ 2x2 switches. So we would make sure that each of the two 8x8s had each line going to a different 2x2 of the 16x16.

Putting it all together, we represent the two 4x4s and the one 8x8 as 2x2s in Figure 8.40. Note that we enforce the criss-cross connection everywhere, and that in each 4x4, the 2x2's have their lines going to different 2x2's in the level above. We repeat the connections that we showed in Figure 8.34: between 1 and 6, 2 and 3, and 7 and 8. They are all still possible, thanks to the way we connected the switches together.

Repeating this symmetry as we move up the tree is pretty straightforward: the important point is that we've found a way to make a tree *scalable*, using only 2x2 switches.

The structure in Figure 8.40 is often called a **fat tree**. Do not be confused by this terminology: a fat tree can scale well, while a tree cannot. With a fat tree, we take the burden off of the higher switches by replacing them with a lot of smaller switches. It's like taking the pressure off of one very capable employee by hiring a number of less capable ones, and spreading the work among them. This is almost necessary, for two reasons.

- Switches just can't handle that much load, and they become expensive very fast as you try to increase their port count. This would be like trying to find an employee that had such a high productivity that he was worth a ton of others.

- Suppose a single root could handle all that load. What would happen if it failed? The entire tree would shut down. By building it with more switches, we're reducing the dependency on any single one of them, and making a single failure that much more tolerable. Just as when we have a lot of employees, losing one of them won't cause the company to collapse.

The Clos Network

We've made some good progress so far. Unfortunately, the fat tree we described in Figure 8.40 isn't perfect. To see why, assume that the current connections are made (1 – 6, 2 – 3, and 7 – 8) as shown. Now, suppose Server 4 needs to communicate with Server 5. So you try to make this happen, but unfortunately, you can't do it. It's impossible to get over to the rightmost middle 2x2 switch, because all paths to it are blocked.

What happened here? The problem is that the current connections are *blocking* us. Does that really matter? Sort of. Remember, within a data center, we really want to make sure all servers can communicate with one another. And sure, if the other connections weren't there, this would be possible. But we would like to have the following property:

- Any unused server can connect to any other *unused* server at any time, *no matter what* the other connections are.

This would mean that the graph is **non-blocking**.

Let's find a way to make this happen with a 4x4 switch. Looking back at Figure 8.37, we see that clearly this won't work, because we don't even have connectivity between the two halves. The top is completely isolated from the bottom.

Thinking of something a little more complex, we come up with the scenario in Figure 8.41. Here, we have have a *two-stage* switch network. Basically, we have taken the right side of the first stage, and hard-wired it to the left side of the second stage. From the top-half, the lower wire from A connects to the bottom, so that 1 and 2 can get to 7 and 8.

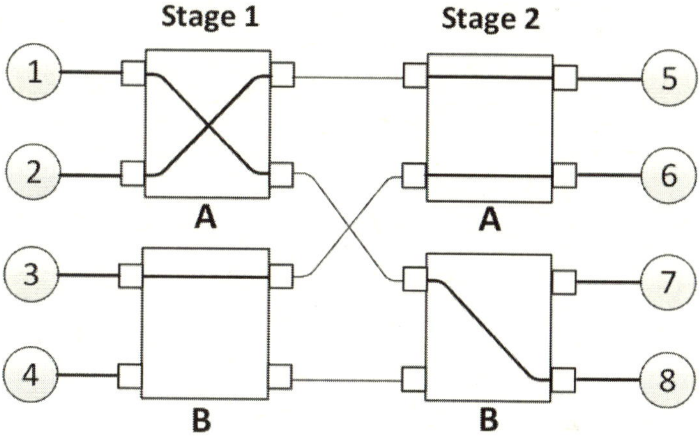

Figure 8.41 Here, we use four 2x2 switches and form a two-stage switch network. Between the two, we have wired the right side of the first stage to the left side of the second. The wiring has been made so the top-left switch can get to the bottom right, and the bottom-left can get to the top-right. You can easily verify that it's possible to get from any port to any other port in this network: from 1 to 8, 2 to 5, and 3 to 6 are highlighted here.

And from the bottom-half, the upper wire from B connects to the top, so that 3 and 4 can get to 5 and 6. In this way, you can verify that it's entirely possible to get from any port to any other. For instance, we can get from 1 to 8, by crossing down in the left switch, going to the bottom half, and then crossing down in the right switch. Similarly, 3 can get to 6, and 2 can get to 5, as we have traced out here. You can verify that it's possible to do this for any two ports.

But this still doesn't give us the non-blocking property. What if port 2 decides it wants to connect to port 7? And what if port 4 decides it wants to connect to port 5? Well, neither of these things can happen, because there's only one path that can satisfy each of these. 2 needs to go to the bottom half, which is already taken up by 1's connection to 8. And 4 needs to go to the top half, another connection that's already occupied.

It turns out that the only way we can solve this is by adding *another* set of switches, in the middle. And in doing so, we have to let A and B in Stage 1 have an additional output port, and in Stage 2 an additional input port. The result we get is shown in Figure 8.42, a three-stage switched network.

We give an example here to illustrate the non-blocking property. The

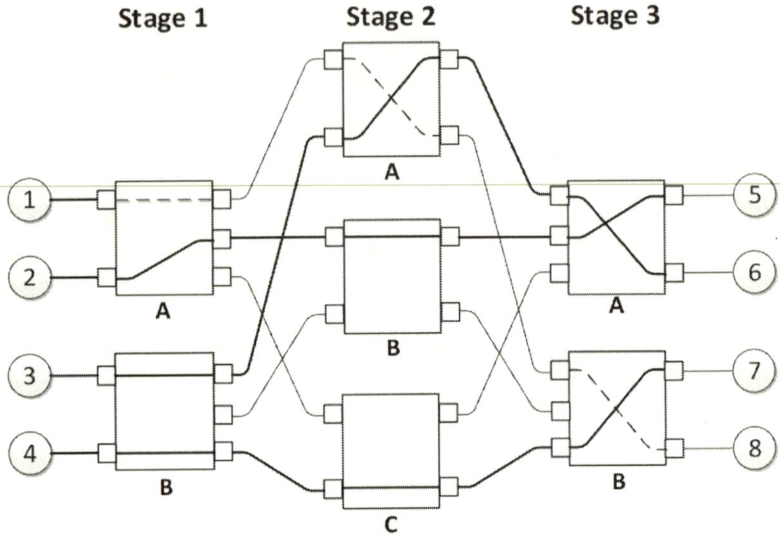

Figure 8.42 Three-stage Clos Network. Here, we have two input switches, three middle switches, and two output switches. The wiring between the stages is always fixed. Here, we have connected 2 and 5, 3 and 6, and 4 and 7 over arbitrary paths. It's still possible to connect 1 and 8. This achieves the non-blocking property.

traffic pattern is such that port 2 needs to be connected to port 5, port 3 to port 6, and port 4 to port 7. At this point, the three pairs have already been connected, as shown, and we need to connect port 1 to port 8. Is this possible? Well, port 2 is already using switch B in the second stage, so we can't go there. And port 4 is already using switch C in the second stage to get to switch B in the third stage. But thankfully, we can still go through port A in the second stage, and make it down to port 8. Had this third middle switch not been there, we would have no way of connecting them. But with the presence of the third middle switch, the network is *non-blocking.*

Figure 8.42 is a specific case of a **Clos Network**, which was formalized in 1953 by Charles Clos. Clos was a telecommunications engineer. If you think about this for a second, it actually makes a lot of sense. Remember, phone networks are all circuit switching, where we *dedicate* connections. So when we are establishing them, we need to be able to *guarantee* we can get from one point to another. If this was a phone network, making sure it's non-blocking would be quite important.

Clos networks, fat trees, and data center networking in general can

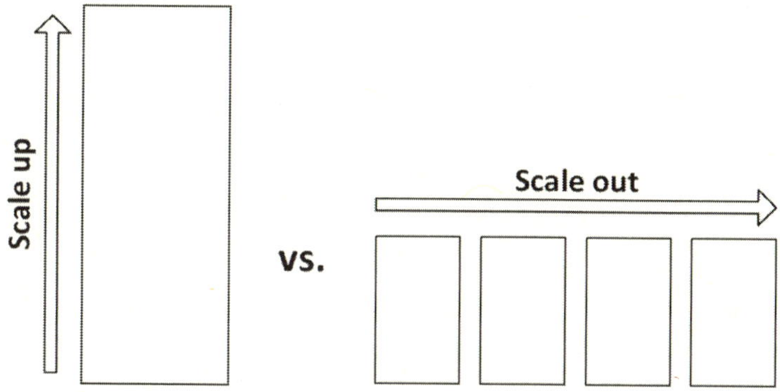

Figure 8.43 When building a data center, it's better to scale *out*, by having a lot of smaller switches, than to scale *up*, by having a few big switches. In fact, the latter is not scalable at all, because after a while, it's not possible to scale up anymore.

get complicated, beyond our scope. But we now have a general idea of how a data center should be built. It's all about *scale*, so we need to make sure we can support an ever-increasing number of servers. Rather than using a tree, we can use a **multi-stage switched network**, like a fat tree. The key idea is that instead of scaling *up* the number of ports per switch, we scale *out* the network by adding more switches. Scale *up*, not *out* (see Figure 8.43). In this way, we use many small switches to make a large switch, just like we use many small lego pieces to build a set. And just like an employee, each is contributing just a small fraction of the total work output, or in this case, the total connectivity.

In this way, we can say that small switches are the "building blocks" of the cloud, making the network not only feasible, but reliable.

8.3 "Achilles' heel"

We just saw how sometimes, it can be difficult for data centers to figure out how to handle the amount of demand they get. Indeed, it's tough to be big, and clearly, the Internet is no exception. The Internet epitomizes the need to be scalable, with the sheer number of people it serves and the amount of devices that are added each year.

It's hard to deny how dependent we've become on the Internet. Every time we update our status on Facebook, post something on Twitter, watch a video on YouTube, or search something on Google, we're sending packets into this deep abyss known as the Internet, and within

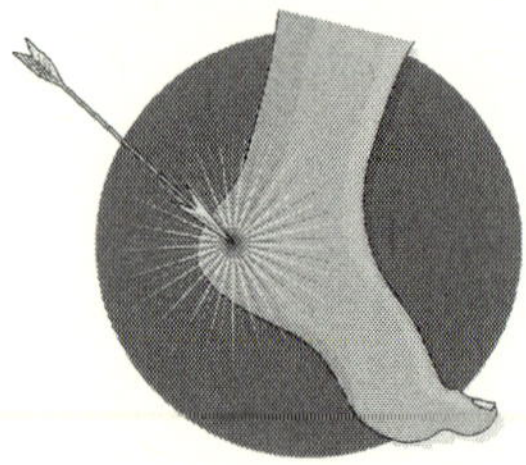

Figure 8.44 An Achilles' heel is known as a deadly weakness in spite of overall strength, that can potentially lead to downfall. Its origin is in Greek mythology, where Achilles' was invincible except for this part of his body, which ultimately caused him to perish. We wonder whether the Internet has an "Achilles' heel" or not.

fractions of a second we can get what we want back. It's almost scary. And the times we are the most frustrated is when the Internet isn't working, during rolling blackouts or when our service providers are working down the street.

During these times, we get a glimpse of how different things would be if we weren't so networked. We wouldn't be able to send emails or enjoy the free WiFi access at Starbucks. We wouldn't be able to do our homework assignments using Wikipedia, or quickly check the definition of a word via Google when we were dying to know. Life would be different.

It's an ominous thought, then: what would happen if someone *destroyed* the Internet? It would be like cutting off our life support. The question we have to ask ourselves is whether or not the Internet has an Achilles' heel (see Figure 8.44), or some central router that someone could find and attack to do significant damage to our networked lives. And if the answer is yes, how easy would it be to find it? Who knows where it is? Can we send the military to protect it, to make sure nobody cuts us off?

Luckily, the answer is *no*. While there are many types of vulnerabilities, the Internet doesn't have a single, highly connected router in the center that one could attack to destroy or severly hinder its function. There's no single "giant" Cisco router to take out (see Figure 8.45). In this section, we investigate the reasons behind this, and we will see how they are related to what we discussed in the last section on hardware limitations.

But first, what was all the hype about in the 1990s, rumoring that such an Achilles' heel exists?

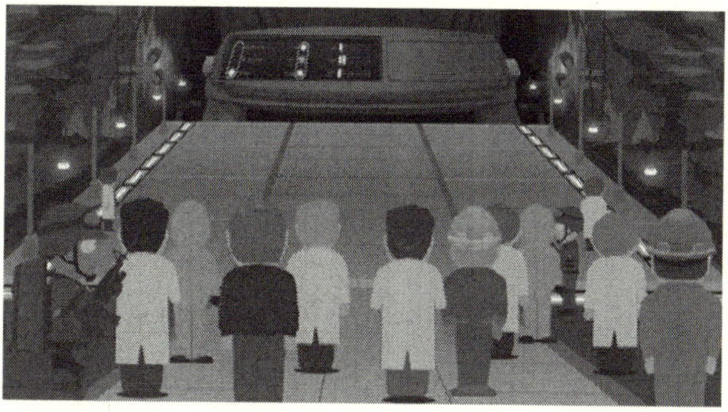

Figure 8.45 In the cartoon show South Park, one episode depicts the entire Internet as a giant Linksys wireless router. Once it ceases to function, everyone loses Internet connection, and chaos begins to ensue. Obviously not at this large of a scale, but we wonder whether it's possible to take out some focal point of the Internet to severely hinder our connections.

8.3.1 Power law distribution

In the first half of the twentieth century, there was a famous American linguist by the name of George Kingsley Zipf. Having devoted his profession to the study of language, he began pondering about words, and our choice of diction both in writing and everyday speech.

Zipf came up with a pretty interesting observation that seems almost uncanny at a first glance. He compiled a lot of texts together, and counted up the number of times he saw each different word. When he ordered them from most frequent to least frequent, he noticed a trend. Specifically, the most frequent word ...

- Came up twice as much as the second most frequent.
- Came up three times as much as the third most frequent.
- Came up four times as much as the fourth most frequent.
- And so on ...

It's remarkable how well this conforms to actuality. If you look at the Brown Corpus, a sample collection of English-language text compiled at Brown University back in the 1960s, you will find that "the" accounts for 7% of all the words, "of" accounts for roughly 3.5% (half of 7%), "and" accounts for about 2.3% (one-third of 7%), and so on.

To summarize, **Zipf's Law** states that the number of times, on average, that a word occurs is its rank's fraction of the number of times the most frequent word appears. We graph this in Figure 8.46; in other words, the second most popular item will appear half the times of the

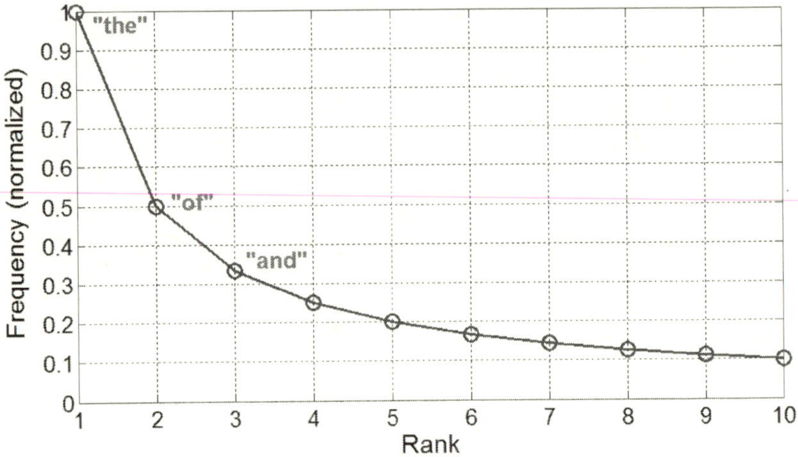

Figure 8.46 When Zipf looked at how often certain words occurred in text, he came up with this "$1/x$" relationship. Specifically, when he added up the frequency of each word, and arranged them from most popular to least popular, he found that the second most popular ("of") came up half as much as the most popular ("the"), the third came up one-third as much, and so on.

most popular, the third will appear one-third of the times, etc. Mathematically,

$$\text{Frequency of a word} = \frac{\text{Frequency of most popular word}}{\text{Rank of word}}.$$

Zipf's findings have become almost like a law in nature. Beyond linguistics, people have applied the same idea to many fields. It has been noted that income distributions conform to it: If we ordered people from richest to poorest, the second richest would have $1/2$ of the money as the richest, and so on. Similarly, the second largest city is $1/2$ the size of the largest, and so on. People have even noticed that the number of requests for files - movies, internet videos, songs, etc. – follows this trend, with the most popular files accounting for most of the requests. Sometimes, the rate of decay is larger or smaller than "$1/x$," and we can control that by adding an exponent to tailor the fit as well as possible.

More generally, Zipf's Law is a special type of **power-law distribution**, where we measure the "popularity" or "frequency" of an item as follows:

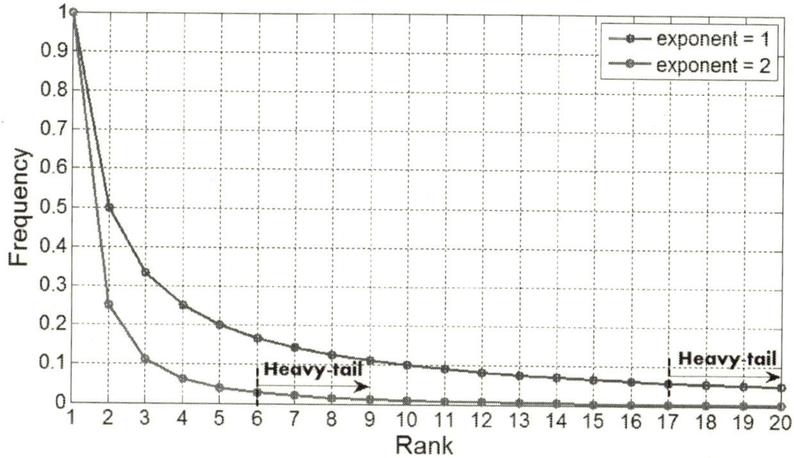

Figure 8.47 Here, we show two properties of power-laws. First, we can change the exponent in Equation (8.3) to control how quickly it decays. For larger exponent, it goes down faster. Second, after a while, the function becomes really small, but it doesn't really go to zero. This is called the "long tail" or "heavy tail."

$$\text{Popularity of item} = \left(\frac{\text{Popularity of most popular}}{\text{Rank of item}}\right)^{\text{exponent}}, \quad (8.3)$$

where that exponent controls the rate of decrease in popularity from item to item. We say that power laws have a **heavy-tail distribution**, because of the long "tail" to the right-hand side, as shown in Figure 8.47.

You're probably saying, this is interesting and all, but how does it apply to the Internet having an Achilles' heel? We'll get to that now. We have to start by taking a look at what the Internet's graph of routers looks like.

Scale-free graphs

The story behind the Achilles' heel argument began in the late 1990s, when people began running tests and measurements to gather data about the complex network of networks. The problem with the Internet is that since it's so big, it's tough to get an exact picture of what it looks like. But people tried, by using tools to trace packets through various paths of the network, and running this many many times to get the best idea possible.

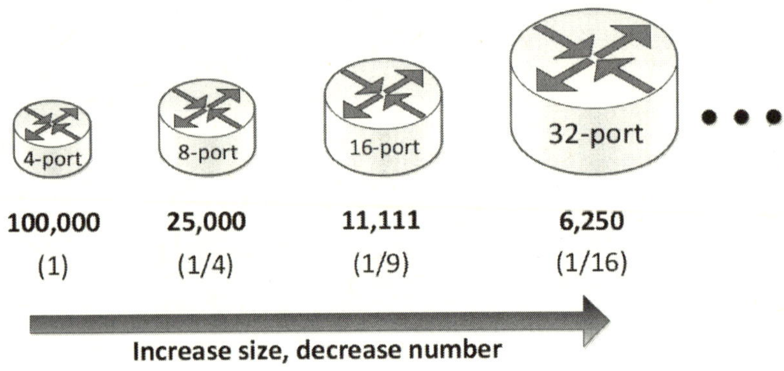

Figure 8.48 Measurements taken on the Internet's graph found that the number of ports that routers have follows a power-law. So the larger the size of the router, the less common it is, and the frequency goes down according to Equation (8.3). The exponent is about 2. Supposing the smallest router is 4-port, and there are 100,000 of them, then we have 25,000 8-port routers $(100,000/4)$, 11,111 16-port $(100,000/9)$, 6,250 32-port $(100,000/16)$, and so on. This is not meant to represent the actual numbers in the Internet.

People came up with an interesting inference result. Like so many other fields and situations, the Internet, too, has a sort of power-law distribution. Specifically, when we look at the *graph of routers*, and we take a look at the number of ports that each router has, it appears to follow the heavy-tail relationship. Routers with high port-degrees are less likely to be found than those with low port-degrees.

As indicated, we should caution on the completeness and validity of such data in the first place, because of the sheer size and intricacy of the Internet. But the idea is that if we took every router, measured the number of ports, and ordered them from smallest to largest, we'd see a power law, with an exponent of larger than two. This means that it cuts more rapidly than $1/x$, and is instead like $1/x^2$, as shown in Figure 8.47. Roughly speaking, suppose a four-port router was the smallest we could see. Then, we might say we had one-fourth as many eight-port routers $(1/2^2 = 1/4)$, one-ninth as many 16-port $(1/3^2 = 1/9)$, one-sixteenth as many 32-port $(1/4^2 = 1/16)$, and so on. Note that we are doubling the router-size here, as if routers increase their port-counts in factors of two (we did this in the last section for switches). These are not meant to represent actual numbers we'd see in the Internet, just to give the idea. We illustrate this in Figure 8.48.

In general, any network whose node degrees follow a power law is called **scale-free**. There are actually a lot of graphs in the Internet

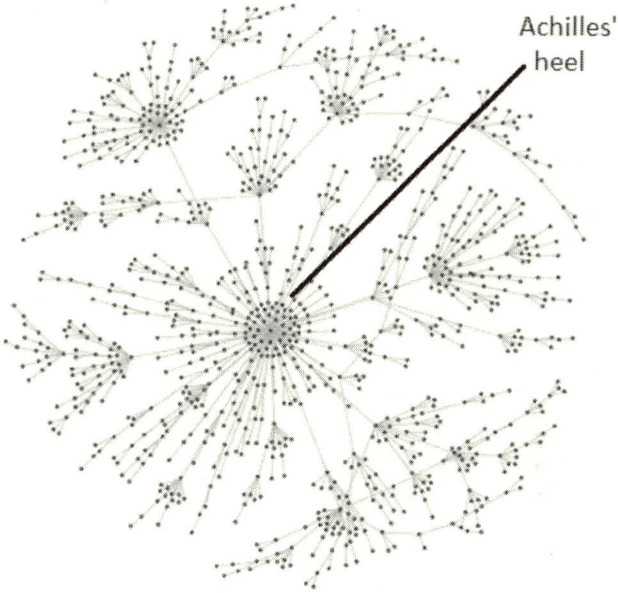

Figure 8.49 This graph is generated by the preferential attachment model. When new nodes are added to the network, they are more likely to connect to those that have a lot of connections. If the Internet were constructed this way, it would be alarming, because the highly-connected nodes at the center (pointed to here) would be an Achilles' heel. Take them out, and we completely partition the network.

that have been reported to follow this, like the graph of webpages. Here, the nodes are webpages and the links URLs, as we saw in Chapter 2 with Google. Scale-free implies that the number of pages with a given number of URLs decreases like a power-law. But in this section, our focus will be in the graph of routers.

We can take away the following from this "scale-free" phenomenon. If we had a graph of the Internet, on the one extreme, we would see *a lot* of routers with *very small* port counts. But at the same time, on the other extreme, we would see *a few* routers with *very high* port counts.

We could infer from this the following: at the edge of the network, close to the end-users, there must be a lot of routers, right? Each of our homes have them, for instance. And each of our home routers don't need that many ports. So then, we'd think that as we move towards the "core" of the Internet, that's where we'd find these few routers with very high port counts. Basically, our intuition is that as we move in a

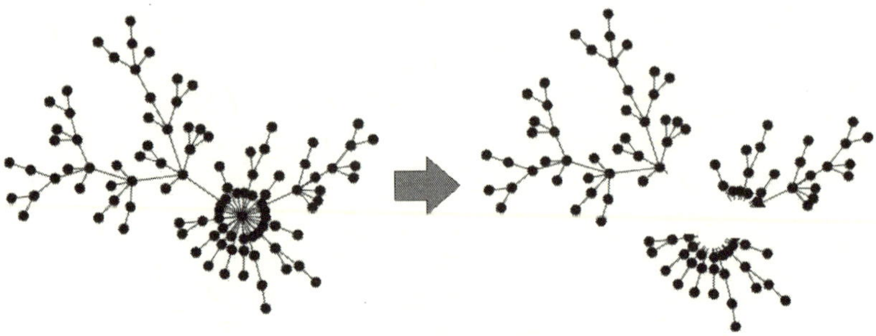

Figure 8.50 If we take the densely-connected nodes out of the graph on the left, we cause the network to partition, or split into pieces. This means that there's no way for the disconnected pieces on the right to communicate.

layer, there's that many more connections to serve, so routers must be building up and taking more connections each.

This is very alarming. If someone randomly chose a router in the network to attack, it's very unlikely that they would get to these large-degree routers at the center, because there are only a few of them. But a *deliberate* attack that specifically removes the most-connected nodes in the center will quickly fragment the network into disconnected pieces. The nodes with large degrees sitting in the center of the network become easy attack targets to break the whole network. These routers, then, must be the Internet's Achilles' heel.

This intuition is supported by the idea of **preferential attachment**, a commonly cited notion in social networks. It states the following:

- When we add a node to a network, it's more likely to connect to nodes that have large degrees than to ones that have small degrees.

In this way, the "rich get richer," because when people enter the network, they will want to connect to the "rich". This often applies in social networks, because that's how videos become viral, and how people become popular on Facebook or Twitter. We show a model generated by preferential attachment in Figure 8.49.

Now, this is truly alarming. If the Internet follows the concept of preferential attachment, then we are certain to have these few "rich" routers that are highly, highly connected. And taking some, or even one, of these routers out could completely partition the Internet, cutting connection off. We show the concept of partitioning in Figure 8.50. It sounds plausible, except for one thing: *it's not true.*

8.3.2 The Internet's reality

There's one major flaw with what we talked about in the last section, which the alarm in the late 90s about "the most connected router in the center of the Internet" didn't realize. Take a look at those highly-connected routers in the middle of the Internet, and look at the number of connections they have. What's wrong with this picture?

The problem is simple: *routers have a maximum capacity.* This is just like what we talked about in the last section on switches, and why we decided to scale out rather than up. A few central routers couldn't possibly do all that work!

Indeed, it's not possible for companies like Cisco and Juniper to make routers that have *both* a large degree *and* a high rate per connection. Each router has a limitation on its total bandwidth, the maximum number of packets it can process each second. So there's an inevitable tradeoff between the number of ports (node degree) and the speed of each port (bandwidth per connection).

We illustrate this in Figure 8.51, where we plot the total bandwidth (top) and the bandwidth per port (bottom) as we increase the number of ports on the router. As we add the first few ports to the router, everything behaves as we would intuitively expect. The total bandwidth increases linearly (top), and each port is able to keep the same bandwidth (bottom). But eventually, we reach the maximum capacity that the router can be built to handle: the technological ceiling. After this point, the total bandwidth stays the same, but we have to divide it among all the ports, so the per-port bandwidth drops quickly. In short, when we add more ports, the speed each port gets will go down.

Once we're at the capacity of the router, there's a simple relationship we can use to relate the bandwidth/port to the number of ports. The total speed needs to stay the same, which must be equal to the product of the capacity per port and the number of ports:

$$\text{Router capacity} = (\text{Capacity per port}) \times (\text{Number of ports}).$$

So then, rearranging this, we see that to keep the router capacity the same, we will inevitably trade-off the capacity per port and the number:

$$\text{Capacity per port} = \frac{\text{Router capacity}}{\text{Number of ports}}.$$

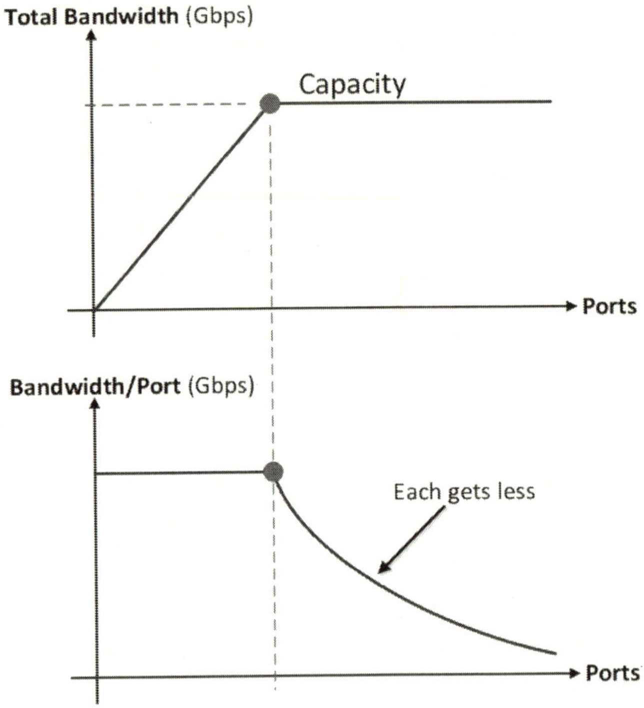

Figure 8.51 Plots of total bandwidth and bandwidth per port, as we increase the number of ports. Initially, the total bandwidth increases linearly, and each port keeps the same bandwidth. But eventually, we hit the capacity that the router can handle, and after this point, the total capacity starts dividing evenly among the number of ports. Each port will then get less. This is why it's impossible to build a router with a large number of ports and a large speed on each port.

Core vs. edge routers

The discussion we had regarding Figure 8.51 implies the following, for routers operating at capacity:

- A large per-port bandwidth means that the number of ports is small.
- A large number of ports means the per-port bandwidth is small.

What can we deduce, then, about the core routers? Well, let's think for a minute, about what happens as we move towards the center of the Internet. One thing we know is that there's a lot of demand coming in at different times, from different people. The Internet relies on the principle of *statistical multiplexing* to smoothen the fluctuation, by pooling demand together. We've discussed this many times.

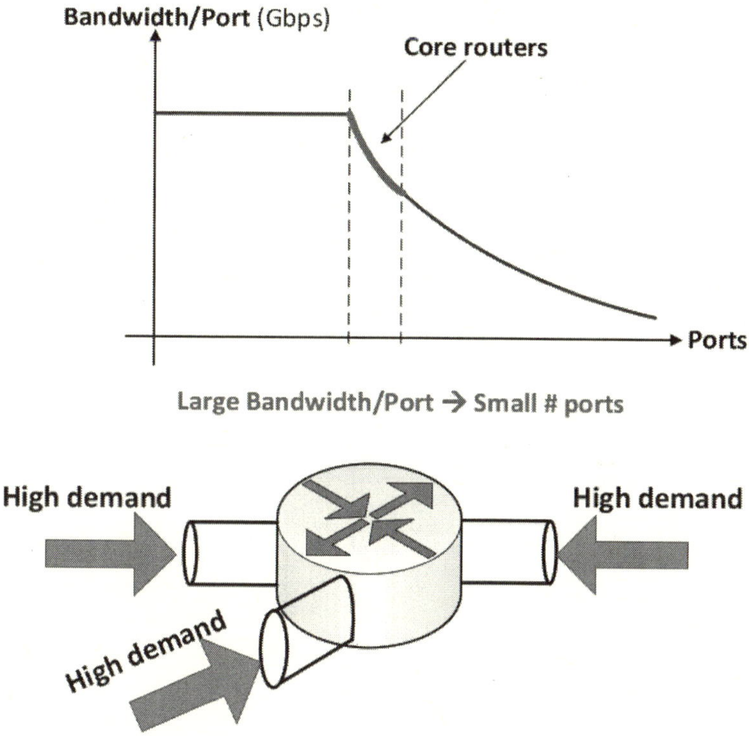

Figure 8.52 From the principle of statistical multiplexing, we know that core routers need to have a high bandwidth per-port, even when operating at full capacity. As a result, they would fit right after the nitch in the graph, as shown, implying that they must have a not-too-large number of ports.

This implies something interesting. If we want to combine a lot of people's demands in the same pipe, we need the Internet's pipes to get *a lot larger* as we move to the core of the network. Indeed, we would need to be able to support all these people's demands peaking at the same time, in the worst case.

So, the routers in the core of the network need to have a large capacity per port. And as a result, they need to have a not-too-large degree. We show this in Figure 8.52.

Well, if the core routers have small degree, how can we fit the reality of the measurements that were taken, which indicated that there were, in fact, some routers with very large degrees? The answer is exactly the opposite of what preferential attachment would predict: these routers must be at the *edge* of the Internet, not at the core.

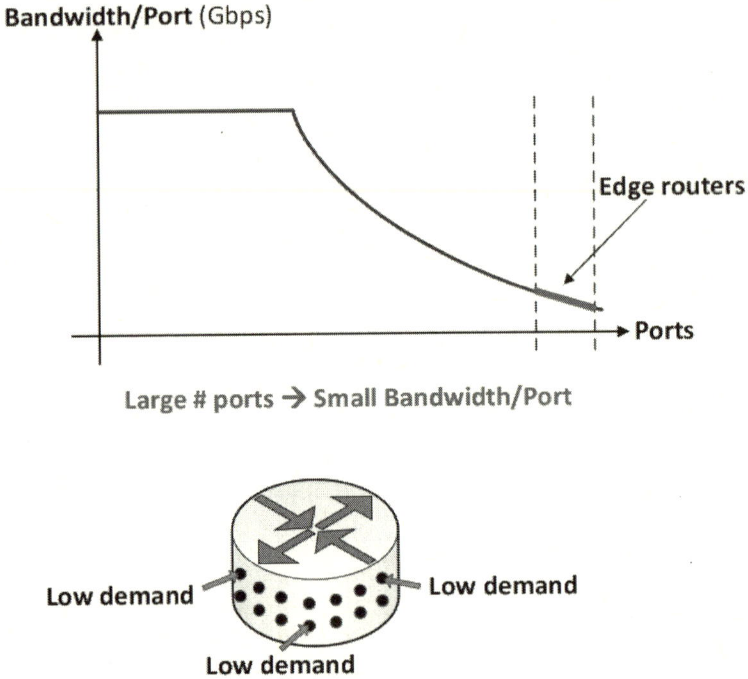

Figure 8.53 If a router has a large number of ports, it must have a small bandwidth capacity on each port. The only place we can achieve this is at the edge of the network, where the demand for each port is smaller. We conclude that some of the routers along the edge create the "heavy tail" in the scale-free network.

This actually makes quite a bit of sense, and conforms well to what we have described. As we increase the number of ports, we inevitably reduce the capacity that each node can handle. So a large in-degree router *must* be near the edge of the Internet, where the amount of demand for the pipes is smaller. Here, we have less users per pipe. We illustrate this in Figure 8.53. In fact, AT&T confirmed this intuition in 2003, when they stated that the maximum degree of a core router in their network was only 68, while the maximum degree of an edge router was 313. That's almost five times as large!

This is not, however, to say that all edge routers have large degrees. In fact, the majority of them *do not*. In order to create the heavy-tail of the scale free network, we only need *a few* routers of large degree. Most edge routers operate well under the capacity required at the core, and with that, most are built for much smaller capacity altogether.

We conclude that while edge-routers have *high* variability in their port numbers, core routers do not. But it's those few at the edge of large degree that skewed the reasoning in the 90s towards believing the Internet had an Achilles' heel.

The "unlikely" Internet

We can then summarize the Internet's router graph as follows:

- Edge routers have much lower demand than core routers. Hence, edge (core) routers have small (large) demand per port.
- Most edge routers have a small number of connections, but a few have a large number of connections.
- Core routers are constrained by their large capacity per port to have small degrees.

We show a picture similar to the *actual* graph of the Internet's routers on the right side of Figure 8.54, and compare it to preferential attachment on the left. For the Internet, there are a few nodes that have high degree right at the edge. On the other hand, for preferential attachment, these high-degree nodes are in the core.

In the Internet, a planned attack on a high-degree node would only affect some small part of the access network, not the whole network. On the other hand, an attack on a medium- to- small- degree node would have a high chance of hitting the access routers, because there are many more of them than there are core routers. So either way we put it, a planned attack would hit the access network, with a very small chance of disrupting the core.

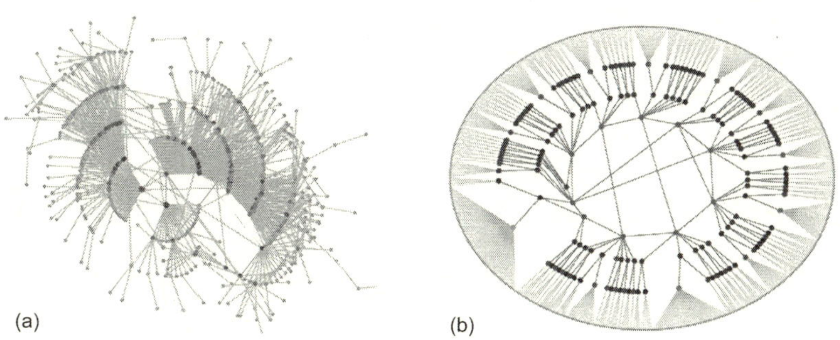

Figure 8.54 Achilles' heel or not, from Alderson et al. (2005): (a) is a typical graph generated from preferential attachment; (b) is a typical graph of the actual Internet's routers. Both satisfy the scale-free condition, but the one on the left has an Achilles' heel whereas the one on the right does not.

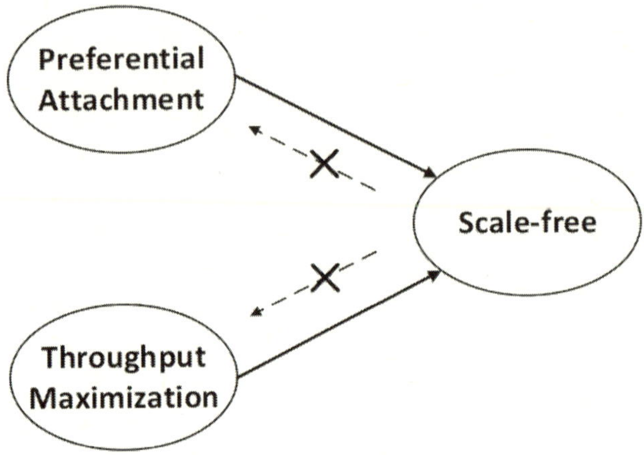

Figure 8.55 While preferential attachment gives us a scale-free network, the reverse is not true. There are many ways to generate a scale-free network; maximizing throughput (i.e., the Internet) itself is one. The biggest mistake in concluding there's an Achilles' heel in the Internet is by making this assumption that scale-free implies preferential attachment, which it does not.

Where did we go wrong, then? The main flaw was as follows: *A power-law degree distribution does not imply preferential attachment.* We jumped to conclusions earlier by assuming that the only way we could get a "heavy tail" of routers with large degree was to have them at the center. But that's not true, and we've shown that the Internet does not do that. The Internet *is* scale-free, but it's *not* vulnerable in the sense of Achilles' heel. We summarize this logical inconsistency in Figure 8.55.

The Internet is sometimes described as "self-organizing." But this self-organization was achieved by smart and efficient design. Indeed, we can say that the Internet is not a very *likely* graph, if we consider all such graphs that can create a power-law. But we didn't get there randomly, we got there by design.

You can see pretty quickly that having high-connected core nodes results in bad performance. A router with a large degree cannot support high-bandwidth links, so if it sits in the core of the network, it becomes a bottleneck for the whole thing. It's exactly like what we discussed in the last section, where we didn't want to have scaled-up root nodes. It simply would not be feasible to handle all this demand with a single switch or router. And if we tried, our performance and vulnerability would suffer greatly.

That's the problem with preferential attachment. Any graph built on this principle will be quite sub-optimal, because the core nodes will (probably) end up having high degree.

In summary, there's no Achilles' heel in the Internet. No need to worry like they did in the 90s, or in South Park. Smart engineering design has carefully avoided it, even if unintentionally.

Summary

Networks are huge, and becoming larger by the day. In this chapter, we surveyed different topics that have emerged from that property. First, we looked at the six degrees of separation in social networks, and analyzed different models that have been proposed to explain the topological and algorithmic "small world" as the networks become large. Then, we studied the concept of scaling out rather than scaling up a cloud datacenter, by relying on a large number of small "building block" switches to face unprecedented data demands. Finally, we deflated the argument that the Internet has an "Achilles' heel" by pointing out the fact that a single router cannot handle the demand that would be required to make it vulnerable. We should consider ourselves lucky that the networks we live and breathe have proven to be scalable to such large sizes.

Index